The Man in the Ice
Volume 4

Sigmar Bortenschlager
Klaus Oeggl (eds.)

The Iceman and his Natural Environment

Palaeobotanical Results

SpringerWienNewYork

Univ.-Prof. Mag. Dr. Sigmar Bortenschlager
Univ.-Prof. Dr. Klaus Oeggl
Institute of Botany, University of Innsbruck, Innsbruck, Austria

This work is sponsored by
Fonds zur Förderung der wissenschaftlichen Forschung, Wien

This work is subject to copyright.
All rights are reserved, whether the whole or part of the material is concerned, specifically those of translation, reprinting, re-use of illustrations, broadcasting, reproduction by photocopying machines or similar means, and storage in data banks.

© 2000 Springer-Verlag/Wien
Printed in Slovenia

Cover: Sheep at pasture on "Hohen Mut", Ötztal, 2.600 m, photo: Prof. Dr. Gernot Patzelt; Dagger of Iceman, photo: RGZM Mainz, Germany.

Typesetting: Thomson Press, India
Printing: Euroadria, SLO-1001 Ljubljana

Printed on acid-free and chlorine-free bleached paper
SPIN: 10556281

With 108 partly coloured Figures

Library of Congress Cataloging-in-Publication Data

The Iceman and his natural environment: palaeobotanical results/Sigmar Bortenschlager, Klaus Oeggl (eds.).
      p. cm. – (The man in the ice, ISSN 0947-3483; v. 4)
  Includes summaries in French, German, and Italian.
  Includes bibliographical references.
  ISBN 3211825502 (acid-free paper)
  1. Ötzi (Ice mummy) 2. Copper age–Italy–Hauslabjoch Pass. 3. Neolithic period–Italy–Hauslabjoch Pass. 4. Human remains (Archaeology)–Italy–Hauslabjoch Pass. 5. Paleobotany–Italy–Hauslabjoch Pass. 6. Paleoethnobotany–Italy–Hauslabjoch Pass. 7. Radiocarbon dating–Italy–Hauslabjoch Pass. 8. Hauslabjoch Pass (Italy)–Antiquities. I. Bortenschlager, Sigmar. II. Oeggl, Klaus. III. Series.

GN778.22.I8 I25 2000
937–dc21

00-038816

ISSN 0947-3483
ISBN 3-211-82660-2 Springer-Verlag Wien New York

Contents

Preface	VII
List of Contributors	IX
Kutschera, W., Golser, R., Priller, A., Rom, W., Steier, P., Wild, E., Arnold, M., Tisnerat-Laborde, N., Possnert, G., Bortenschlager, S. and Oeggl, K.: Radiocarbon dating of equipment from the Iceman	1
Bortenschlager, S.: The Iceman's environment	11
Birks, H. H.: The amount of CO_2 in the air breathed by the Iceman	25
Oeggl, K. and Schoch, W.: Dendrological analyses of artefacts and other remains	29
Oberhuber, W. and Knapp, R.: The bow of the Tyrolean Iceman: A dendrological investigation by computed tomography	63
Pfeifer, K. and Oeggl, K.: Analysis of the bast used by the Iceman as binding material	69
Dickson, J. H.: Bryology and the Iceman: Chorology, Ecology and Ethnobotany of the Mosses *Neckera complanata* Hedw. and *N. crispa* Hedw.	77
Oeggl, K.: The diet of the Iceman	89
Rott, E.: Diatoms from the colon of the Iceman	117
Aspöck, H., Auer, H., Picher, O. and Platzer, W.: Parasitological examination of the Iceman	127
Tessadri, R.: Vivianite from the Iceman of the Tisenjoch (Tyrol, Austria): Mineralogical-chemical data	137
Peintner, U. and Pöder, R.: Ethnomycological remarks on the Iceman's fungi	143
Schedl, W.: Contribution to insect remains from the accompanying equipment of the Iceman	151
Antonini, S., Luciani, S., Marota, I., Ubaldi, M. and Rollo, F.: Compilation of DNA sequences from the Iceman's grass clothing	157
Oeggl, K., Dickson, J. H. and Bortenschlager, S.: Epilogue: The search for explanations and future developments	163

Preface

Before the discovery of the "Man in the Ice" in September 1991, little was known about the Neolithic period in the Central Alps. Suddenly and without precedent, here was the very well preserved corpse of a man who had lived more than 5,000 years ago with his clothing and equipment almost intact. The discovery was not just deservedly a world-wide sensation but a unique opportunity for the scientific community to investigate the life and death of a human from such very ancient times. It opened up wholly new horizons in prehistoric research, and with the help of a full range of modern research techniques an attempt was made in a multidisciplinary approach to extract as much information as possible about the life of Neolithic people in the Alpine region. The very unusual location and circumstances of the discovery caught the imagination of the general public and also raised numerous questions relating to the find. Some answers have already been provided through archaeological, medical and scientific research, but many questions still remain.

Immediately after the discovery of the Iceman, the Department of Botany at Innsbruck University initiated a detailed archaeobotanical research programme in support of the archaeological investigations with the aim of reconstructing the Iceman's environment and way of life. This archaeobotanical research is based on two independent data sources and thus provides a more global view of the unparalleled find. On the one hand the botanical and zoological remains that were excavated or collected from the site of the find have been subjected to full analysis. They are a source of information on the availability and use of plants and animals in the Neolithic, fundamental data for the reconstruction of the wooden artefacts found at the site, and indirect insights into the Iceman's environment. The data obtained from the on-site finds, however, represent no more than an excerpt from the life of the Iceman. They only permit reconstruction of a limited area within a short period of the Neolithic. The results of the analyses performed on the finds from the site therefore have to be placed in a wider spatial and temporal context. On the other hand the reconstruction of the Iceman's life-style as derived from data obtained from the site has to be correlated with independent data sets. For that purpose pollen analysis was performed on peat-bogs in the vicinity of the find and further afield to obtain precise data on the vegetation cover and climate in the Neolithic. Pollen analysis was performed for a vertical transect extending from the timber line almost up to the nival zone. The results of the analyses reveal changes to the vegetation patterns caused by pastoral farming long before the time of the Iceman. Many additional and highly specific questions have arisen in the course of this research, only some of which are addressed in this volume.

A programme of archaeobotanical investigations into the Iceman is of course one thing, and the resources needed to perform the research work another. In this context a debt of gratitude is owed to all the organisations and individuals whose financial and personal assistance have contributed to the success of the programme. In particular we would like to thank the Austrian Science Foundation, the Norwegian Research Council and the Grolle Olsen Fund, the Scottish Universities' Carnegie Trust, the University of Glasgow, the Royal Society of London, the University of Innsbruck, and the Austrian Academy of Sciences. Their financial support has been essential for the research work in all its depth. The Roman Germanic Central Museum in Mainz has made extensive photographic records available and thus contributed to the outstanding standard of documentation of the finds. Professor Jim Dickson of Glasgow University has played a key role in the genesis of this volume, given the various contributions the necessary stylistic polish and generally contributing to the readability of the final product. Last but not least, the Springer publishing house has taken on the task of publishing the results of these archaeobotanical investigations to the same standard as the three previous volumes.

Innsbruck, December 1998

Sigmar Bortenschlager
Klaus Oeggl

List of Contributors

Antonini Dr. Silvia, Universitá degli Studi di Camerino, Dipartimenti di Biologia moleculare, cellulare e animale, Via Camerini, I-62032 Camerino, Italia

Arnold Dr. Maurice, Laboratoire des Science du Climat et de l' Environnement (LSCE), Unité Mixte de Recherche CEA-CNRS, Avenue de la Terasse, F-91198 Gif-sur-Yvette Cedex, France

Aspöck Univ. Prof. Dr. Horst, Klinisches Institut für Hygiene der Universität Wien, Abteilung für Medizinische Parasitologie, Kinderspitalgasse 15, A-1095 Wien, Austria

Auer Univ. Prof. Dr. Herbert, Klinisches Institut für Hygiene der Universität Wien, Abteilung für Medizinische Parasitologie, Kinderspitalgasse 15, A-1095 Wien, Austria

Birks Dr. Hilary H., Botanical Institute, University of Bergen, Allégaten 41, N-5007 Bergen, Norway

Bortenschlager Univ.-Prof. Dr. Sigmar, Institut für Botanik der Universität Innsbruck, Sternwartestrasse 15, A-6020 Innsbruck, Austria

Dickson Univ.-Prof. Dr. James H., Division of Environmental and Evolutionary Biology, University of Glasgow, G12 8QQ, United Kingdom

Golser Dipl.-Ing. Dr. Robin, Institut für Radiumforschung und Kernphysik der Universität Wien, Vienna Environmental Research Accelerator, Währinger Strasse 17, A-1090 Wien, Austria

Knapp Dr. Rudolf, Universitäts-Kliniken Innsbruck, Abteilung für Röntgendiagnostik und Computertomographie, Anichstraße 35, A-6020 Innsbruck, Austria

Kutschera Univ.-Prof. Dr. Walter, Institut für Radiumforschung und Kernphysik der Universität Wien, Vienna Environmental Research Accelerator, Währinger Strasse 17, A-1090 Wien, Austria

Luciani Dr. Stefania, Universitá degli Studi di Camerino, Dipartimenti di Biologia moleculare, cellulare e animale, Via Camerini, I-62032 Camerino, Italia

Marota Dr. Isolina, Universitá degli Studi di Camerino, Dipartimenti di Biologia moleculare, cellulare e animale, Via Camerini, I-62032 Camerino, Italia

Oberhuber Dr. Walter, Institut für Botanik der Universität Innsbruck, Sternwartestrasse 15, A-6020 Innsbruck, Austria

Oeggl Univ.-Prof. Dr. Klaus, Institut für Botanik der Universität Innsbruck, Sternwartestrasse 15, A-6020 Innsbruck, Austria

Peintner Dr. Ulrike, Institut für Mikrobiologie der Universität Innsbruck, Technikerstraße 25, A-6020 Innsbruck, Austria

Pfeifer Mag. Klaus, Institut für Botanik der Universität Innsbruck, Sternwartestrasse 15, A-6020 Innsbruck, Austria

Picher Wiss. Oberrat Dr. phil. Otto, Klinisches Institut für Hygiene der Universität Wien, Abteilung für Medizinische Parasitologie, Kinderspitalgasse 15, A-1095 Wien, Austria

Platzer Univ. Prof. Dr. Werner, Institut für Anatomie der Universität Innsbruck, Müllerstraße 59, A-6020 Innsbruck, Austria

Pöder Univ.-Prof. Dr. Reinhold, Institut für Mikrobiologie der Universität Innsbruck, Technikerstraße 25, A-6020 Innsbruck, Austria

Possnert Dr. Göran, Uppsala University, The Svedberg Laboratory, Box 533, S-75121 Uppsala, Sweden

Priller Dipl. Phys. Dr. Alfred, Institut für Radiumforschung und Kernphysik der Universität Wien, Vienna Environmental Research Accelerator, Währinger Strasse 17, A-1090 Wien, Austria

Rollo Univ.-Prof. Dr. Franco, Universitá degli Studi di Camerino, Dipartimenti di Biologia moleculare, cellulare e animale, Via Camerini, I-62032 Camerino, Italia

Rom Mag. Werner, Institut für Radiumforschung und Kernphysik der Universität Wien, Vienna Environmental Research Accelerator, Währinger Strasse 17, A-1090 Wien, Austria

Rott Univ.-Prof. Dr. Eugen, Institut für Botanik der Universität Innsbruck, Sternwartestrasse 15, A-6020 Innsbruck, Austria

Schedl Univ.-Prof. Dr. Walter, Institut für Zoologie und Limnologie der Universität Innsbruck, Technikerstr. 25, A-6020 Innsbruck, Austria

Schoch Werner, Labor für Quartäre Hölzer, Tobelhof 13, CH-8134 Adliswil, Switzerland

Steier Mag. Peter, Institut für Radiumforschung und Kernphysik der Universität Wien, Vienna Environmental Research Accelerator, Währinger Strasse 17, A-1090 Wien, Austria

Tessadri Dr. Richard, Institut für Mineralogie und Petrographie der Universität Innsbruck, Innrain 52, A-6020 Innsbruck, Austria

Tisnerat-Laborde Dr. Nadine, Laboratoire des Science du Climat et de l'Environnement (LSCE), Unité Mixte de Recherche CEA-CNRS, Avenue de la Terasse, F-91198 Gif-sur-Yvette Cedex, France

Ubaldi Dr. Massimo, Universitá degli Studi di Camerino, Dipartimenti di Biologia moleculare, cellulare e animale, Via Camerini, I-62032 Camerino, Italia

Wild Mag. Dr. Eva, Vienna Environmental Research Accelerator, Institut für Radiumforschung und Kernphysik der Universität Wien, Währinger Strasse 17, A-1090 Wien, Austria

Radiocarbon dating of equipment from the Iceman

W. Kutschera[1], R. Golser[1], A. Priller[1], W. Rom[1], P. Steier[1], E. Wild[1], M. Arnold[2], N. Tisnerat–Laborde[2], G. Possnert[3], S. Bortenschlager[4], and K. Oeggl[4]

[1] Vienna Environmental Research Accelerator, Institut für Radiumforschung und Kernphysik, Universität Wien
[2] Laboratoire des Sciences du Climat et de l'Environnement (LSCE), Unité Mixte de Recherche CEA-CNRS, Gif-sur-Yvette
[3] The Svedberg Laboratory, Uppsala University
[4] Institut für Botanik, Universität Innsbruck

1. Introduction

The Iceman was discovered on September 19, 1991 at the "Tisenjoch", a usually glacier-covered mountain pass at 3210 m altitude located at the Italian/Austrian border in the Ötztal Alps. Shortly thereafter, AMS measurements of ^{14}C in bone and tissue of the Iceman revealed that he had died somewhere between 5300 and 5100 years ago (Bonani et al., 1992; Hedges et al., 1992; Bonani et al., 1994; Prinotoh–Fornwagner and Niklaus, 1994). Besides the body of the Iceman, numerous pieces of equipment of the Iceman and other materials associated with the finding place were recovered (Lippert, 1992; Bagolini et al., 1996). A small but representative fraction have now been radiocarbon dated at three different AMS laboratories. The present paper summarizes the results of these measurements (see also Rom et al., 1999).

Although the method of ^{14}C dating is well established by now, we briefly describe its basic principle in Sect. 2. The AMS measuring technique is described in Sect. 3 referring mainly to the newly established AMS facility in Vienna (Kutschera et al., 1997). In Sect. 4 we present the ^{14}C data measured at the three different AMS laboratories. In Sect. 5 we discuss possible implications of the results.

2. Radiocarbon dating

The idea of using the radioactive decay of ^{14}C ($t_{1/2}$ = 5730 y) as a clock to determine the age of carbon-containing materials was developed about 50 years ago by Willard Libby at the University of Chicago (Libby, 1946; Anderson et al., 1947; Arnold and Libby, 1949). Like many great ideas, the principle of ^{14}C dating is simple:

Primary cosmic rays (mainly protons) produce secondary neutrons through nuclear reactions on atomic nuclei of the atmosphere. These neutrons convert ^{14}N nuclei of the atmosphere (78% of the atmosphere consist of N_2) into ^{14}C by emitting a proton (^{14}N + n → ^{14}C + p). Atmospheric chemistry converts these cosmogenic ^{14}C atoms into $^{14}CO_2$ molecules, where they become a minute part ($\sim 10^{-12}$) of the global atmospheric CO_2 pool. The overwhelming part of CO_2 contains two stable isotopes of carbon, ^{12}C (98.9%) and ^{13}C (1.1%). Since about 20% of the atmospheric CO_2 exchanges annually with the biosphere and the hydrosphere (Graedel and Crutzen, 1994; Levin 1994; Levin, 1998), the result is a well-mixed global inventory of ^{14}C. The global balance between ^{14}C production and decay (the so-called secular equilibrium) creates a very small but distinct isotopic ratio of ^{14}C/^{12}C = 1.2×10^{-12} within this "live" carbon pool. Assuming that one knows the equilibrium ratio for live carbon at all times, a measurement of this ratio in an ancient object today tells us how much time has passed since the object was last part of the live carbon pool. Although the original assumption of an approximately constant ^{14}C/^{12}C ratio (Arnold and Libby, 1949) proved not to be acceptable (de Vries, 1958) due to fluctuations in the ^{14}C production and in the exchange processes between the different global reservoirs, the natural ^{14}C/^{12}C variations can now be accurately traced back 12 000 years by measuring ^{14}C in tree rings, absolutely dated by dendrochronology (Taylor et al., 1996, see also references therein). An absolute age determination within this time range is therefore possible by converting the measured, so-called "^{14}C age" (given in years before present, 0 B.P. = 1950), into a calendar date using standard calibration curves (Stuiver et al., 1993). This transformation sometimes leads to a considerable loss in precision as compared to the ^{14}C age, depending on the natural ^{14}C/^{12}C fluctuations of atmospheric CO_2 in the particular time period to be dated.

Various attempts are under way to extend the ^{14}C/^{12}C calibration beyond the range of dendrochronology by comparing ^{14}C with other absolute dating methods. The most recent calibration of ^{14}C (Stuiver et al., 1998) extends the time range to 24 000 years ago, using uranium–thorium dating of corals and periodic layers of marine sediments (varves) as an absolute time reference.

In modern times, ^{14}C/^{12}C ratios in atmospheric CO_2 have been influenced by anthropogenic contributions. Starting around 1900, the release of CO_2 from fossil-fuel burning added "dead" carbon (having no measurable ^{14}C content), which decreased the natural ^{14}C/^{12}C ratio by a few percent in the first 50 years of this century. From 1950 on, atmospheric nuclear weapons testing led to a sharp increase in ^{14}C by supplying anthropogenic neutrons in addition to the cosmogenic ones converting more ^{14}N into ^{14}C (see above). At the time of the Nuclear Test Ban Treaty in 1963, the ^{14}C/^{12}C ratio increased by almost a factor of two over values of 1950 (Levin et al., 1985). Since then the ^{14}C/^{12}C ratio in atmospheric CO_2 has been decreasing steadily through exchange of CO_2 with the biosphere and the ocean. In 1996 it is still about 10% above the natural level (Levin and Kromer, 1997). Currently, it continues to decrease at a rate of about 1% per year. Since the precision of ^{14}C/^{12}C ratio measurements with modern AMS facilities is better than 1%, it is possible to date very recent material with a precision of a few years using this ^{14}C "bomb peak" (Wild et al., 1998).

3. ^{14}C Measurements with AMS

The ^{14}C/^{12}C content in carbon can be measured in two ways: either by counting the beta particles (e^-) from the radioactive decay of ^{14}C (^{14}C $\rightarrow$ ^{14}N + e^- + $\bar{\nu}_e$) with a proportional counter or a liquid scintillation detector to determine the specific ^{14}C decay rate (decays per minute per gram carbon), or by a ^{14}C/^{12}C isotopic ratio measurement with AMS. It can easily be shown that counting ^{14}C atoms directly rather than waiting for their infrequent decay is a much more sensitive method. While 1 milligram of modern carbon with a ^{14}C/^{12}C ratio of 1.2×10^{-12} contains 6×10^7 ^{14}C atoms, ^{14}C decays only with a rate of about 1 decay per hour because of the long half-life. In contrast, approximately 1% of the sixty million ^{14}C atoms can be measured with AMS in 1 hour. The much higher detection sensitivity of AMS allows one to perform ^{14}C dating measurements with milligram quantities of carbon, whereas gram amounts are typically required for beta counting. However, AMS requires measuring ^{14}C/^{12}C isotopic ratios in the range of 10^{-12} (modern sample) to 10^{-15} (50 000-year-old sample). AMS is capable to provide the necessary selectivity and precision to measure these extremely small isotope ratios. Most AMS facilities performing ^{14}C measurements reach an overall precision of $\sim 0.5\%$ (e.g., Rom et al., 1998), corresponding to a ^{14}C age uncertainty of ± 40 years.

As mentioned above, there are two stable isotopes of carbon: ^{12}C, which comprises 98.9% of carbon, and ^{13}C with an isotopic abundance of 1.1%. The ^{13}C/^{12}C isotopic ratio, which is also measured at most AMS laboratories, provides important information on mass fractionation, i.e., mass-dependent effects in natural processes (e.g., biochemical processes), chemical processing of the sample, and the measurement procedure. For a correct age determination from the measured ^{14}C/^{12}C ratios, these mass-dependent effects have to be taken into account.

In Fig. 1 we show a schematic view of the new AMS system in Vienna based on a 3-MV Pelletron tandem accelerator (Kutschera et al., 1997). Essentially all AMS systems use tandem accelerators (Kutschera, 1997), which provide a particularly useful selectivity through the use of negative ions. Taking the VERA facility of Fig. 1 as a guideline, the principle of ^{14}C measurement with AMS is briefly described.

Aliquots of about 1 mg of solid carbon, produced from the original sample material by combustion to CO_2 and subsequent catalytic reduction to graphite (e.g., Wild et al., 1998), are mounted into the multi-sample ion source. Negative ions of carbon of the order of 20 μA are produced by sputtering the carbon sample with a Cs ion beam (Middleton, 1984; Middleton, 1989). The pre-accelerated C^- ions first pass through an electrostatic analyser for energy selection, and are then separated by the injection magnet according to their mass. Mass-12 ($^{12}C^-$) and mass-13 ($^{13}C^-$ + $^{12}CH^-$) negative ions are measured as electrical (ion) currents in Faraday cups indicated on the figure. Mass-14 negative ions ($^{14}C^-$ + $^{13}CH^-$ + $^{12}CH_2^-$ + $^7Li^{2-}$ + $\cdots$) are selected for injection into the tandem accelerator operated at a positive voltage of 2.7 MV. It is most important for a ^{14}C detection that ^{14}N, the omnipresent stable isobar of ^{14}C, does not form negative ions and is therefore not present in the mass-14 ions. However, the rare $^{14}C^-$ ions are nevertheless deeply buried in the background of $^{13}CH^-$ + $^{12}CH_2^-$ molecular ions which are typically a billion (10^9) times more abundant than $^{14}C^-$.

After injection into the tandem accelerator, the negative ions are accelerated to the positively charged terminal acquiring an energy of 2.7 MeV. At the terminal, the negative ions pass at high speed ($\sim 2\%$ of the velocity of light, i.e. it would take them only 2 seconds to fly from Vienna to San Francisco) through a gas canal where several electrons are stripped off, breaking molecules apart. After being stripped to multiply charged

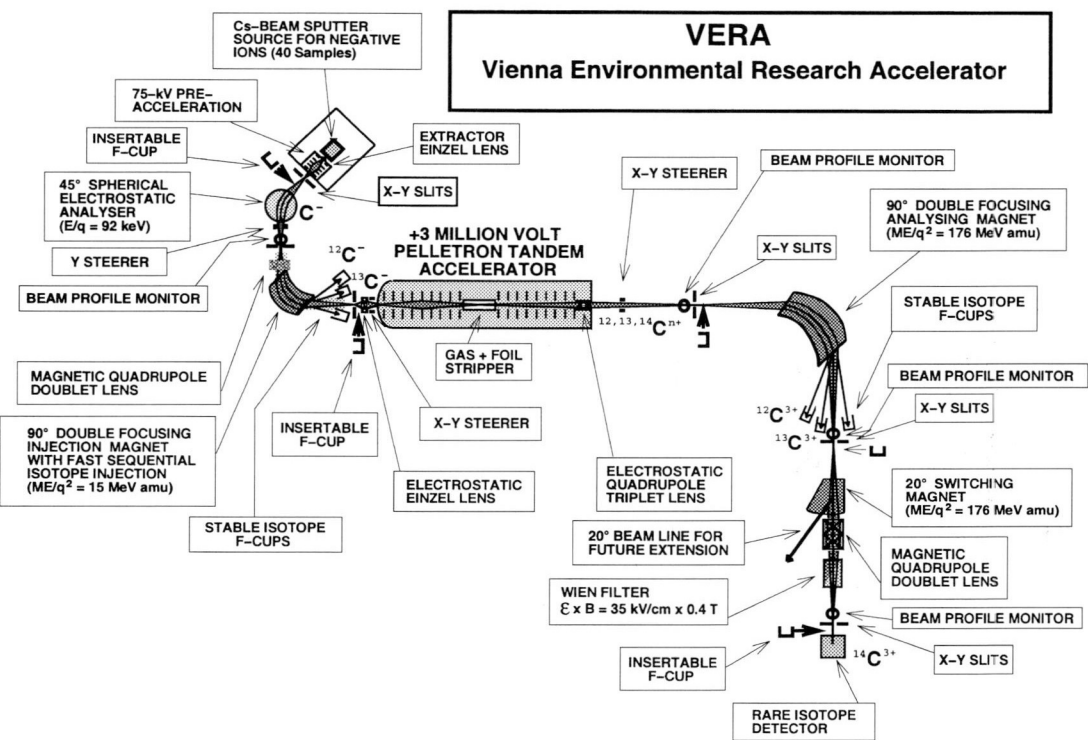

Fig. 1. Schematic layout of VERA showing the essential features of the system

positive ions, the ions obtain another boost in energy by being pushed away from the positive voltage of the terminal (tandem principle).

Selecting $^{14}C^{3+}$ ions with the analyzing magnet at the high energy side ensures that no mass-14 molecules survive the stripping process. However, since the magnet only selects ions with a particular magnetic rigidity, i.e. a particular momentum-to-charge ratio, a certain fraction of $^{12}C^{3+}$ and $^{13}C^{3+}$ ions matches this ratio by acquiring the right momentum through higher-order processes (e.g., additional charge exchange processes along the acceleration path). After the analyzing magnet these ions have different velocities and can therefore be removed by the Wien Filter, a velocity filter which selects only ions with the same velocity as $^{14}C^{3+}$. Finally, an energy spectrum of the ions passing through the Wien filter is measured with the rare isotope detector (a silicon surface barrier detector). Typical $^{14}C^{3+}$ counting rates for modern carbon samples are 50 to 100 ions per second.

In order to measure both $^{14}C/^{12}C$ and $^{13}C/^{12}C$ ratios after the accelerator with high precision, a fast sequential isotope injection scheme (Suter et al., 1984) is employed. With the magnetic field of the injector magnet remaining at a fixed value, the energy of the C^- ions is varied in such a way as to obtain equal magnetic rigidities for all three carbon isotopes. This is done by applying a variable voltage to the electrically insulated magnet vacuum chamber. For typical ^{14}C measurements (Priller et al., 1997) the system is set to inject $^{12}C^-$ for 0.5 ms followed by $^{13}C^-$ for 1.5 ms and $^{14}C^-$ for 100 ms. Including waiting periods between the isotopes the total cycle time for the three isotopes is 120 ms. On the high-energy side, the short current pulses of $^{12}C^{3+}$ and $^{13}C^{3+}$ are measured in offset Faraday cups after the analyzing magnet (see Fig. 1). The fast cycling of isotopes through the machine allows one to monitor isotope ratio measurements with high time resolution, thus leading to a higher achievable precision. Once the whole system is properly tuned for measuring ^{14}C, the sample measurements proceed in a fully automated (soft-ware controlled) manner. Each sample in the 40-position sample wheel of the ion source is measured for 3 minutes, and this is repeated five to fifteen times depending on the radiocarbon age (and the corresponding ^{14}C counting rate).

4. ^{14}C Measurements of the Iceman samples

The various specimens collected at the Iceman site were distributed to the three AMS laboratories at

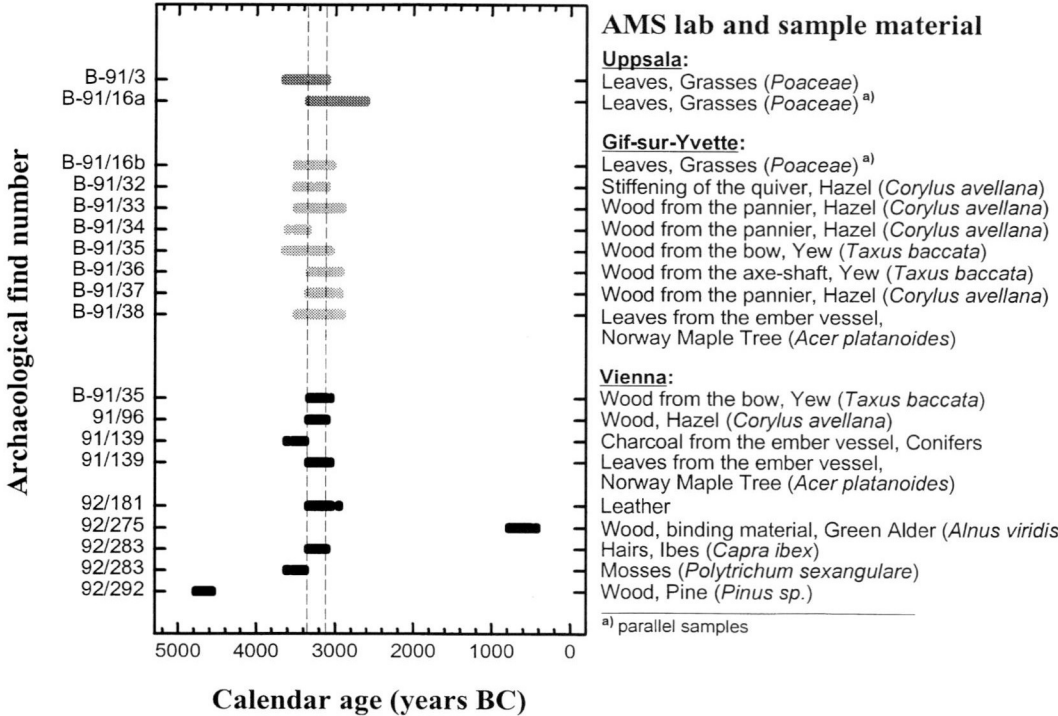

Fig. 2. Summary of samples from the Iceman's equipment. Horizontal bars indicate 95% confidence ranges (2σ, see Table 1). The dashed vertical lines show the 2σ range obtained from tissue and bone samples measured at the AMS laboratories in Zurich and Oxford (Bonani et al., 1992; Bonani et al., 1994)

Uppsala, Gif-sur-Yvette, and Vienna. The measurements at VERA were the most recent ones. They will be described in some detail in Sect. 4.1., whereas the procedures used at Uppsala and Gif-sur-Yvette are described in somewhat more condensed form in Sects. 4.2 and 4.3, respectively. Table 1 summarizes the results of all three laboratories.

4.1. Measurements at Vienna

In most cases the original sample material was first cleaned in an ultrasonic bath to remove adherent particles. The so-called ABA (acid-base-acid, sometimes also referred as AAA acid-alkaline-acid) method was applied as a further pretreatment. An amount of 10 mg pretreated sample material was then transferred to a quartz tube containing 1 g CuO. Some silver wire was added as a binder for sulphur and halogens and the tubes were evacuated and sealed with a glass-blowing torch. For complete combustion of the sample material to CO_2 the sealed samples were heated in a muffle furnace at 900 °C for four hours. The catalytic reduction of CO_2 to elemental carbon was adopted from Vogel et al. (1984) according to the reaction:

$$CO_2 + 2H_2 \xrightarrow{Fe, 580\,°C} C + 2H_2O$$

This "graphitisation" technique is now a standard method used by many AMS ^{14}C laboratories and was also chosen for the VERA lab (Wild et al., 1998). An amount of the catalyst-carbon mixture corresponding to ~1 mg carbon was pressed into the 1-mm holes of the aluminum target holders with a recess of 0.5 mm. From one graphitisation process up to three aliquots of samples can be produced.

When possible, triplicates of each sample material were produced for the AMS measurement. Chemistry blanks were prepared by processing "dead" graphite in the same way as the Iceman samples. For the ^{14}C normalisation, the IAEA standard materials C-6 sucrose, C-5 wood, and C-3 cellulose (Rozanski, 1992) were used. Using several standards provides a high degree of quality control. The loading of the 40-position wheel consisted of nine Iceman samples (6 triplicates, 3 duplicates), three standard samples (triplicate C-6, duplicate C-5, triplicate C-3), duplicate chemistry blanks, and three machine blanks (unprocessed "dead" graphite). As a quality control of the set up we used the ratios of standards. Chemistry blanks gave a ^{14}C age of 49 800 yr B.P., and machine blanks 63 600 yr B.P. From the raw data, mean values were calculated according to the formulae given in Priller et al. (1997). For each individual target of a particular sample material the unweighted mean of several 3-minutes runs (9–16) and its standard deviation were calculated. Then the weighted mean of all targets of the same material was calculated, and the external standard deviation (scatter) and the internal standard deviation. For the final value of each sample material, the larger of the two uncertainties was adopted. The resulting ^{14}C/^{12}C ratio was then corrected for mass fractionation using the measured stable isotope ^{13}C/^{12}C ratio, the measured chemistry blank contribution was subtracted, and finally the ratios were normalized according to standard procedures (Stuiver and Polach, 1977) to calculate the uncalibrated ^{14}C ages. The results are summarized in Table 1.

4.2. Measurements at Uppsala

The two grass samples (Ua-2373 and Ua-2374) measured at the AMS facility of Uppsala (Possnert, 1984) were subjected to the following pretreatment procedure:

1) mechanical cleaning to remove visible particles etc.
2) 6–8 hours application of 1% HCl below boiling
3) washing in distilled water
4) 6–8 hours application of 1% NaOH below boiling
5) washing in distilled water.

The insoluble fraction was combusted to CO_2 with CuO by heating it for 10 minutes to 800 °C. The CO_2 was then graphitized to elemental carbon with H_2 at 750 °C using Fe as catalyst. One pretreatment was done for each sample but two independent combustion-graphitization-AMS measurements were performed. Since the material consisted of grass which was not homogenized, different blades were used in the two analyses from the same sample. The results are summarized in Table 1, where the weighted averages of the ^{14}C ages are also reported.

4.3. Measurements at Gif-sur-Yvette

The ^{14}C measurements of samples from the Iceman's equipment were performed at the Gif-sur-Yvette tandetron AMS facility (Arnold et al., 1987). All samples were divided into two or three subsamples to which two different pretreatment procedures were applied:

i) an ABA method similar to the one applied in Vienna, and

ii) a cellulose extraction method for the wood samples replacing the last acid step in i) by a bleaching in $NaClO_2$ at 80 °C. Each subsample was also combusted and graphitized separately using Fe as a catalyst. The carbon-catalyst mixture yielded two to three targets with 1 mm diameter. The results of the AMS measurements are summarized in Table 1. Whenever three subsamples were treated and analyzed separately, the weighted average of the ^{14}C age is also reported. The confidence of the weighting procedure and the associated error was checked with the so-called χ^2 test.

5. Discussion of the results

In Table 1 all ^{14}C ages were converted into calibrated calendar dates using the OxCal program from Oxford (Bronk Ramsey, 1995), which is available on the Internet. Samples from the same archaeological find (identified by the number in the first column of Table 1) measured at different AMS labs agree well within the respective 2σ uncertainties. Figure 2 shows that most dates of the Iceman's equipment fall into the time range determined from the ^{14}C dating of tissue and bone of the Iceman himself, from 3350 to 3120 B.C. (Bonani et al., 1992; Hedges et al., 1992; Bonani et al., 1994). There are, however, two wood specimens which clearly deviate from this time period. One is pointing to a much earlier time from 4790 to 4550 B.C., i.e., to the transition of the Mesolithic to the Neolithic Period, and the other one to a much later time from 790 to 480 B.C., i.e., within the Hallstatt period.

Most of the ^{14}C data fall well into the general time range when the Iceman lived. The two samples deviating from this period indicate that the "Tisenjoch", a high-altitude mountain pass in the Alps where the Iceman was found, was used also at other times. The radiocarbon date measured on a pine twig belongs to the early Neolithic. This is a new aspect because within the investigation area archaeological finds of this cultural epoch are missing up to now. Probably it was brought to the findspot by humans still living in mesolithic tradition. Several mesolithic dwelling sites at high altitudes are known from the valleys sloping north and south from the finding place (Niederwanger, pers. comm.; Leitner, 1996), indicating that the highland zones were of considerable interest long before the Iceman's lifetime. Also surprising is the date measured on a wooden artifact made of green alder (*Alnus viridis*). This piece of wood shows clear working traces (sample no.: 92/275; Oeggl and Schoch, this volume) and reveals that it was brought to the site by prehistoric men during the Hallstatt Period. This is remarkable because it is the first artefact from the Iron Age in the entire

Table 1. Summary of ^{14}C dating results for samples taken from the Iceman's equipment

Archaeological identification number	Laboratory numbers [a]	Specification	Species	Detailed localization	Dry weight [b] (g)	^{14}C age [c] (yr. B.P.)	$\delta^{13}C$ [d] (‰)	Calibrated age ranges [e] (yr. B.C.)	Fraction of age range [f]
B-91/3	Ua-2373 average	Leaves	Grasses (Poaceae)	Grasses from the filling of his left shoe	0.13	4620 ± 75 4605 ± 70 4612 ± 51	−26.0 −25	3650–3100	1.00
B-91/16a	Ua-2374 average	Leaves	Grasses (Poaceae)	Grasses from the cape	0.07	4250 ± 70 4450 ± 75 4343 ± 100	24.5 −25	3350–2600	1.00
B-91/16b	GifA91413	Leaves, parallel sample to no. B-91/16a	Grasses (Poaceae)	Grasses from the cape	0.09	4550 ± 60	−22	3500–3450 3380–3030	0.44 0.96
B-91/32	GifA93033 GifA93034 GifA94367 average	Stiffening of the quiver	Hazel (Corylus avellana)	Inner parts of the notch	0.04	4690 ± 70 4620 ± 60 4460 ± 80 4605 ± 40	−29 −27 −21 −26	3510–3410 3390–3300 3240–3100	0.34 0.43 0.23
B-91/33	GifA93035 GifA93036 GifA94368 average	Wood from the pannier	Hazel (Corylus avellana)	Hardwood from the broken end of the hazel stem	0.05	4430 ± 60 4680 ± 60 4500 ± 80 4540 ± 70	−30 −30 −19 −26	3500–2900	1.00
B-91/34	GifA93038 GifA93039 GifA94369 average	Wood from the pannier	Hazel (Corylus avellana)	Wood from the broken end of the hazel stem	0.06	4710 ± 70 4670 ± 60 4480 ± 90 4645 ± 40	−32 −32 −27 −30	3620–3590 3520–3340	0.02 0.98
B-91/35	GifA93040 GifA93041 GifA94370 average	Wood from the bow	Yew (Taxus baccata)	Splinter from the broken end	0.13	4540 ± 70 4700 ± 70 4530 ± 80 4595 ± 40	−27 −27 −23 −26	3510–3420 3380–3270 3240–3100	0.22 0.43 0.35
B-91/36	GifA93043 GifA93044 GifA94371 average	Wood from the axe-shaft	Yew (Taxus baccata)	Wood from the inner sides of the split branch	0.04	4440 ± 60 4500 ± 70 4450 ± 70 4460 ± 40	−31 −25 −27 −28	3340–3020 2990–2920	0.87 0.13
B-91/37	GifA93045 GifA93046 GifA94372 average	Wood from the pannier	Hazel (Corylus avellana)	Splinter from the broken end	0.21	4430 ± 70 4540 ± 50 4420 ± 70 4480 ± 40	−32 −28 −31 −30	3350–3030 2970–2930	0.97 0.03

Table 1. (continued)

Archaeological identification number	Laboratory numbers [a]	Specification	Species	Detailed localization	Dry weight [b] (g)	^{14}C age [c] (yr. B.P.)	δ^{13}C [d] (‰)	Calibrated age ranges [e] (yr. B.C.)	Fraction of age range [f]
B-91/38	GifA93047	Leaves from the ember vessel	Norway Maple Tree (*Acer platanoides*)	Found nearby the quiver	0.08	4540 ± 70	−28	3500−2900	1.00
B-91/35	VERA0050	Wood from the bow, parallel sample to no. B-91/35	Yew (*Taxus baccata*)	Recovered during sample washing		4500 ± 30	−24.9 ± 12	3340−3090 3060−3040	0.98 0.02
91/96	VERA0051	Wood	Hazel (*Corylus avellana*)	Recovered during sample washing	0.06	4520 ± 30	−27.2 ± 1.2	3350−3090	1.00
91/139	VERA0053	Charcoal from the ember vessel	Conifers	Recovered during from the meltwater sample washing channel	0.27	4690 ± 40	−23.0 ± 1.9	3630−3580 3540−3360	0.12 0.88
91/139	VERA0049	Leaves from the ember vessel, parallel sample to no. B-91/38	Norway Maple Tree (*Acer platanoides*)	Recovered during sample washing from the meltwater channel	0.06	4510 ± 40	−27.2 ± 1.8	3350−3080 3060−3040	0.97 0.03
92/181	VERA0054	Leather	–	Recovered during sample washing	0.07	4480 ± 40	−22.1 ± 1.4	3350−3030 2970−2930	0.97 0.03
92/275	VERA0048	Wood, binding material	Green Alder (*Alnus viridis*)	Found in the western part of the southern rock-rib	0.04	2500 ± 40	−24.9 ± 1.8	790−480 450−410	0.96 0.04
92/283	VERA0056	Hairs	Ibex (*Capra ibex*)	Sediment from the gully	0.49	4510 ± 30	−22.2 ± 1.7	3350−3090	1.00
92/283	VERA0055	Mosses	*Polytrichum sexangulare*	Sediment from the gully	0.08	4700 ± 40	−23.0 ± 1.6	3630−3570 3540−3360	0.17 0.83
92/292	VERA0052	Wood	Pine (*Pinus sp.*)	Found on the southern rock-rib	0.02	5820 ± 40	−21.0 ± 1.5	4790−4550	1.00

a) The initial letters of the laboratory numbers identify the AMS facilities of Uppsala (Ua), Gif-sur-Yvette (GifA), and Vienna (VERA).
b) The total dry weight of the distributed sample material is given. For individual samples 3 to 10 mg of the respective material were used.
c) The ^{14}C age is calculated from the measured ^{14}C/^{12}C ratio with a standardized prescription (Stuiver and Polach, 1977). It is assumed that the ^{14}C/^{12}C ratio in the atmosphere has a fixed value was the same at all times. The age is calculated from the radioactive decay law using the so-called Libby half-life of 5568 yr, and is reported in years before present (B.P., i.e. before 1950). Errors are given as one standard deviation (1σ, 68% confidence).
d) The δ^{13}C values of −25 for the second subsample of Ua-2373 and Ua-2374 are estimated, not measured. For all δ^{13}C values where no errors are given, the errors are estimated to be 3‰.
e) Calibration by OxCal v2.18. The 95% confidence ranges [2σ] are given. B.C. = Before Christ.
f) Relative probability of finding the true age in the respective time range. The absolute probability is obtained by multiplying with 0.95 (2σ range).

Ötztal region. Whereas the Iron Age is well documented with archaeological finds in the Vinschgau (Leitner, 1986), artefacts from the Ötztal were still lacking up to now. ^{14}C dating of more material recovered from the site would be desirable to reveal a clearer picture of the history of this unique location in prehistoric times.

Abstract

We present the results of ^{14}C dating measurements of various samples from the equipment of the Iceman, performed with Accelerator Mass Spectrometry (AMS) at the AMS facilities of Uppsala, Gif-sur-Yvette and Vienna. All dates except two overlaps with the time period of 3350–3120 B.C., determined previously from tissue and bone material of the Iceman himself at the AMS facilities of Zurich and Oxford. The two exceptional dates give time periods of 4790–4550 B.C. and 790–480 B.C., respectively. These two data measured on wooden artefacts indicate that the finding place of the Iceman at the "Tisenjoch" was visited by prehistoric men at considerably earlier and even later times than the Iceman lived.

Zusammenfassung

Die Ergebnisse der C$_{14}$-Messungen verschiedener Proben von den Ausrüstungsgegenständen des Mannes im Eis, die mit der Beschleuniger-Massen-Spekrometrie (AMS) an den AMS-Institutionen von Uppsala, Gif-sur-Yvette und Wien durchgeführt wurden, werden vorgestellt. Alle Daten, bis auf zwei, überlappen mit der Zeitspanne von 3350–3120 v. Chr., die vorher schon an Gewebe und Knochenmaterial vom Körper des Mannes im Eis an den AMS-Möglichkeiten in Zürich und Oxford bestimmt wurden. Die zwei abweichenden Daten liefern eine Zeitspanne von 4790–4550 v. Chr. und 790–480 v. Chr. Sie zeigen an, daß der Fundort wahrscheinlich schon lange vor und auch noch lange nach der Zeitperiode des Mannes im Eis aufgesucht wurde.

Résumé

Le présent article reproduit les résultats des mesures au carbone 14 exécutées sur différents échantillons provenant de l'équipement de l'Homme des glaces. La méthode employée a été la spectrométrie de masse par accélérateur pratiquée dans les instituts AMS d'Uppsala, de Gif-sur-Yvette et de Vienne. Toutes les dates, à l'exception de deux, recoupent la période comprise entre 3 350 et 3 210 av. J.-C., période déjà établie auparavant pour les échantillons de tissus et d'os prélevés sur le corps de l'Homme des glaces et analysés par les installations AMS de Zurich et d'Oxford. Les deux dates divergentes signalent une période comprise entre 4 790–4 550 av. J.-C. et 790–480 av. J.-C. Elles indiquent que le site en question avait sans doute été fréquenté bien avant et encore bien après le temps de vie de l'Homme des glaces.

Riassunto

Vengono presentati i risultati delle misurazioni C14, eseguite su alcuni campioni di oggetti dall'equipaggiamento, tramite AMS presso gli Istituti di AMS di Uppsala, Gif-sur-Yvette e Vienna. Tutti i dati, ad eccezione di due, coincidono con la datazione 3350–3120 a.C., già determinata sulla base di campioni di tessuto ed ossa della mummia presso gli istituti di Spettrometria AM di Zurigo e di Oxford. Le due datazioni divergenti forniscono i periodi 4790–4550 a.C. e 790–480 a.C. Essi indicano che il sito di ritrovamento veniva frequentato probabilmente già molto prima ed anche molto dopo la vita della mummia.

References

Anderson E. C., Libby W. F., Weinhouse S., Reid A. F., Kirschenbaum A. D. and Grosse A. V. (1947) Natural Radiocarbon from Cosmic Radiation. Phys. Rev. 72: 931–936.

Arnold J. R. and Libby W. F. (1949) Age determinations by Radiocarbon Content: Checks with Samples of Known Age. Science 110: 678–680.

Arnold M., Bard E., Maurice P. and Duplessy J. C. (1987) ^{14}C Dating with the Gif-sur-Yvette Tandetron Accelerator: Status Report. Nucl. Instr. and Meth. B 29: 120–123.

Bagolini B., Dal Ri L., Lippert A. und Nothdurfter H. (1996) Der Mann im Eis: Die Fundbergung 1992 am Tisenjoch, Gem. Schnals, Südtirol. In: Spindler K., Rastbichler-Zissernigg E., Wilfling H., zur Nedden D. und Nothdurfter H. (eds.): Der Mann im Eis. Neue Funde und Ergebnisse. The Man in the Ice, vol. 2, Springer Verlag, Wien: 3–23.

Bonani G., Ivy S., Niklaus Th. R., Suter M., Housley R. A., Bronk C. R., van Klinken G. J. and Hedges R. E. M. (1992) Altersbestimmung von Milligrammproben der Ötztaler Gletscherleiche mit der Beschleuniger-Massenspektrometrie- Methode (AMS). In: Höpfel F., Platzer W. und Spindler K. (eds.): Der Mann im Eis, Bd. 1. Veröffentlichungen der Universität Innsbruck 187: 108–116.

Bonani G., Ivy S., Hyjdas I., Niklaus Th. R. and Suter M. (1994) AMS ^{14}C Age Determination of Tissue, Bone and Grass Samples from the Ötztal Iceman. Radiocarbon 36/2: 247–250.

Bronk Ramsey C. (1995) Radiocarbon calibration and Analysis of Stratigraphy: The OxCal Program. Radiocarbon 37(2): 425–430; available also on internet: *http://units.ox.ac.uk/departments/rlaha/oxcal/oxcal_h.html.*

Graedel Th. E. und Crutzen P. (1994) Chemie der Atmosphäre: Bedeutung für Klima und Umwelt. Spektrum Akademischer Verlag, Heidelberg: 511 pp.

Hedges R. E. M., Housley R. A., Bronk C. R. and van Klinken G. J. (1992) Radiocarbon Dates from the Oxford AMS System: Archaeometry Datelist 15. Archaeometry 34(2): 337–357.

Kitagawa H. and van der Plicht J. (1998) Atmospheric Radiocarbon Calibration to 45 000 yr B.P.: Late Glacial Fluctuations and Cosmogenic Isotope Production. Science 279: 1187–1190.

Kutschera W. (1997) Conference Summary: Trends in AMS. Nucl. Instr. and Meth. B 123: 594–598.

Kutschera W., Collon P., Friedmann H., Golser R., Hille P., Priller A., Rom W., Steier P., Tagesen S., Wallner A., Wild E. and Winkler G. (1997) VERA: A New AMS Facility in Vienna. Nucl. Instr. and Meth. B 123: 47–50.

Leitner W. (1995) Der "Hohle Stein" - eine steinzeitliche Jägerstation im hinteren Ötztal. (Archäologische Sondagen 1992/93). In: Spindler K., Wilfling H., Rastbichler-Zissernig E., zur Nedden D. und Nothdurfter, H. (eds.):

Der Mann im Eis: Neue Funde und Ergebnisse. The Man in the Ice, vol. 2, Springer Verlag, Wien: 209–213.

Levin I., Kromer B., Schoch-Fischer H., Bruns M., Münnich M., Berdau D., Vogel J. C. and Münnich K. O. (1985) 25 Years of Tropospheric ^{14}C Observations in Central Europe. Radiocarbon 27(1): 1–19.

Levin I. (1994) The Recent State of Carbon Cycling Through the Atmosphere. In: Zahn R., Pedersen T. F., Kaminski M. A. and Labeyrie L. (eds.): NATO ASI Series, Springer Verlag, Berlin, vol. I, 17: 3–13.

Levin I. (1998) Institut für Umweltphysik, Universität Heidelberg (private communication).

Levin I. and Kromer B. (1997) Twenty Years of Atmospheric $^{14}CO_2$ Observations at Schauinsland Station, Germany. Radiocarbon 39(2): 205–218.

Libby W. F. (1946) Atmospheric Helium Three and Radiocarbon from Cosmic Radiation. Phys. Rev. 69: 671.

Lippert A. (1992) Die erste archäologische Nachuntersuchung am Tisenjoch. In: Höpfel F., Platzer W. und Spindler K. (eds.): Der Mann im Eis, Bd. 1. Veröffentlichungen der Universität Innsbruck 187: 234–244.

Middleton R. (1984) A Versatile High Intensity Negative Ion Source. Nucl. Instrum. and Meth. 214: 139–150.

Middleton R. (1989) A Negative Ion Cookbook, University of Pennsylvania, PA, USA. (unpublished).

Possnert G. (1984) AMS with the Uppsala EN Tandem Accelerator. Nucl. Instr. and Meth. B 5: 159–161.

Priller A., Golser R., Hille P., Kutschera W., Rom W., Steier P., Wallner A. and Wild E. (1997) First Performance Tests of VERA. Nucl. Instr. and Meth. B 123: 193–198.

Prinoth-Fornwagner R. and Niklaus Th. R. (1994) The Man in the Ice: Results from Radiocarbon Dating. Nucl. Instr. and Meth. B 92: 282–290.

Rom W., Golser R., Kutschera W., Priller A., Steier P. and Wild E. (1998) Systematic Investigations of ^{14}C Measurements at VERA. Radiocarbon 40(1): 255–263.

Rom W., Golser R., Kutschera W., Priller A., Steier P. and Wild E. M. (1999) AMS ^{14}C Dating of Equipment from the Iceman and of Spruce Logs from the Prehistoric Salt Mines of Hallstatt. Radiocarbon 41(2): 183–197.

Rozanski K., Stichler W., Gonfiantini R., Scott E. M., Beukens R. P., Kromer B. and van der Plicht J. (1992) The IAEA ^{14}C Intercomparison Exercise 1990. Radiocarbon 34(3): 506–519.

Stuiver M. and Polach H. (1977) Discussion Reporting of ^{14}C Data. Radiocarbon 19(3): 355–363.

Stuiver M., Long A. and Kra R. S. (eds.) (1993) Calibration 1993. Radiocarbon 35(1): 1–244.

Stuiver M., Reimer P. J., Bard E., Beck J. W., Burr G. S., Hughen K. A., Kromer B., McCormac G., van der Plicht J. and Spurk M. (1998) INTCAL 98 Radiocarbon Age Calibration, 24 000–0 cal BP. Radiocarbon 40(3): 1041–1083.

Suter M., Balzer R., Bonani G. and Wölfli W. (1984) A fast Beam Pulsing System for Isotope Ratio Measurements, Nucl. Instr. and Meth. B 5: 242–246.

Taylor R. E., Stuiver M. and Reimer P. J. (1996) Development and Extension of the Radiocarbon Time Scale: Archaeological Applications. Quart. Sci. Rev. 15: 655–668.

Vogel J. S., Southon J. R., Nelson D. E. and Brown T. A. (1984) Performance of Catalytically Condensed Carbon for Use in Accelerator Mass Spectrometry, Nucl. Instr. and Meth. B 5: 289–293.

Vries H. de (1958) Variations in Concentration of Radiocarbon with Time and Location on Earth. Proceedings, Nederlandsche Akademie van Wetenschappen. Series B 61: 96–102.

Wild E., Golser R., Hille P., Kutschera W., Priller A., Puchegger S., Rom W. and Steier P. (1998) First ^{14}C Results from Archaeological and Forensic Studies at the Vienna Environmental Research Accelerator. Radiocarbon 40(1): 273–281.

Zissernig E. (1992) Der Mann vom Hauslabjoch: Von der Entdeckung bis zur Bergung. In: Höpfel F., Platzer W. und Spindler, K. (eds.): Der Mann im Eis, Bd. 1. Veröffentlichungen der Universität Innsbruck 187: 445–254.

The Iceman's environment

S. Bortenschlager

Institut für Botanik, Universität Innsbruck

The dating of the Iceman to 3300–3350 calendar years BC by Bonani et al. (1992) places this find in the closing stages of the Atlantic chronozone. The Atlantic chronozone encompasses the period 8000–5000 ^{14}C years BP (Mangerud et al., 1974) which includes the start of the Neolithic period, i.e., the beginning of agriculture in the Alps, indeed in the entire Central European region.

The start of the Neolithic period, however, also marks a decisive phase in the development of the vegetation in the post-glacial. From this time onwards man began to have an intensive influence on the natural vegetation cover, which is clearly seen in pollen analyses.

This influence was a result, on the one hand, of forest clearance for agriculture and, on the other hand, of an increase in the number of livestock and the corresponding need for supplies of fodder plants. The sources of fodder for the animals – sheep, goats and even already cattle – were firstly the dried foliage ("Laubheu") from branches cut from deciduous trees, especially elm (*Ulmus*) and ash (*Fraxinus*) growing in the valleys. This served above all as winter fodder. Secondly, the alpine pastures above the forest limit were utilised for summer grazing. Because of the dense forests in the valleys, pastures at those lower altitudes scarcely existed then.

1. Development of the vegetation after the end of the last Glacial period

Recolonisation of the formerly ice-covered areas began with the retreat and disintegration of the glaciers at the end of the last ice age. Derived from a cold-steppe vegetation that was present in the foothills ("Alpenvorland") of the Alps, the alpine valleys very quickly became colonised by a herbaceous plant cover, exemplified for the Tyrol by the pollen profile (Figs. 2, 3) from the Lanser See (Bortenschlager S., 1984). The dominants were species of mugwort (*Artemisia*), goosefoot (*Chenopodium*), pinks (Caryophyllaceae), joint pine (*Ephedra*), the daisy family (Asteraceae) and other indicator species of an open vegetation. From about 14.000 ^{14}C years, commencing on the edaphically more favourable places, dwarf-shrubs and shrubs began to spread, of which the main species were juniper (*Juniperus*), dwarf birch (*Betula nana*) and sea-buckthorn (*Hippophaë rhamnoides*). This shrub vegetation initiated, in practice, the virtual re-afforestation of the entire Alps. At about 13.000 ^{14}C years, the start of the Bölling chronozone, the onset of a marked climatic amelioration saw the colonisation of tree birches (*Betula*), Scots pine (*Pinus sylvestris*), dwarf mountain pine (*Pinus mugo*), Arolla pine (*Pinus cembra*) and larch (*Larix*). During the Bölling chronozone in the Central Alps, these species grew at an altitude approaching 1900 m and in the Alleröd chronozone approaching 2100 m. They all suffered setbacks during the Older and Younger Dryas chronozones. At the end of the Late-glacial, at about 10.000

^{14}C a BP	Age	Chronozone	Archaeologic Classification	Cal a AD/BC
			Modern Times	1.500
		Subatlantic	Middle Age	500
			Roman Times	0
2.500			Iron Age	800
	Post-glacial Flandrian Holocene	Subboreal	Bronze Age	2.400
5.000			Neolithic	3.900
		Atlantic		5.800
			Mesolithic	
8.000				7.050
9.000		Boreal		
		Preboreal		7.850
10.000				9.150
11.000		Younger Dryas		
11.800	Late-glacial Weichselian Pleistocene	Alleröd		
12.000		Older Dryas		
13.000		Bölling	Paleolithic	

Fig. 1. Summary of chronozones and archaeologic periods

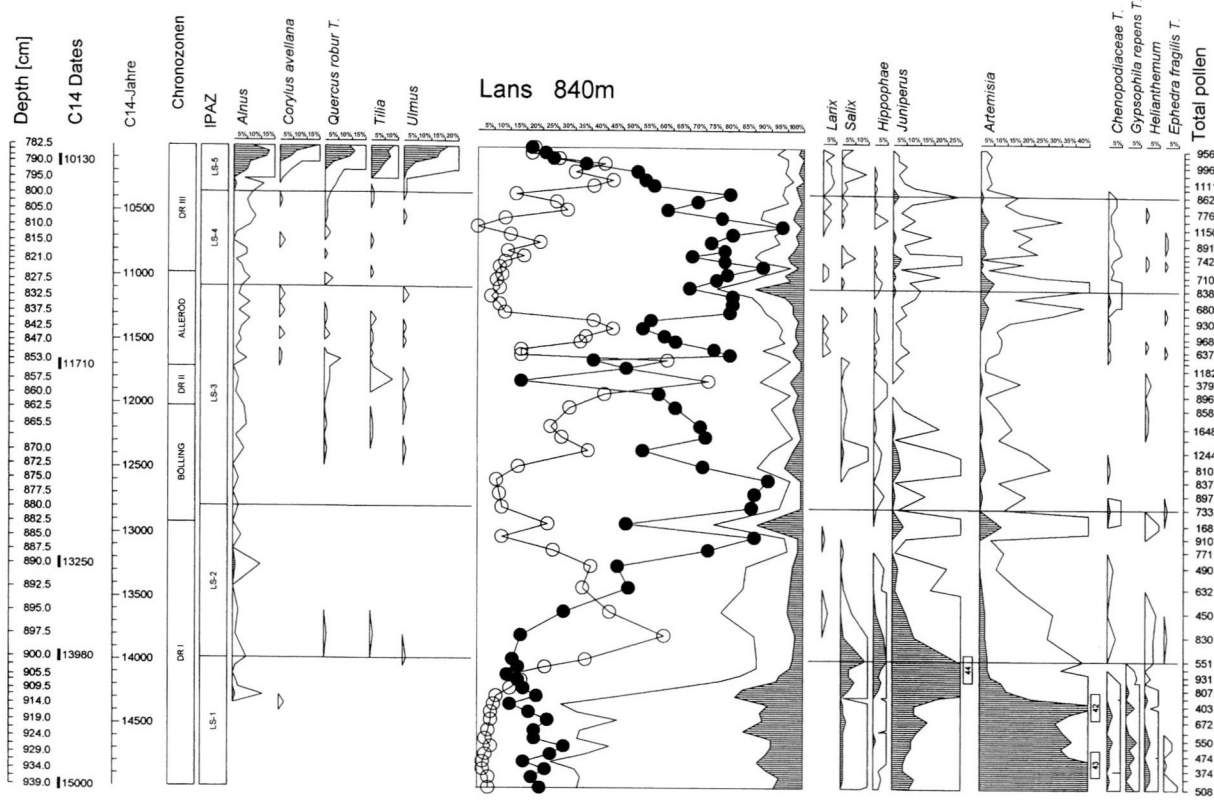

Fig. 2. Simplified relative pollendiagram from the "Lanser See" showing the initial phase of reforestation in the Tyrol

^{14}C years, and shortly afterwards, the final improvement in the climate took place, with temperatures reaching present-day values. During the initial stage of the post-glacial, the Preboreal chronozone, the above-mentioned species formed the forest limit at roughly its present-day altitude. Throughout the remainder of the post-glacial the climatic parameters, and thereby the forest limit, fluctuated to only a minor degree around the present-day mean values.

The subsequent constituents of the forest cover immigrated into the Alps during the Post-glacial, at rates dependent on the distance to be covered from their glacial refugia, their dispersal ability and also in relation to certain competitive factors.

In the North and South Tyrol, the Iceman's habitat, the mixed deciduous woodland (EMW), represented by elm (*Ulmus*), lime (*Tilia*), oak (*Quercus*), ash (*Fraxinus*) and also hazel (*Corylus*), arrived from the south. These tree species had already reached the vicinity of Bozen and Brixen in the final stages of the late-glacial (Seiwald, 1980; Wahlmüller, 1990; Kompatscher, 1997), and at the start of the post-glacial had crossed the Brenner pass (see Figs. 2, 3) and occurred en masse in the vicinity of Innsbruck already in the Preboreal chronozone (Bortenschlager S., 1984). From eastward, from the Laibach basin, spruce (*Picea*) immigrated at intermediate altitudes, on the one hand along the southern margin of the Alps, via the Drau valley and the Pustertal, into the South Tyrol, and, on the other hand, around the eastern end of the Eastern Alps, to reach the North Tyrol via the Alpine foothills during the second half of the Preboreal chronozone (Lang, 1994). In the period of time from the Boreal to the first half of the Atlantic, which corresponds to roughly the major part of the Mesolithic period (Fig. 1), stable vegetation conditions were able to develop. The altitudinal zonation of the forests became established: mixed deciduous forest in the valleys, mixed deciduous and coniferous forest at intermediate altitudes, and subalpine coniferous forest up to the forest limit. The mixed deciduous forest was composed of oak (*Quercus*), elm (*Ulmus*), lime (*Tilia*), ash (*Fraxinus*), maple (*Acer*) and hornbeam (*Carpinus*). Manna ash (*Fraxinus ornus*) and hop hornbeam (*Ostrya carpinifolia*) only spread into the South Tyrol at a later time (Kompatscher, 1997). In accordance with the habitat conditions, the riverine woodlands

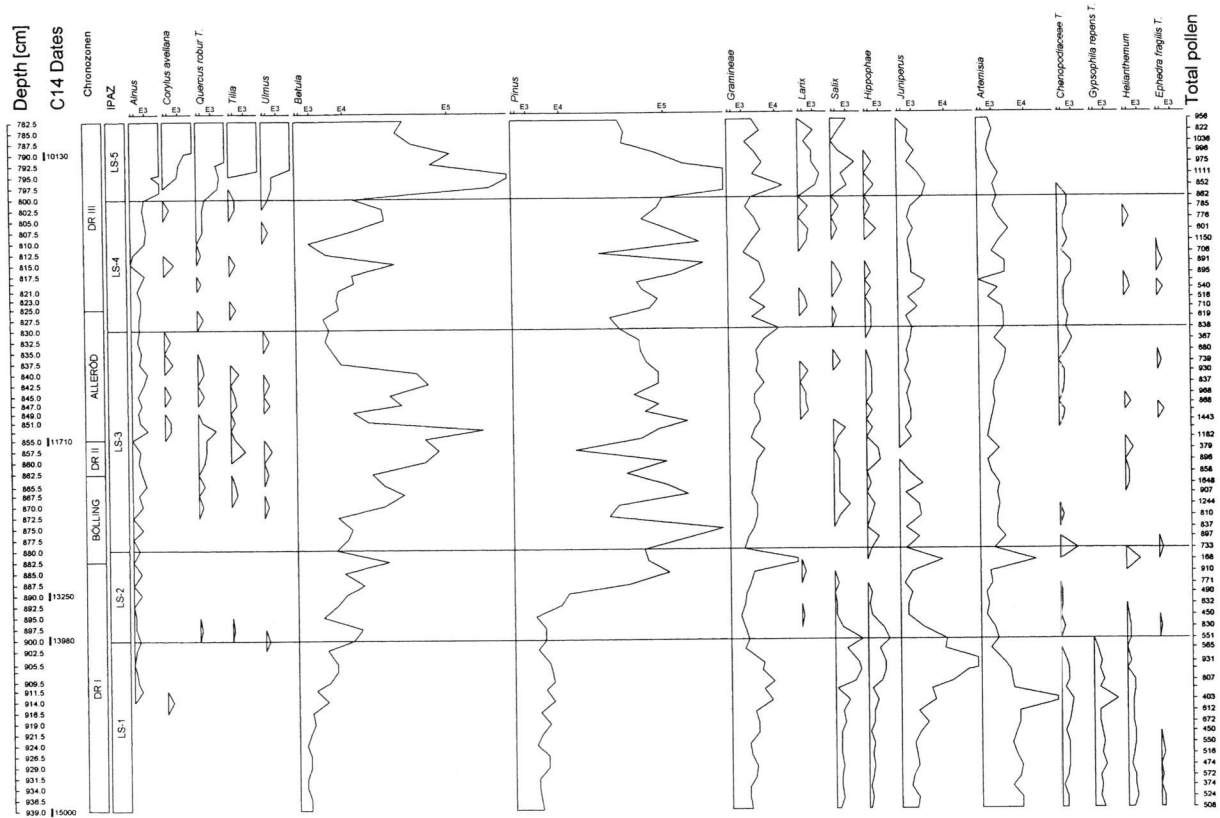

Fig. 3. Simplified pollenconcentration diagram (grains/cm² year) from the "Lanser See" showing the initial phase of reforestation in the Tyrol

comprised willow (*Salix*) and alder (*Alnus*). The mixed forest at intermediate altitudes was composed of spruce (*Picea*), maple (*Acer*), elm (*Ulmus*), alder (*Alnus*), hazel (*Corylus*) and birch (*Betula*). The subalpine coniferous forest had roughly the same composition as it has today; Arolla pine (*Pinus cembra*), dwarf mountain pine (*Pinus mugo*), Scots pine (*Pinus sylvestris*) and larch (*Larix*) made up the upper part of this zone and formed the forest limit. Spruce (*Picea*) was similarly strongly represented in the lower part of the subalpine coniferous forest.

The Mesolithic people lived among this natural and stable vegetation. They sought out open, and sparsely wooded sites, since they were food gatherers and predominantly hunters, which were at that time mainly to be found around and above the tree limit. Thanks to the investigations made by Bagolini et al. (1984, 1992) and Broglio (1994), a large number of such Mesolithic hunters' campsites are known from the South Tyrol. An equivalent site in the East Tyrol was located by Stadler (1991) and investigated by Oeggl and Wahlmüller (1994, 1997). More recently, also in the North Tyrol, such hunting camps at high altitudes have been located. The site "Hohler Stein" has been investigated by Leitner (1995) and a Mesolithic site at Ullafelsen in the Fotschertal has been even more intensively investigated by Schäfer (1997), and also further sites in the northern limestone Alps, in the Karwendel. The influence of the Mesolithic people on the vegetation was relatively slight and was restricted probably to occasional fires lit to open up the plant cover. This was clearly demonstrated by Oeggl and Wahlmüller (1997), but an unequivocal distinction between fires from natural causes and deliberately caused ones, however, cannot be drawn. Consequently a direct human influence on the forest is still uncertain during this period.

A restructuring of the forests by a fresh wave of immigrant species occurs during the second half of the Atlantic chronozone. Fir (*Abies*) arrives from the southwest and beech (*Fagus*) from the southeast; both are shade-casting species which can out-compete spruce once dense stands have been formed. Mixed forest of fir, beech and spruce began to be formed at intermediate altitudes. This changing composition of the forest in the Tyrolean region occured for the most part also at the beginning of the Neolithic period. Whether or not a direct relationship between the start of agriculture and this restructuring of the forest existed cannot yet be unequivocally established. In some cases, a relationship between the spread of the beech and utilisation of its fruits – beechmast – has been postulated.

2. The influence of the Neolithic people on the vegetation in the valleys

The start of the Neolithic period is by definition synchronous with the start of agriculture. This start, in the Alps, is placed more or less in the final stages of the Atlantic chronozone, in the period 6000–5000 ^{14}C years BP. The first finds of pollen grains of cereals (Cerealia) have been made during this period of time in a large number of pollen profiles from sites in the Tyrol. The much-discussed decline in the pollen values for certain constituents of the mixed deciduous forest – especially the elm (*Ulmus*), the ash (*Fraxinus*) and the lime (*Tilia*) – are very clearly visible in the profiles from Lindenmoos (Bortenschlager S., 1984) and Giering (Bortenschlager I., 1976, Bortenschlager I. and S., 1981) and are an indication of human influence on the natural vegetation. The combination, declines in the pollen of the representatives of the mixed deciduous forest, increases in those of the light-demanding species birch (*Betula*) and hazel (*Corylus*), in conjunction with an increased representation of arboreal-tree pollen (NAP) and the first finds of cereal pollen, had already been characterized by Iversen (1956) as the "Stone Age Landnam Phase". The Neolithic farmers were obliged to clear areas in the valleys on which to grow their cereal crops. These clearances, however, were so small-scale that there was still no lasting effect on the valley vegetation. In addition, the regular alternation of the cultivated patches, induced by soil exhaustion, led to there being no really long-lasting changes in the vegetation. On the other hand, the utilisation of the trees for leaf-fodder had a more marked effect. This leaf-hay was necessary for the overwintering of their animals and was the only possibility of maintaining a reduced number of beasts. There were virtually no meadows in the valleys at that time and only the wetlands in the valley bottoms were available for pastures and possibly for hay making.

Lopping of the trees for leaf fodder produced more marked changes in the pollen diagrams than the actual effect it had on the tree populations. The declines in the pollen of elm (*Ulmus*), ash (*Fraxinus*) and lime (*Tilia*) would seem to indicate a complete disappearance of these species, as exemplified by the pollen diagram for Giering (Bortenschlager I., 1976), for the regions to the north of the main alpine chain, and by that for Tammerle Moos (Wahlmüller, 1990) for those to the south of it. When

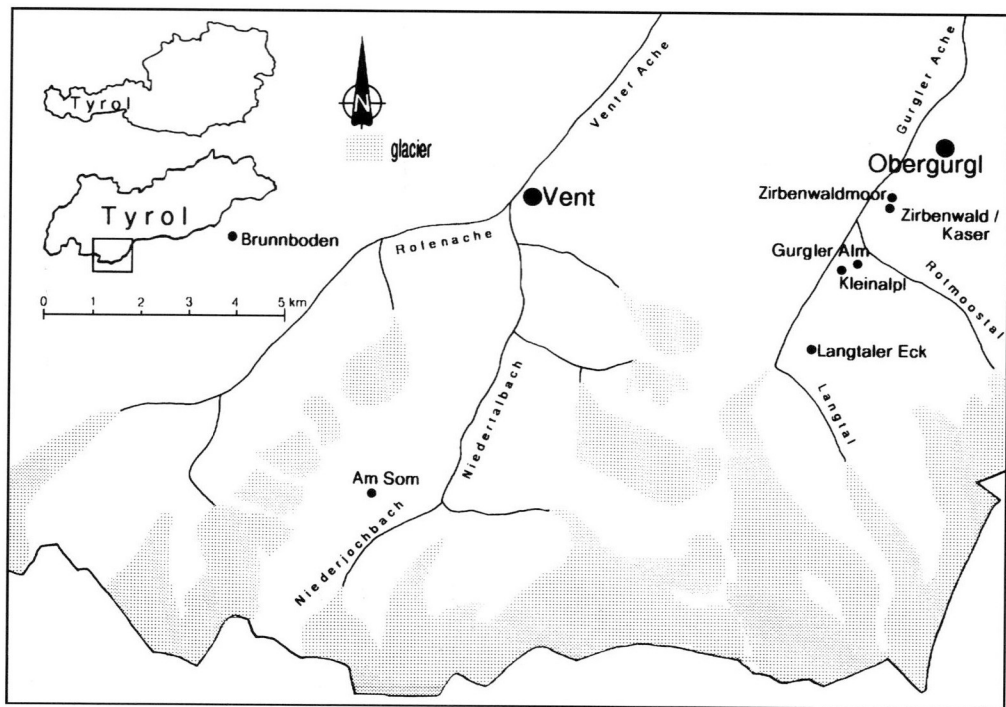

Fig. 4. Location of the sites investigated in the Vent/Obergurgl area

this utilisation ceased, the values of the pollen curves relatively soon returned to their previous levels, showing that a return to the natural forest cover had occurred. A more intensive and longer lasting effect on the vegetation first arose following the introduction of mining, whether for copper or for salt, in the later stages of the Neolithic period, as well as with the intensification of agriculture in the Bronze Age. The problem that faced the Neolithic population was that of land clearance. "Slash and burn" was the method adopted to clear patches for cereal growing. S. Jørgensen and J. Troels-Smith, with their experimental archaeology trials, provided a most successful example of such methods (Iversen, 1973). This method of clearance, however, was far too time-consuming for creating pasturage and instead the naturally forest-free areas above the tree limit were utilized.

3. The influence of the Neolithic people on the montane vegetation

The herb-rich alpine meadows and grasslands were, and still are, ideal pastures for farm stock. Even today thousands of sheep are driven annually from the valleys on the south side of the alpine chain, across the glaciers, to the upland pastures above the forest limit on the northern side. In autumn they return by the same routes to the southern valleys. Nowadays these sheep can be kept during the winter on the hay made during the summer, but in the Neolithic period hay making in the uplands was still unknown. First, during the Bronze Age there is indirect evidence from irrigation systems, that hay making was practiced (Patzelt, 1996, 1997).

The natural species composition of the upland pastures naturally began to change under the influence of such annual grazing. These changes can be clearly seen in the pollen diagrams. Both the abundance and the dominance of the plant species are affected. Those species that can better withstand the grazing pressure, or have developed better resistance mechanisms to cope with such pressure, are successful. These are species that have a growth-form of rosettes, or adpressed leaves, such as the plantains (*Plantago*), arnica (*Arnica*) or alpine lovage (*Ligusticum*), or those that possess unpalatable substances, in particular the gentians (*Gentiana*). Likewise, some species increase in frequency due to the favourable effect, for them, of the increased nitrate supply from the animal droppings, i.e., on the resting places, characterised by docks (*Rumex*), goosefoots (*Chenopodium*) and nettles (*Urtica*).

This phenomenon was first identified for the inner Ötztal in the pollen diagram from the Rofenberg (Bortenschlager S. et al., 1992; Bortenschlager S., 1993), which had already been analyzed in 1973. This profile, following the discovery of the Iceman, became extremely important, since the two sites are mutually visible in the field. In the pollen diagram, the clearly marked increase in the pasture indicator species, when the values were combined into a single pollen curve, was pointed out at that time.

The following pollen types have been grouped together as pasture indicators in all the pollen diagrams under discussion:

Apiaceae	*Ligusticum*	*Rhinanthus*
Chenopodiaceae	*Lotus*	Rosaceae
Fabaceae	*Plantago*	*Rumex*
Filipendula	Ranunculaceae	*Urtica*
Gentianaceae		

The Asteraceae pollen type was omitted, since this includes both pasture indicator species and others that belong to the natural vegetation and therefore they cannot be separated and shown selectively in the pollen diagrams.

To amplify the available evidence in support of the above-mentioned phenomenon, further pollen diagrams from sites in the inner Ötztal are presented here, five of which are now published in full for the first time.

Profile "Brunnboden" 2640 m. above s.l.

The Brunnboden Moor is a spring-fed bog that slopes very gently towards the southwest. It is situated in a wholly open landscape and the plant cover is for the most part composed of sedges (Cyperaceae) – *Carex fusca*, *Trichophorum* – with a few herbs and mosses. This same composition of the vegetation extends down throughout the entire profile depth of 192,5 cm, on to the basal gravel.

The pollen diagram reflects the vegetational development typical for these altitudes. Non-tree pollen predominates from the start, of which the main contribution is sedge pollen (Cyperaceae). At about 170 cm depth the tree pollen proportion, pine (*Pinus*) and spruce (*Picea*), markedly increases; but also the value for grasses (Gramineae). This initial marked change in the pollen spectra probably occurred at about 7490–7290 calender years BC and is a reflection of a climatic improvement – the end of the Venediger fluctuation (Patzelt and Bortenschlager, 1973). From this time onwards, the tree-pollen values vary between 5 and 15%, indicative for this locality of a complete absence of forest cover. Some small fragments of wood recovered from the peat column proved to be of dwarf-shrubs (Ericaceae), juniper (*Juniperus*) and willow (*Salix*).

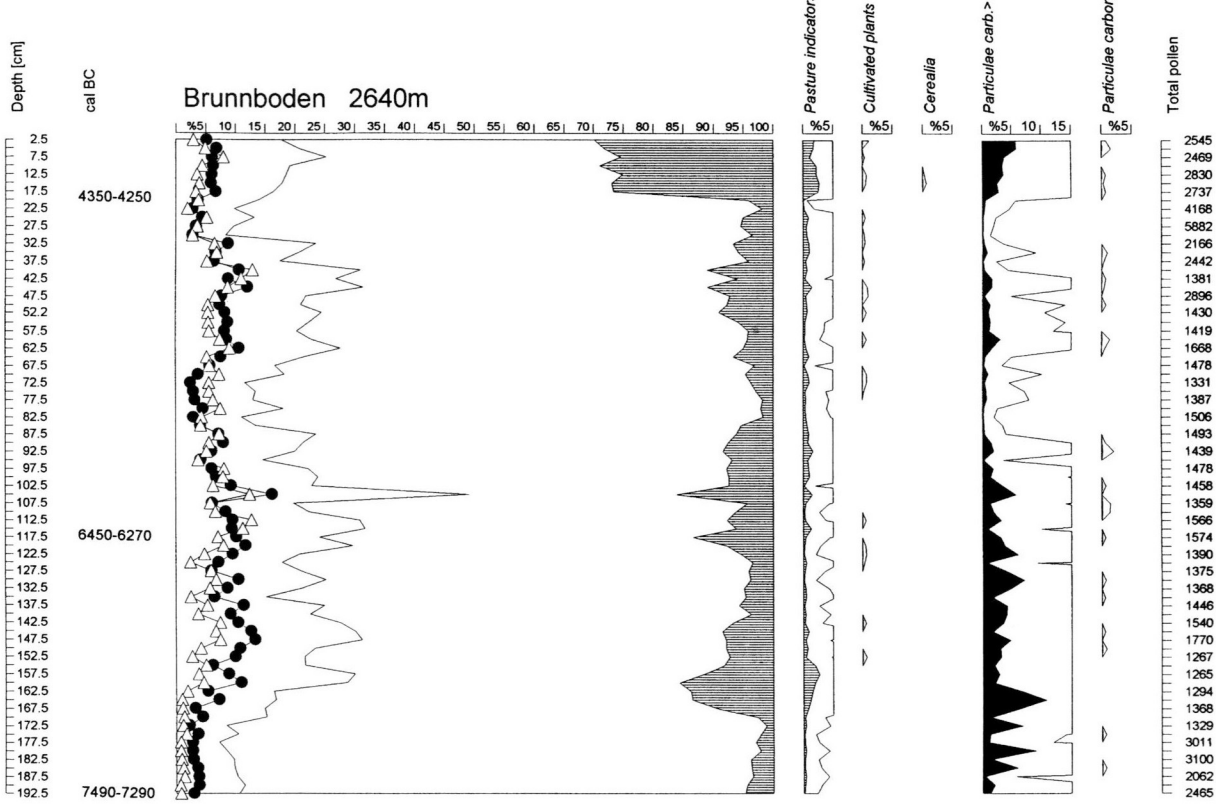

Fig. 5. Simplified relative pollendiagram "Brunnboden"; for details see the complete relative and pollenconcentration diagram in the Appendix

The next decisive and clear change in the pollen spectrum occurs at about 20 cm depth. Grass (Gramineae) pollen suddenly increases to 25% and those non-tree pollen comprising the pasture indicator species' curve show a similar two- to three-fold increase. This change was dated to 4350–4250 calender years BC. Whereas in the lower part of the profile, from 192,5 to 20 cm depth, the average rate of peat accumulation was of the order of 5 cm per 100 years, a value similar to those established for other moors at this altitude (Bortenschlager S., 1984), from 20 cm depth to the present-day surface there appears to have been practically no peat accumulation at all. The explanation would seem to be the 5500 years of continuous grazing of the bog surface. The normal annual growth of the bog vegetation was simply removed by the grazing sheep and thus unavailable for peat formation.

Profile "Am Soom" 2620 m. above s.l.

This profile occupies a shallow shelf on a south-east facing hillside (Fig. 4). The bog surface has been affected by solifluction many times in the past. The heaviest deposition of soliflucted material occurs at a depth of 145–120 cm, causing an obvious hiatus in the peat profile, though of only a very short duration in time. The vegetation cover of the locality at the present-day is an alpine grassland, in which sedges (Cyperaceae) are predominant. Due to the situation and position of the moor, the source of the regional pollen component has been the Vent valley and this is appreciably more obvious than in the Brunnboden peat profile. On the other hand, the local conditions have not followed the normal pattern, due to interruptions by soil slippage and the deposition of water-borne material. The organic sections of the peat profile are composed of sedge peat, which is disrupted at a depth of 145–120 cm by a flooding horizon of coarse sediment that contains no pollen. Similar sandy and gravelly material, flooding horizons, form interruptions in the peat profile at 75–65 cm and at the base at 195–180 cm depth. The peat profile is capped on the surface by a deposit of sand and humus 20 cm thick.

The peat accumulation rate of the organic parts of the profile, on a basis of ^{14}C-datings, lies in the region of 2–3 cm per 100 years, a value comparable with those found for other moors at similar altitudes.

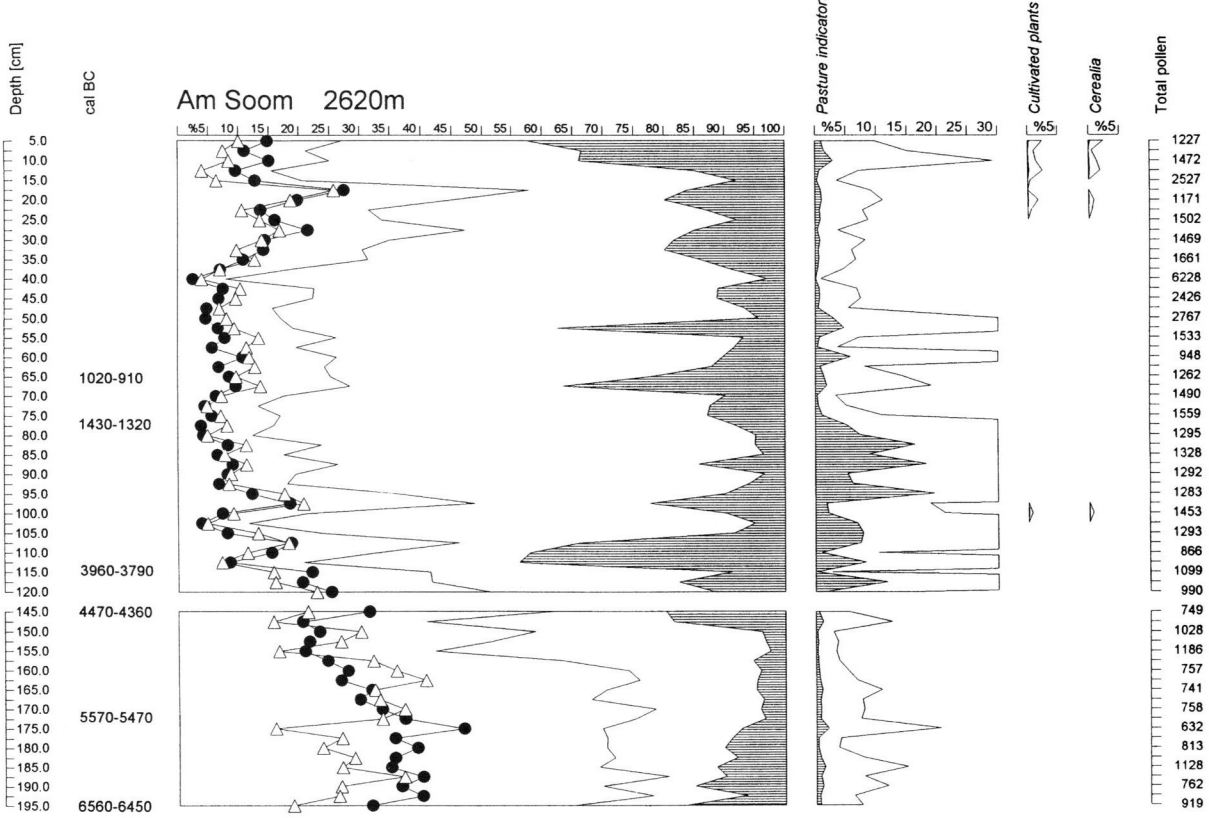

Fig. 6. Simplified relative pollendiagram from the "Am Soom"; for details see the complete relative and pollenconcentration diagram in the Appendix

The vegetational development seen in the pollen diagram from this profile is basically the same as that for Brunnboden, although the local developments are more strongly influenced by the regional component, because of the site position and the local wind conditions that determine the pollen source. On these accounts, the curve for the ratio arboreal/non-arboreal pollen fluctuates more strongly here and the proportion of tree pollen is greater. Nevertheless, on the whole, the high proportion of non-tree pollen supports the view that this area was also unforested. Pine (*Pinus*) and spruce (*Picea*) exhibit pollen values of more or less similar magnitude, although the dominance of one or the other changes at times. As in the Brunnboden-profile, the dominance of pine (Pinus) pollen changes to that of spruce (*Picea*) in the older section, about 4470–4360 calender years BC. There is a reflection here of the influence of a climatic deterioration on the position of the forest limit, which was formed here by Arolla pine (*Pinus cembra*). Spruce (*Picea*), which was growing at a lower altitude, was less strongly affected by the change.

Immediately following the hiatus induced by the soil slippage – the cause of which is unknown – the pasture indicator species' pollen values increase strongly also in this profile and this is similarly interpreted as a influence of animal pasturage. The continuous dominance of spruce (*Picea*) from this time onwards could be interpreted as caused by the influence of the Neolithic settlers on the forest limit. Forest clearance by means of fire could be held responsible. At the appreciably lower-lying site, at that time below the forest limit, of the Gurgler Alm (Vorren et al., 1993), the pollen profile shows an increase in the density of particulate charcoal from this point in time onwards.

Profile "Langtalereck" 2420 m. above s.l.

The Langtalereck-profile was excavated from a rock basin in the valley-side trough below the orographic right-hand-side moraine of the Gurgler glacier (Fig. 4). It was situated beyond the direct influence of the glacier throughout the post-glacial maximum of the Gurgler glacier. Only at sometime after Christ era the moor was buried by minerogenous material and peat accumulation brought to halt. At the present time, the surroundings are occupied by alpine grassland and

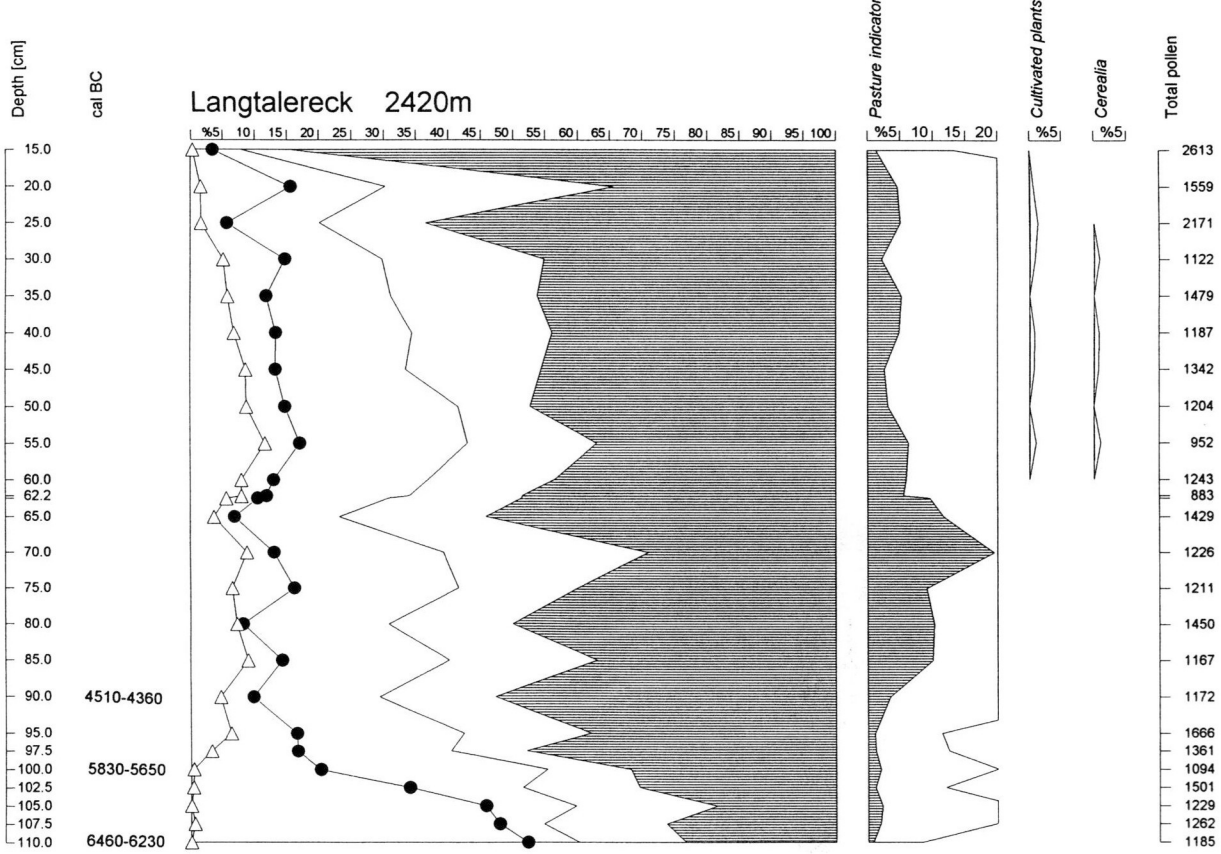

Fig. 7. Simplified relative pollendiagram from the "Langtalereck"; for details see the complete relative and pollenconcentration diagram in the Appendix

snowbed vegetation; in the wettest parts sedges (Cyperaceae) are predominant, with damp habitats occupied by *Carex fusca* and *Trichophorum*, though grasses (Gramineae) – *Nardus* and *Deschampsia* – are also present. The snow bed communities are very nicely characterised in the pollen diagram by the snowbell (*Soldanella*) and in part also by the Asteraceae-type pollen. The entire peat profile of 110 cm is composed of sedge peat (Cyperaceae), to which the above-mentioned sedges contribute the major part. The top of the peat column is covered by a 15 cm thick layer of recent morainic material.

The basal peat dating of 6460–6230 calender years BC also in this case documents the start of peat formation in a climatically favourable period following the Venediger fluctuation. The picture provided by the pollen spectra, in contrast, differs from those hitherto discussed. Pine (*Pinus*) values are high at the start and remain dominant throughout the time of peat formation. This is correlated with the proximity of the site to the forest limit, which lies at about 2300–2350 m altitude hereabouts. At this site, the local pollen sources depress the values for the regional pollen derived from the sub-alpine forest. The Arolla pine (*Pinus cembra*), the Scots pine (*Pinus sylvestris*) and the dwarf mountain pine (*Pinus mugo*) are the predominant tree pollen representing the forest element. The composition of the non-tree pollen spectra differs in a similar way. Whereas in the foregoing two pollen profiles the main constituent was sedge (Cyperaceae) pollen, at this site the grasses (Gramineae) come to the fore as the local element. The cause is the grassland cover in the immediate vicinity of the site. The small size and relative unattractiveness of the site as pasturage have allowed a largely uninterrupted peat growth to take place.

Profile "Gurgler Alm" 2240 m. above s.l.

The pollen diagram published by Vorren et al. (1993) has been re-drawn so that it is directly comparable with those discussed so far. The site (Fig. 4) lies below the potential forest limit, although at the present time there is no forest cover. The peat column is wholly composed of sedge (Cyperceae) peat, the growth of which com-

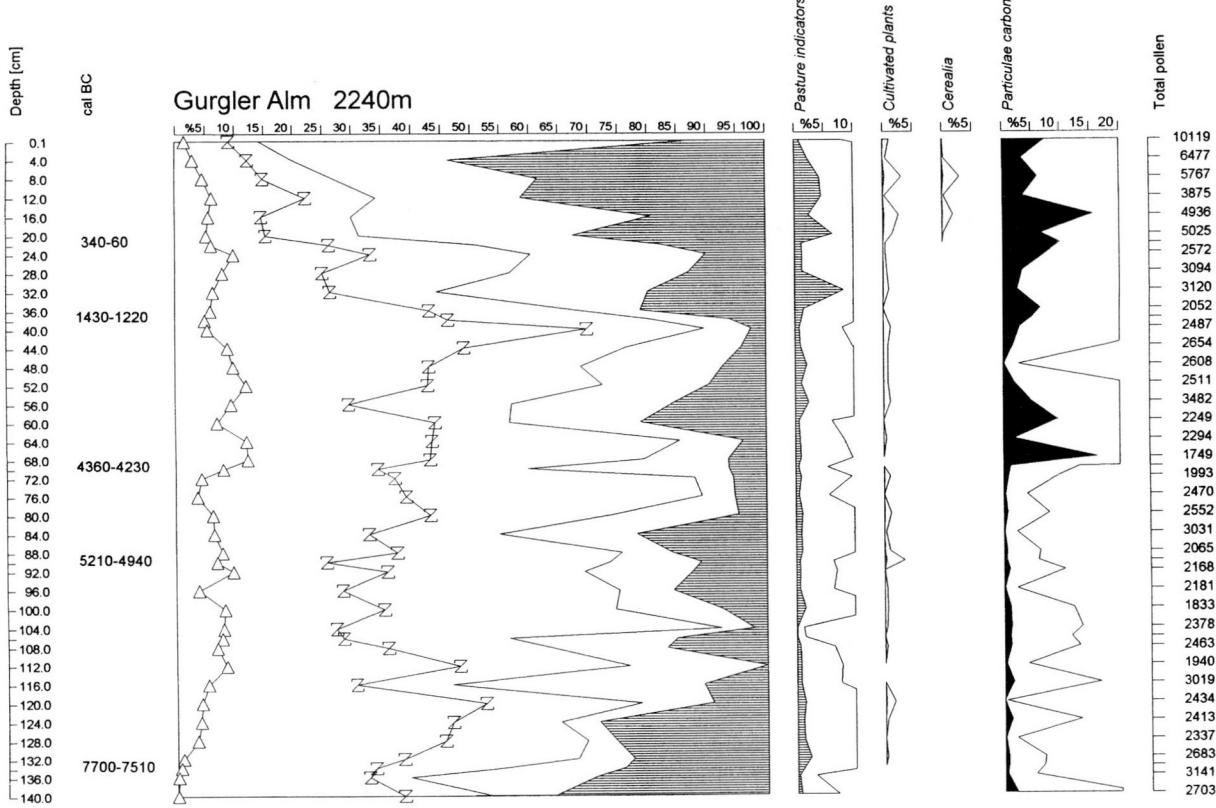

Fig. 8. Simplified relative pollendiagram from the "Gurgler Alm" (after K. D. Vorren et al., 1993)

menced just before 7700–7510 calender years BC. The pollen spectra are in conformity with the site situation, within the potential forest limit when peat growth started, at the end of the Venediger fluctuation, the non-tree pollen values lie at over 40%, indicating that the area was not forest-covered at that time. At about 7490–7290 calender years BC, the tree pollen proportion rises to over 70%, i.e., a forest cover had become established. Arolla pine (*Pinus cembra*) is the predominant tree, but spruce (*Picea*), with values of up to 10%, ocurred sporadically. Periods of increased non-tree pollen values indicate climatic oscillations, especially glacier maxima, as has been documented by Patzelt (1996).

A special feature of this pollen diagram is the curve for charcoal particles. The first marked increase occurs at a depth of 68 cm, dated to 4360–4230 calender years BC, and followed by a clear increase in the pollen of pasture indicator species. These two features taken together are interpreted as evidence that the Neolithic population attempted to increase the pasturage area at the expense of the forest limit, by fire clearance; perhaps as a reaction to the subsequent climatic deterioration, which is documented by the high non-tree pollen values. This climatic deterioration diminished the available pasturage at higher altitudes. The final forest clearance at this site, however, first occurred at the turn of the last century.

Profile "Kleinalpl" 2205 m. above s.l.

This peat profile shows marked alternations between sedge (Cyperaceae) peat and fluvioglacial solifluced sediments, clay and silt or even sand, throughout the 80 cm depth. The site (Fig. 4) lay beyond, but still within reach of, deposited minerogenous material from the orographic right-hand marginal moraine of the Gurgler glacier, on the margin of the alm pastures of the Gurgler Alm. The minerogenous intercalations are directly related to former maxima of the Gurgler glacier and could be assigned fixed dates in the profile. The pollen spectra of the peat layers indicate that in the vicinity of the site there was a forest cover, at least to the turn of the century, although the vegetation at the site itself was predominantly of alpine grassland. The

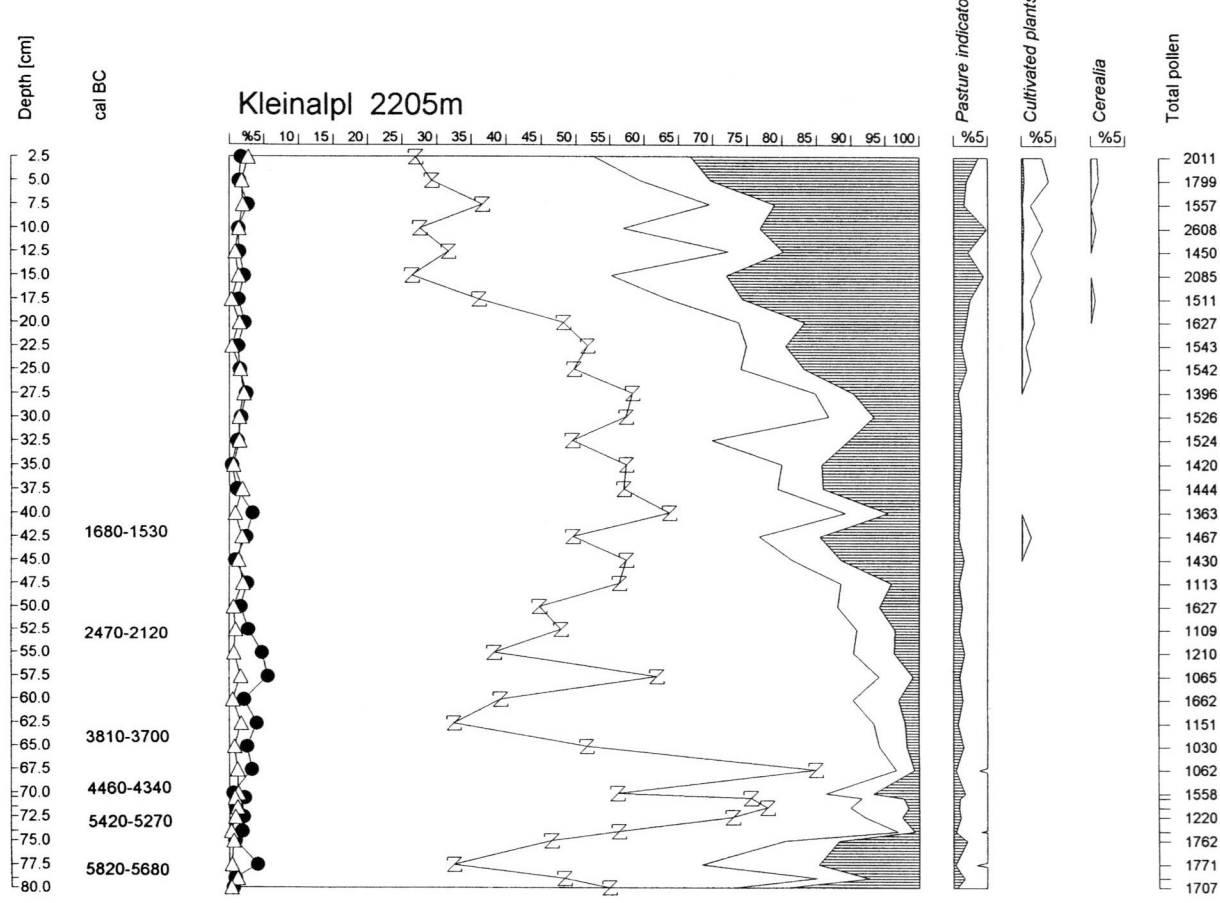

Fig. 9. Simplified relative pollendiagram from the "Kleinalpl"; for details see the complete relative and pollenconcentration diagram in the Appendix

forest cover consisted of Arolla pine (*Pinus cembra*); spruce (*Picea*) and Scots pine (*Pinus sylvestris*) were not present, their pollen was derived from distant transport of the regional component from lower altitudes.

The increase in the pasture indicator species' pollen is relatively minor here, at 67 cm depth. An opening-up of the forest cover, in parallel with the increase in the pasture indicator species' pollen, is supported by the fall in the pollen of the Arolla pine and a contemporaneous and massive increase in alder (*Alnus*) pollen. This phenomenon, however, is superimposed in the younger part of the profile by a glacial maximum, revealed in the stratigraphy by a layer of sand and clay. An exact relationship between these facts is difficult to determine. The increase in the pollen of the pasture indicator species, in combination with the decline in Arolla pine pollen and the increase in the shrub alder (*Alnus viridis*) at about 3810–3700 calender years BC is, nevertheless, unequivocally attributable to the start of the pastoral economy.

Profile "Zirbenwald auf der Kaser" 2085 m. above s.l.

This profile was excavated in a sedge moor (Cyperaceae – *Trichophorum* and *Carex fusca*) located in a shallow depression just above the Gurgler pine forest (*Pinus cembra*) and lying well within the potential forest zone (Fig. 4). It is underlain by late-glacial ground moraine. Charcoal was present at the moraine/peat transition and another marked charcoal horizon was encountered at a depth of 28–25 cm. At the present time, the moor lies in an unforested area of dwarf-shrub heath, dominated by alpenrose (Rhododendron) and bilberry (*Vaccinium* sp.). The nearest stand of trees, those of the Gurgler pine forest (*Pinus cembra*) are about 150 m away. At present practically only Arolla pine (*Pinus cembra*) and scattered dwarf mountain pines (*Pinus mugo*) are present. From the lowest dated peat layer, 4230–4000 calender years BC at 40 cm depth, upwards, spruce (*Picea*) plays virtually no role in the profile.

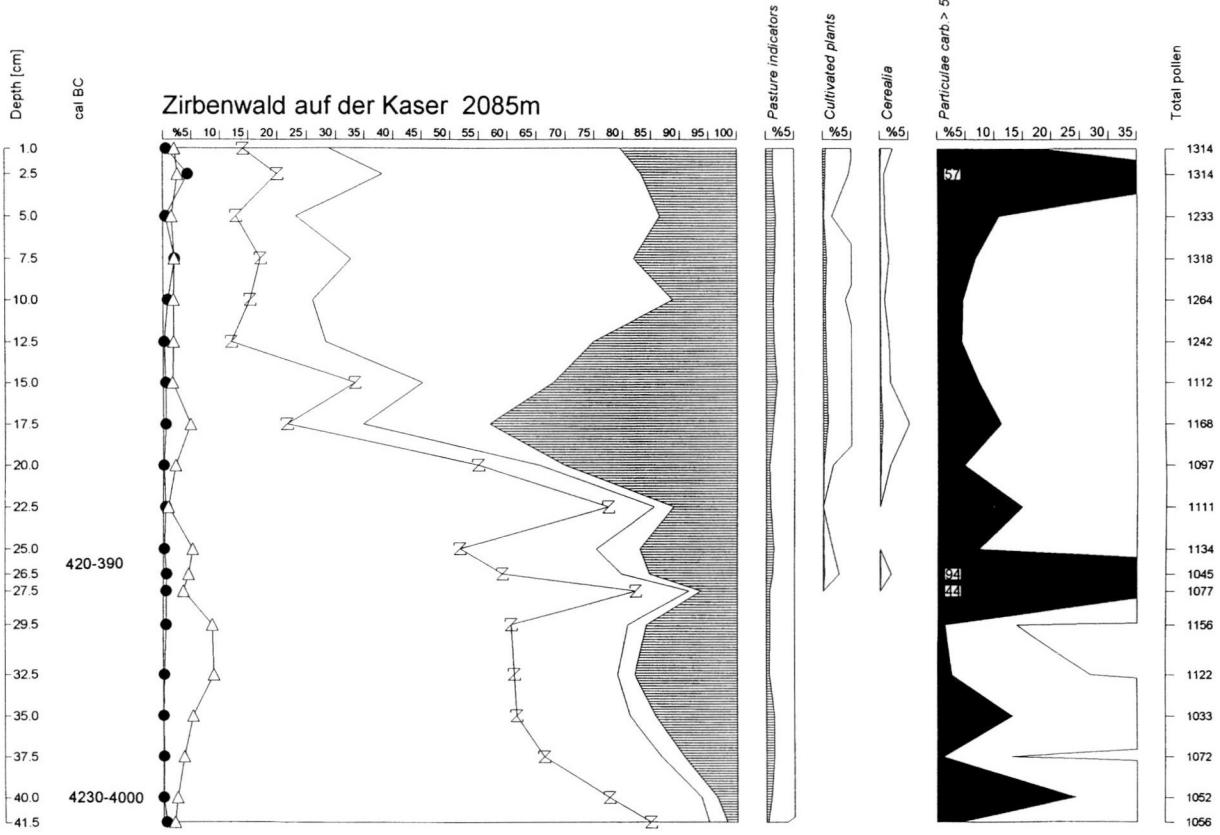

Fig. 10. Simplified relative pollendiagram from the "Zirbenwald auf der Kaser"; for details see the complete relative and pollen-concentration diagram in the Appendix

Rybnicek et al. (1977), found in the nearly Zirbenwald moor that there was a clear decrease in Arolla pine (*Pinus cembra*) pollen, coincident with a birch (*Betula*) pollen peak and accompanied, or followed by higher values of spruce (*Picea*) and alder (*Alnus*) pollen and dated to 4790–4775 calender years BC. These major changes in the tree pollen spectra were followed by markedly higher values for the pollen of grasses (Gramineae), as well as of those of certain of the pasture indicator species, viz gentian (*Gentiana*), daisy family (Asteraceae) and cow-wheat (*Melampyrum*). The occurrence of charcoal in this profile was not investigated. Taken as a whole, the above facts reflect the initial forest clearance episode here.

The basal date for the profile "Auf der Kaser" places the start of peat growth in the closing stages of the Atlantic chronozone. Relatively high values of Arolla pine (*Pinus cembra*) pollen are accompanied by a minor peak of charcoal particles and a low peak for pasture indicator pollen, both of which should be synchronous with the findings from the Zirbenwald moor. The start of the pastoral economy in the Neolithic period is also here in evidence, although not to the same degree as in the naturally unforested region above the present forest limit. Only at the turn of the century, with the occurrence of a second charcoal horizon, is there evidence for a more intensive utilisation of this area. The charcoal particle curve indicates an intensive forest clearance, by fire, in two stages, in the course of which the forest cover of this region was reduced to small islands of trees. In parallel with this development, the pollen representation of pasture indicator species markedly increases and the occurrence of cereal (Cerealia) pollen – distant transport from lower altitudes – provides evidence for a more intensive utilisation of nature by man.

4. Conclusions

The six pollen diagrams presented in this paper document the development of the vegetation from at least the younger Atlantic chronozone up to the present day. They together represent a transect at about 550 m extent, from the Arolla pine forests of the subalpine forest zone, though the forest limit, to an altitude in the alpine grasslands which was never forest-covered throughout the post-glacial period. Despite these differences, in regard to vegetation cover, position and site altitude, a marked change in the non-tree pollen spectra in the last stages of the Atlantic chronozone could be detected in all the profiles. This change consisted of both relative and absolute increases in the pollen of those plant species that are grouped together as pasture indicators. In parallel with this change, in all the profiles that were derived from sites situated within the subalpine forests and for which charcoal particle counts were carried out, i.e., the profiles for the Gurgler Alm and the Zirbenwald, a marked increase in charcoal particles was noted. Even in the Brunnboden-profile, the highest-lying of all the sites presented here, a similar increase was noted. Whereas in the two first-mentioned profiles the charcoal particles were relatively large (up to more than $100\,\mu$ in diameter), those in the Brunnboden-profile were mostly smaller-sized. The charcoal particles found in the Zirbenwald-and Gurgler Alm-profiles maybe considered as being autochthonous, whereas those in the Brunnboden-profile will have been derived by distant transport from lower levels, below the forest limit. Taken together, these two findings are considered as evidence that the alpine grassland has been utilised intensively as summer pasturage since the beginning of the Neolithic period. The climatic fluctuation, termed the Rotmoos fluctuation, which dates from the last stages of the Atlantic chronozone, though it restricted the opportunities for grazing, perhaps at the same time led to a greater grazing intensity, which was one reason why additional pasturage had to be created. This was achieved by the use of fire. The forest limit was lowered anthropogenically.

A further intensification of such usage thereafter occurred in the Bronze Age, with the construction of irrigation channels for the meadows, which provide evidence for the start of hay making, and this enabling permanent settlements to be established (Patzelt, 1996, 1997).

Finally, it can be stated that the alpine grasslands above the forest limit have been utilised as pasturage for more than 5000 years. This usage has become more and more intensive, in a series of stages, up to the present day. Forest clearance in all these areas began more or less contemporaneously, whereby the forest limit was lowered following such clearance by burning. Permanent occupation at these altitudes, however, is likely only at the end of the Bronze Age. What is certain, however, is that use and settlement of the inner Ötztal occurred from higher altitudes downwards and from the southern side of the main alpine chain northwards.

Abstract

On a basis of six pollen diagrams, evidence is presented that the alpine grasslands of the inner Ötztal have been utilised as pastures since the start of the Neolithic period. This usage as pasture led to a restructuring of the plant communities, whereby all species considered as pasture indicators increased in abundance. In parallel, and almost simultaneously, an increased intensity of forest clearance by fire is evidenced by an increase in charcoal particles in the pollen diagrams. This occasioned a depression of the forest limit for the purpose of increasing the available pasturage. Utilisation of the alpine grassland took place from the southern side of the main alpine chain by the Neolithic population living in those valleys. The pollen diagrams also show that such utilisation took place in a series of waves and increased steadily up to recent times. The Iceman provides a proof for the start of this usage of the alpine grasslands in the Neolithic period.

Zusammenfassung

Durch die Datierung des Eismannes durch Bonani et al. (1992) auf 3300 bis 3350 Kalenderjahre v. Chr. wurde der Fund in das ausgehende Atlantikum gestellt. In das Atlantikum, das den Zeitraum 8000 BP – 5000 BP (Mangerud et al., 1974) umfaßt, fällt der Beginn des Ackerbaus im Alpenraum, ja im gesamten mitteleuropäischen Raum.

Dieser Beginn des Neolithikums markiert aber auch einen entscheidenden Punkt in der Vegetationsentwicklung des Postglazials. Ab diesem Zeitpunkt nimmt der Mensch intensiver, und durch die Pollenanalyse nachweisbar, Einfluß auf die ursprüngliche Vegetation. Diese Einflußnahme erfolgt einerseits durch Rodungen für den Ackerbau und andererseits durch die Intensivierung der Viehhaltung und der dafür notwendigen Inanspruchnahme von Futterpflanzen. Als Futterquelle für das Vieh – Schafe, Ziegen und auch schon Rinder – diente einerseits Laubheu, das durch Schneiteln von Laubbäumen, vor allem Ulme (Ulmus) und Esche (Fraxinus), in den Tallagen gewonnen wurde. Es diente vor allem als Winterfutter. Andererseits wurden die inneralpinen Rasen über der Waldgrenze als Sommerweiden genutzt. Weideflächen in den Tallagen waren bei der damals noch dichten Bewaldung kaum vorhanden.

Résumé

La datation de l'Homme des glaces par Bonani et coll. (1992), le faisant remonter à 3 300–3 350 av. J.-C., le place ainsi dans l'Atlantique final. C'est à l'époque de l'Atlantique, comprise entre 8 000 et 5 000 B.P. (Mangerud et coll., 1974), que nous situons les débuts de l'agriculture dans les Alpes, voire dans toute la région d'Europe centrale.

Ce début du néolithique marque également un tournant décisif de l'évolution du monde végétal dans l'ère postglaciaire. C'est le moment à partir duquel l'Homme commence à exercer une influence plus intensive – mise en évidence par l'analyse pollinique – sur la végétation primitive. Cette intervention se réalise d'une part sous forme de défrichements effectués en vue de la mise en culture des terres, d'autre part au moyen d'une intensification de l'élevage avec le recours aux plantes fourragères qu'elle nécessite. La nourriture du bétail – ovins, caprins et, déjà, bovins – était constitué en partie de feuillages séchés, obtenus par la taille en têtard d'arbres feuillus, dont notamment l'orme (*Ulmus*) et le frêne (*Fraxinus*). Ces feuilles récoltées dans les vallées servaient surtout de fourrage d'hiver. De plus, les pelouses centralpines situées au-dessus de la limite forestière étaient exploitées comme pâturages d'été. En effet, rares étaient à cette époque-là les pâturages dans les vallées, encore largement couvertes par des boisements.

Riassunto

Con la datazione 3300–3350 a. C., effettuata da Bonani et. al. nel 1992, il periodo suggerito era il tardo Atlantico. In tale era dell'Atlantico che comprende, secondo Mangerud et al. (1974) i millenni tra l'8000 ed il 5000 BP l'uomo introdusse l'agricoltura non solo nell'area alpina, ma anche in tutta l'Europa centrale.

Tale inizio del Neolitico segna tuttavia anche una svolta nell'evoluzione vegetale postglaciale. Da allora l'uomo interviene più intensamente sulla vegetazione originale. Ciò è dimostrabile tramite l'analisi dei pollini. L'intervento dell'uomo avvenne per mezzo del dissodamento da un lato, della zootecnia dall'altro. Le sue attività erano l'agricoltura con la produzione foraggera. Il foraggio, prevalentemente invernale, per gli allevamenti ovini, caprini e bovini era costituito da fogliame tagliato da olmi e frassini di fondovalle. In estate venivano utilizzati prati oltre il limite arboreo dato che il fondovalle era ancora fortemente boscoso e quindi pressoché privo di pascoli.

References

Bagolini B. et al. (1984) Pian dei Laghetti - S. Martino di Castrozza. Preistoria Alpina 20: 39–52.

Bagolini B. und A. Pedrotti (1992) Vorgeschichtliche Höhenfunde in Trentino-Südtirol und im Dolomitenraum vom Spätpaläolithikum bis zu den Anfängen der Metallurgie. In: Höpfel F., Platzer W. und Spindler K. (eds.): Der Mann im Eis, Bd. 1. Veröffentlichungen der Universität Innsbruck 187: 359–377.

Bonani G. et al. (1992) Altersbestimmung von Milligrammproben der Ötztaler Gletscherleiche mit der Beschleunigermassenspektrometrie-Methode (AMS). In: Höpfel F., Platzer W. und Spindler K. (eds.): Der Mann im Eis, Bd. 1. Veröffentlichungen der Universität Innsbruck 187: 108–116.

Bortenschlager I. (1976) Beiträge zur Vegetationsgeschichte Tirols. II. Kufstein - Kitzbühel - Paß Thurn. Ber. nat.-med. Ver. Innsbruck 63: 105–137.

Bortenschlager I. und Bortenschlager S. (1981) Pollenanalytischer Nachweis früher menschlicher Tätigkeit in Tirol. Veröffentlichungen des Museum Ferdinandeum, Innsbruck 61: 5–12.

Bortenschlager S. (1984) Die Vegetationsentwicklung im Spätglazial: das Moor beim Lanser See III, ein Typprofil für die Ostalpen. Diss. Botanicae 72: 71–79.

Bortenschlager S. (1984) Beiträge zur Vegetationsgeschichte Tirols I. Inneres Ötztal und unteres Inntal. Ber. nat.-med. Ver. Innsbruck 71: 19–56.

Bortenschlager S. (1993) Das höchst gelegene Moor der Ostalpen "Moor am Rofenberg" 2760 m. Diss. Botanicae 196: 329–334.

Bortenschlager S., Kofler W., Oeggl K. und Schoch W. (1992): Erste Ergebnisse der Auswertung der vegetabilischen Reste vom Hauslabjoch. In: Höpfel F., Platzer W. und Spindler K. (eds.): Der Mann im Eis, Bd. 1. Veröffentlichungen der Universität Innsbruck 187: 307–312.

Broglio A. (1994) Mountain sites in the context of the North-East Italian Upper Palaeolithic and Mesolithic. Preistoria Alpina 28(1): 293–310.

Iversen J. (1956) Forest clearance in the Stone Age. Scientific American 194: 36–41.

Iversen J. (1973) The development of Denmark's nature since the Last Glacial. Geology of Denmark III: pp. 126.

Kompatscher K. (1997) Pollenanalytische Untersuchungen zur spät- und postglazialen Vegetationsentwicklung des Überetsch (Südtirol/Italien). Diplomarbeit Universität Innsbruck, pp. 98.

Lang G. (1994) Quartäre Vegetationsgeschichte Europas - Methoden und Ergebnisse. Fischer, Jena, pp. 462.

Leitner W. (1995) Der "Hohle Stein" - eine steinzeitliche Jägerstation im hinteren Ötztal, Tirol (Archäologische Sondagen 1992/93). In: Spindler K. et al. (eds.): Der Mann im Eis - Neue Funde und Ergebnisse. The Man in the Ice, vol. 2. Springer-Verlag, Wien: 209–213.

Mangerud J. et al. (1974) Quaternary stratigraphy of Norden, a proposal for terminology and classification. Boreas, Oslo 3: 109–128.

Oeggl K. and Wahlmüller N. (1994) Vegetation and climate history of a high alpine mesolithic camp site in the Eastern Alps. Preistoria Alpina 28(1): 71–82.

Oeggl K. und Wahlmüller N. (1997) Die Waldgrenze in den Zentralalpen während des Mesolithikums. In: Alpine Vorzeit in Tirol - Begleitheft zur Ausstellung: 29–44.

Patzelt G. (1996) Modellstudie Ötztal - Landschaftsgeschichte im Hochgebirgsraum. Mitt. Österr. Geogr. Ges. Wien 138: 53–70.

Patzelt G. (1997) Die Ötztalstudie - Entwicklung der Landnutzung. In: Alpine Vorzeit in Tirol - Begleitheft zur Ausstellung: 46–62.

Patzelt G. und Bortenschlager S. (1973) Die postglazialen Gletscher- und Klimaschwankungen in der Venedigergruppe (Hohe Tauern, Ostalpen). Z. Geomorph. N. F. 16: 25–72.

Patzelt G., Bortenschlager S. und Poscher G. (1996) Exkursion A1 Tirol: Ötztal - Inntal. Exkursionsführer DEUQUA. pp. 23.

Rybnicek K. und Rybnickova, E. (1977) Mooruntersuchungen im oberen Gurgltal, Ötztaler Alpen. Folia Geobot. Phytotax. 12: 245–291.

Schäfer D. (1997) Mittelsteinzeitliche Fundplätze in Tirol. In: Oeggl K., Patzelt G., Schäfer D.: Alpine Vorzeit in Tirol - Begleitheft zur Ausstellung: 7–25.

Seiwald A. (1980) Beiträge zur Vegetationsgeschichte Tirols IV: Natzer Plateau - Villanderer Alm. Ber. nat.-med. Ver. Innsbruck 67: 31–72.

Stadler H. (1991) Eine mesolithische Freilandstation auf dem Hirschbichl im Defereggental, Gem. St. Jakob/Osttirol. Archäologie Österreichs 2(1): 23–26.

Vorren K.-D., Morkved B. and Bortenschlager, S. (1993) Human impact on the Holocene forest line in the Central Alps. Vegetation History and Archaeobotany 2: 145–156.

Wahlmüller N. (1990) Spät- und postglaziale Vegetationsgeschichte des Tschöggelberges (Südtirol). Ber. nat.-med. Ver. Innsbruck 77: 7–16.

The pollenanalyses were done by:

Bortenschlager Sigmar	Lans
Kofler Werner	Brunnboden
	Kleinalpl
	Zirbenwald auf der Kaser
Kofler Werner, Wahlmüller Notburga	Langtalereck
Kucher Wolfgang	Am Soom
Vorren Karl Dag, Moerkved Brunhild	Gurgler Alm

The amount of CO_2 in the air breathed by the Iceman

Hilary H. Birks

Botanical Institute, University of Bergen

1. Introduction

The amount of CO_2 in the atmosphere is steadily rising at the present day due to human activities. In pre-industrial times, before 1800, the concentration was ca. 280 ppmv (parts per million by volume, or mixing ratio). Since then it has risen at an increasing rate, as shown by ice-core measurements (Friedli et al., 1986) and by direct atmospheric measurements since 1958 (Keeling et al., 1995; Mauna Loa Data, 1997). Air bubbles trapped in polar ice preserve a very long-term record of CO_2 concentrations, and direct measurements on this air have shown that CO_2 concentrations were some 100 ppmv lower in glacial periods, at ca. 180 ppmv (Barnola et al., 1987; Raynaud et al., 1993). The concentration of CO_2 rose at the end of the last glaciation and was ca. 270 ppmv in the CO_2 record from the Taylor Dome ice core at 5–6000 yr BP (Indermühle et al., 1999).

Woodward (1987) showed that the density of stomata on leaves was sensitive to CO_2 concentration. He showed, using herbarium material collected over the last 200 years, that stomatal density (the number of stomata per mm^2) decreased linearly with increasing CO_2 concentration over the same time period. Other factors have long been known to affect stomatal density (e.g. Salisbury, 1927), such as light and humidity, but in spite of the variability caused by these other environmental factors, the relationship with CO_2 is still strongly present in many, although not all, species (Woodward and Kelly, 1995).

Salix herbacea (dwarf willow) is a widespread species in the European mountains, preferring habitats where snow lies late in the season. An excellent relationship has been found between its stomatal density and CO_2 concentration (Beerling et al., 1995). Hattam (1997) used the decrease in CO_2 partial pressure with altitude to provide different CO_2 absolute volume concentrations (ppm) in the natural environment, and carefully characterized the relationship. The regression statistics she obtained indicate a strong statistical relationship between stomatal density and CO_2 concentration, with $R^2_{adj} = 0.59$, indicating that 59% of the variance in stomatal density can be attributed to the influence of CO_2 concentration. This is a high percentage for a natural system.

This strong relationship justified the use of stomatal density of fossil *Salix herbacea* leaves to reconstruct CO_2 in the past. The statistical method of inverse regression and calibration was used by Beerling et al. (1995) and Rundgren and Beerling (1999) to reconstruct CO_2 concentrations from fossil leaves during the late-glacial to Holocene transition and during the Holocene, respectively.

Fossil *Salix herbacea* leaves at Tisenjoch were collected by Klaus Oeggl from where the Iceman was lying. This locality is some 200 m higher than the present altitudinal limit of 3000 m for *Salix herbacea* in the Alps (Braun-Blanquet and Rübel, 1932–1935), suggesting that the climate some 5000 years ago was at least 1.0 °C warmer than ca. 1850. Since then, the annual mean air temperature of Austria has increased by ca. 1 °C (Grabherr et al., 1995), resulting in the ice-melting that exposed the Iceman. The upper plant limits increase much more slowly. The high elevation of *Salix herbacea* 5000 years ago is consistent with the glacial evidence for a warmer climate at that time discussed by Baroni and Orombelli (1996). The fossil leaves can be used to reconstruct CO_2 concentration in the same way as for the late-glacial and Holocene reconstructions.

2. *Salix herbacea* leaves from under the Iceman

One whole and several large fragments of *Salix herbacea* leaves were available. Their stomatal density was measured nondestructively, by viewing untreated leaves with a Zeiss Axioscope microscope at 400× magnification, using fluorescent light with an emission wavelength of 420 nm and an excitation wavelength of 365 nm (blue). All the stomata in a field of view were counted for 10 fields (4 in one case) of view over each surface of 5 leaves, and on one surface of a further 2 leaf fragments ($\Sigma = 114$ counts). The mean stomatal density is 169 mm^{-2}.

3. Enlarged calibration set

Because the data set of Hattam (1997) only reached an altitude of 2860 m and the Iceman was found at 3210 m, additional high-altitude leaves were obtained for the calibration set of stomatal density and CO_2 concentrations. Klaus Oeggl collected leaves at 2900 m from Am Bild, Niedertal, Vent, close to the Iceman locality, and Frank Graf collected leaves at 2950 m on Flüela Schwarzhorn, Switzerland. Leaves from 3000 m altitude were provided by Zürich ETH herbarium from Almagell, near Zwischbergenpass, ct. VS, Switzerland (1887) and 2950 m from Mutthorn, Lauterbrunnental, ct. BE, Switzerland (1916). For purposes of the calibration set, the modern sea-level value used was 360 ppm, reduced for altitude using the physical relationship given in Jones (1992). For 1887 and 1916 the appropriate CO_2 values from the Siple ice-core (Friedli et al., 1986) were adjusted for the altitudes of the collections.

The enlarged calibration set consisted of Hattam's original 24 collections, plus 2 modern and 2 herbarium high-altitude collections, and the 16.000 year old "full-glacial" value used by Beerling et al. (1995) ($\Sigma = 29$ collections). The regression statistics showed a much improved statistical relationship between stomatal density and CO_2 concentration, with $R^2_{adj} = 0.75$, indicating that 75% of the variance in stomatal density is explained by CO_2 concentration.

4. Results

The mean stomatal density of 169 mm^{-2} of the "Iceman" leaves results in a reconstructed CO_2 absolute volume concentration of 218 ± 20 ppm when the calibration set is used in an inverse regression and calibration. This is an estimate of the concentration of CO_2 breathed by the Iceman in his last breath. Supposing that the Iceman came from a village in Vinschgau (Val Venosta) (Dickson et al., 1996), the CO_2 concentration at this altitude of ca. 800 m breathed by his contemporaries would have been around 290 ppm. This estimate is based on the inferred value for 3210 m adjusted to an elevation of 800 m (Jones, 1992).

When the Taylor Dome (Indermühle et al., 1999) ice-core CO_2 absolute volume concentration between 5–6000 yr BP is adjusted to 3200 m the value is ca. 187 ppm, with a very small standard error. This is somewhat lower than the reconstruction from the Iceman *Salix herbacea* leaves. If the CO_2 concentration 5–6000 years ago from the Holocene leaf sequence of Rundgren and Beerling (1999) is similarly adjusted to 3200 m, the result is ca. 192 ppm. Even without their 95% confidence interval of ca. 90 ppm, this falls close to the Iceman reconstruction of 218 ± 20 ppm. These comparisons illustrate that the stomatal density method of reconstructing CO_2 concentration gives reasonable estimates considering that it is a biological method with many inherent and environmental sources of variability. This study has demonstrated how the method can be used to reconstruct CO_2 concentrations in local situations of interest to environmental archaeologists.

Abstract

The stomatal density of *Salix herbacea* leaves is inversely proportional to the absolute volume concentration (ppm) of CO_2 in the air. The relationship was quantified with a modern calibration data-set from leaves collected between 40 and 2860 m a.s.l. covering a range of CO_2 concentration from 360–200 ppm. Because the Iceman lay at 3210 m a.s.l., the calibration set was extended by new collections from as high as possible in the Alps, 2900 and 2950 m, and herbarium collections at 3000 m. *Salix herbacea* leaves from under the Iceman are assumed to be also ca. 5000 years old. Their stomatal density was measured, and the atmospheric CO_2 concentration breathed by the Iceman was reconstructed to 218 ± 20 ppm. His likely contemporaries at ca. 800 m a.s.l. in Val Venosta were breathing air with ca. 290 ppm CO_2. These values compare quite well with contemporary measurements from an Antarctic ice core and reconstructions from a Holocene sequence of fossil *Salix herbacea* leaves.

Zusammenfassung

Die Stomatadichte der Blätter von *Salix herbacea* ist verkehrt proportional zum absoluten CO_2 Gehalt der Luft (ppm). Das genaue Verhältnis wurde anhand eines modernen Eichdatensatzes mit einem Bereich von 360–200 ppm CO_2 quantifiziert. Die Eichdaten stammen von Blättern, die auf Höhen zwischen 40 und 2.860 m.ü.M. gesammelt wurden. Da die Fundstelle des Eismanns aber auf 3.210 m lag, wurden diese Daten mit neuen Ergebnissen von möglichst hohen Berglagen – 2.900 und 2.950 m – sowie älteren Herbarien von 3.000 m Höhe ergänzt. Es wird angenommen, daß die *Salix herbacea* Blätter, die unter dem Eismann lagen, ebenfalls ca. 5.000 Jahre alt sind. Die Dichte ihrer Stomata wurde gemessen, und der CO_2 Gehalt der Luft, die der Eismann atmete, als 218 ± 20 ppm bestimmt. Seine Zeitgenossen im Vinschgau, auf 800 m.ü.M., atmeten eine Luft mit ca. 290 ppm CO_2. Diese Werte stimmen gut mit zeitgleichen Messungen aus einem antarktischen Eiskern und den Rekonstruktionen einer holozänen Sequenz fossiler *Salix herbacea* Blätter überein.

Résumé

La densité stomatique des feuilles de *Salix herbacea* est inversement proportionnelle à la concentration (ppm) en CO_2 de l'atmosphère. Le rapport exact a été quantifié à l'aide d'un ensemble de données d'étalonnage constitué à partir de feuilles récoltées entre 40 et 2 860 m d'altitude et comprenant une plage de concentrations en CO_2 de 360 à 200 ppm. L'Homme des glaces ayant été découvert à 3 210 m d'altitude, l'ensemble d'étalonnage a été étendu par des collectes supplémentaires, effectuées aussi haut que possible dans les Alpes – entre 2 290 et 2 950 m – et d'herbiers anciens collectés à une altitude de 3 000 m. Les feuilles de *Salix herbacea* découvertes sous le corps de la momie ont elles aussi un âge supposé d'environ 5000 ans. Leur densité stomatique a été mesurée, et la teneur en CO_2 de l'air respiré par l'Homme des glaces établie à 218 ± 20 ppm. Ses contemporains vivant à 800 m d'altitude au Vinschgau (Val Venosta) ont respiré un air dont la teneur en CO_2 était de 290 ppm. Cette concentration en CO_2 est proche de celle mesurée dans les bulles d'air piégées dans des noyaux de glace antarctique pendant l'Holocène.

Riassunto

La densità stomatica di foglie di *Salix herbacea* è inversamente proporzionale alla concentrazione (ppm) di CO_2 nell'aria. Il rapporto è stato quantificato con una serie calibrata di dati su foglie raccolte tra i 40 e i 2860 m s.l.m. con un tasso di concentrazione di CO_2 da 360 a 200 ppm. Essendo il sito di ritrovamento della mummia a quota 3210 m s.l.m. la serie calibrata di dati è stata completata da nuove raccolte di foglie a quote più alte possibile nell'area alpina (2900, 2950 m) e da un antico erbario a quota 3000 m. Foglie di *Salix herbacea* da siti al di sotto di quello dell'uomo dell'Hauslabjoch sono ritenute risalire a 5000 anni fa. È stata misurata la loro densità stomatica e la concentrazione di CO_2 nell'aria respirata dall'uomo del ghiacciaio si assume sia stata di 218 ± 20 ppm. I suoi contemporanei a quota 800 m s.l.m. in Vinschgau (Val Venosta) respiravano aria con 290 ppm, livello simile a quello di concentrazione di CO_2 misurata in bolle d'aria incluse in nuclei di ghiaccio nell'Antartide durante l'Olocene.

Acknowledgements

Klaus Oeggl supplied me with the Iceman leaves and with a modern collection of *Salix herbacea* from near the Iceman locality. Frank Graf collected leaves from the Flüela Schwarzhorn, Switzerland. Curator Matthias Baltisberger gave me leaves from specimens in the Zürich herbarium. John Birks did the statistical calculations and Walter Kutschera made useful comments. The stomatal density measurements were made on the Zeiss Axioscope belonging to the Institute of Fish and Marine Biology, University of Bergen. During the work, I was supported by the Norwegian Research Council and the Grolle Olsen Fund. To all these people and institutions I am very grateful.

References

Barnola J. M., Raynaud D., Korotkevich U. S. and Lorius D. (1987) Vostok ice core provides a 160,000 year record of atmospheric CO_2. Nature 329: 408–414.

Baroni C. and Orombelli G. (1996) The alpine "Iceman" and Holocene climatic change. Quaternary Research 46: 78–83.

Beerling D. J., Birks H. H. and Woodward F. I. (1995) Rapid late-glacial atmospheric CO_2 changes reconstructed from the stomatal density record of fossil leaves. Journal of Quaternary Science 10: 379–384.

Braun-Blanquet J. und Rübel E. (1932–1935) Flora von Graubünden. Veröff. Geobot. Inst. Rübel 7: 1695 pp.

Dickson J. H., Bortenschlager S., Oeggl K., Porley R. and McMullen A. (1996) Mosses and the Tyrolean Iceman's southern provenance. Proceedings of the Royal Society of London B 263: 567–571.

Friedli H., Lötscher H., Oeschger H., Siegenthaler U. and Stauffer B. (1986) Ice core record of the $^{13}C/^{12}C$ ratio of atmospheric CO_2 in the past two centuries. Nature 324: 237–238.

Grabherr G., Gottfried M., Gruber A. and Pauli H. (1995) Patterns and current changes in alpine plant diversity. Ecological Studies 113: 167–181 (F. S. Chapin and C. Körner, eds.).

Hattam C. (1997) How does stomatal density of *Salix herbacea* L. vary with atmospheric CO_2 concentration? Thesis, University of Wales, Bangor.

Indermühle A., Stocker T. F., Joos F., Fischer H., Smith H. J., Wahlen M., Deck B., Mastrioanni D., Tschumi J., Blunier T., Meyer R. and Stauffer B. (1999) Holocene carbon-cycle dynamics based on CO_2 trapped in ice at Taylor Dome, Antarctica. Nature 398: 121–126.

Jones H. G. (1992) Plants and Microclimate. A quantitative approach to environmental physiology, 2nd ed., Cambridge University Press, Cambridge.

Keeling C. D., Whorf T. P., Wahlen M. and van der Plicht J. (1995) Interannual extremes in the rate of rise of atmospheric carbon dioxide since 1980. Nature 375: 666–670.

Mauna Loa Data (1997) Internet. http://cdiac.esd.ornl.gov/trends_html/trends/tables/sio-maun.htm.

Raynaud D., Jouzel J., Barnola J. M., Chappellaz J., Delmas R. J. and Lorius C. (1993) The ice record of greenhouse gases. Science 259: 926–934.

Rundgren M. and Beerling D. (1999) A Holocene CO_2 record from the stomatal index of subfossil *Salix herbacea* L. leaves from northern Sweden. The Holocene 9: 509–513.

Salisbury E. J. (1927) On the causes and ecological significance of stomatal frequency, with special reference to the woodland flora. Philosophical Transactions of the Royal Society of London B 216: 1–66.

Woodward F. I. (1987) Stomatal numbers are sensitive to increases in CO_2 from pre-industrial levels. Nature 327: 617–618.

Woodward F. I. and Kelly C. K. (1995) The influence of CO_2 concentration on stomatal density. New Phytologist 131: 311–327.

Dendrological analyses of artefacts and other remains

K. Oeggl[1] and W. Schoch[2]

[1] Institut für Botanik, Universität Innsbruck
[2] Labor für Quartäre Hölzer, Adliswil

1. Introduction

When the "Man in the Ice" was discovered in September 1991 at the head of the Ötztal, all attention was concentrated on the exposed corpse frozen into the ice. The unusual features of the discovery strongly suggested a corpse of someone who had perished relatively recently. The associated artefacts, therefore, attracted no attention and were first noticed during a more careful examination of the find site. These finds, then, enabled a provisional estimate of the age of the corpse, whereby it then became clear that this was the body of a prehistoric man, together with his entire equipment, something that in view of the remoteness of the site had not previously been taken into consideration.

The sensational discovery was made at the Tisenjoch, an east-west orientated rock saddle at the head of the Ventertal, a side valley of the Ötztal (Fig. 1). Nowadays, a rock scramble route runs along the crest of the ridge. This pass is seldom used now as a route to get from the Ötztal into the head of the Schnalstal, a side valley of the Vinschgau. Instead, the usual route leads above the Niederjoch, a little further eastward, a pass which has been used since the dawn of history and is still so used today by shepherds each year to drive their sheep from the Vinschgau to their summer pastures on the grassy heaths at the head of the Ötztal. It is only when the Niederjoch is blocked by snow cornices that the shepheards make the detour to the narrow Tisenjoch at the end of the Tisental (Lippert and Spindler, 1991).

The site lies on the north side of the Tisenjoch on a rock shelf that is covered by ice and snow throughout the year. The underlying rock-strata, a schistose-gneiss, are exposed in a few places on near vertical rock faces and on wind exposed edges. The rock strata have an almost vertical strike and run across the shelf from north to south. Because of erosion in cracks and crevices, chamber-like depressions had been created. In one of these rocky depressions, ca. 100 m away from the rocky ridge on the Tisenjoch (Fig. 2), the Iceman thawed out in September 1991. This depression runs northeast-southwest with a length of 30 m and a breadth of 5–10 m. On average, the depression is from 2,5 to 4 m deep, sunk into the rock (Lippert, 1992a). The corpse itself was discovered lying face-downwards over a boulder in the southern part of this shallow gully (Fig. 2). The upper part of the body was naked, the legs were covered by leather. On the rock surface beneath his body there was a network, which later turned out to be a cloak. The other artefacts were found in the immediate vicinity of the body. The remains of the first container made of birch bark lay about 1 m away, southwest of the body. This tubular piece of birch bark had been flattened and in its original state would have been 20 cm long and ca. 10–15 cm wide. Grass, or hay and leaves, still with discernible venation, were found inside it (Koler, cited in Zissernig, 1992). The remains of a second birch bark container were later found beside the pack-frame on the rock-shelf southeast of the body. In the neighbouring depression to eastward about 20 m away from the site of the body, the base of one of them was found. The remains of a pack-frame, a U-shaped bent piece of solid wood with two wooden lathes (Fig. 2), were found on the rock-shelf 4 m south of the body. Right beside the pack-frame the hafted axe was lying on hide remains in a crevice. In front of it, the bow stuck in the ice and, unfortunately, became broken during a rescue attempt by tourists. On the opposite side of the rock depression, about 2.5 m west of the above mentioned remains, the quiver containing 14 arrows was found, hard frozen onto a stone slab. In the meltwater issuing from this vicinity a grooved piece of wood was discovered, which later proved to be a part of the stiffening of the quiver (Lippert, 1992a,b; Zissernig, 1992).

Immediately in the same year following the discovery of the body an excavation started though it had to be abandoned after three days because of bad weather. In the course of this later investigation the grass mat (cloak), nets and ropes made from plant fibres were rescued. In addition, other remains of botanical interest, a length of twig (sample no. B-91/23) and in the vicinity of the pack-frame a tiny splinter of wood (sample no. B-91/22) came to light (Lippert and Spindler, 1991).

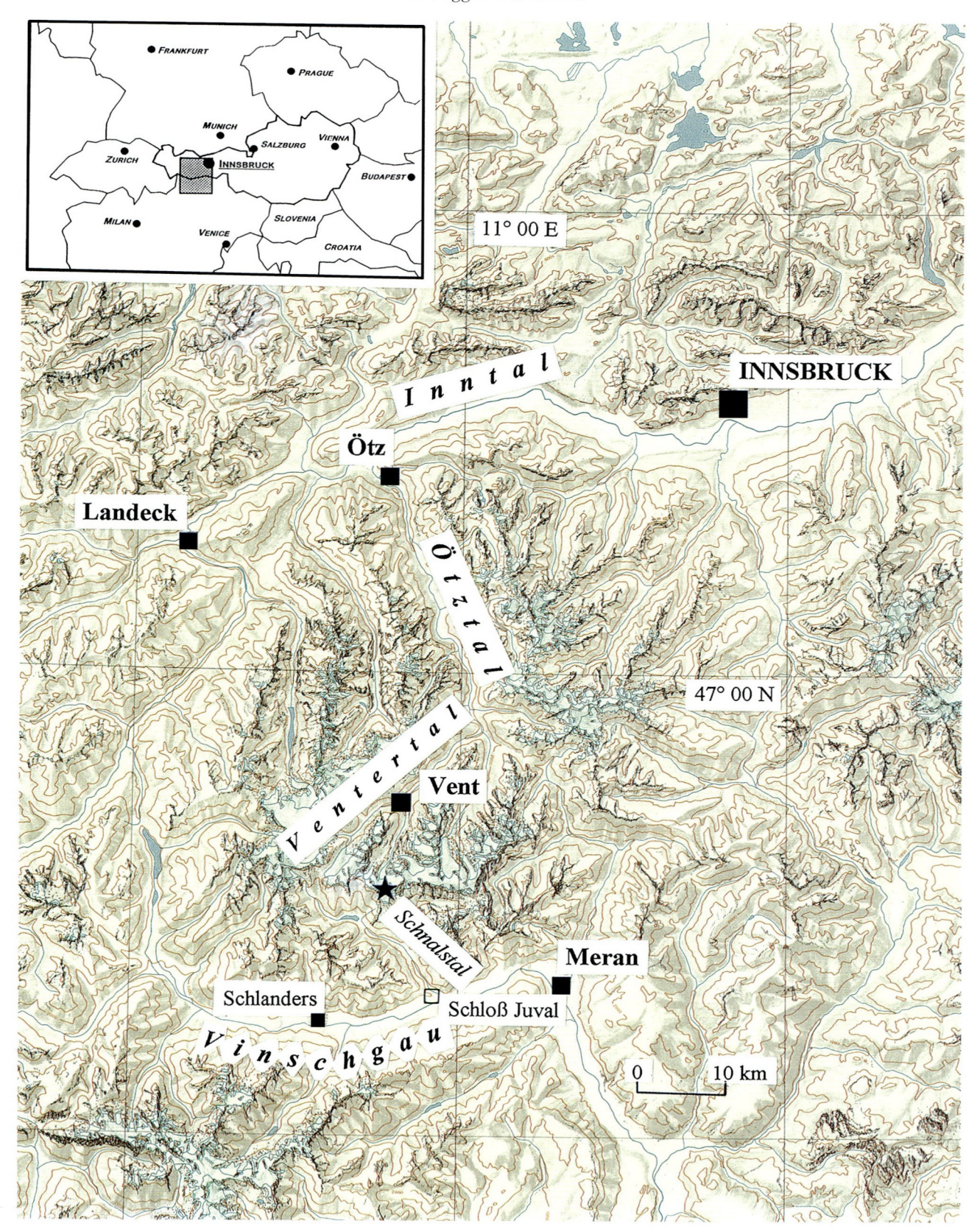

Fig. 1. The area in which the Iceman was found. The findspot is marked with an asterisk (map production: Atlas of the Tyrol, design by Ernest Troger, Institut für Landeskunde der Universität Innsbruck)

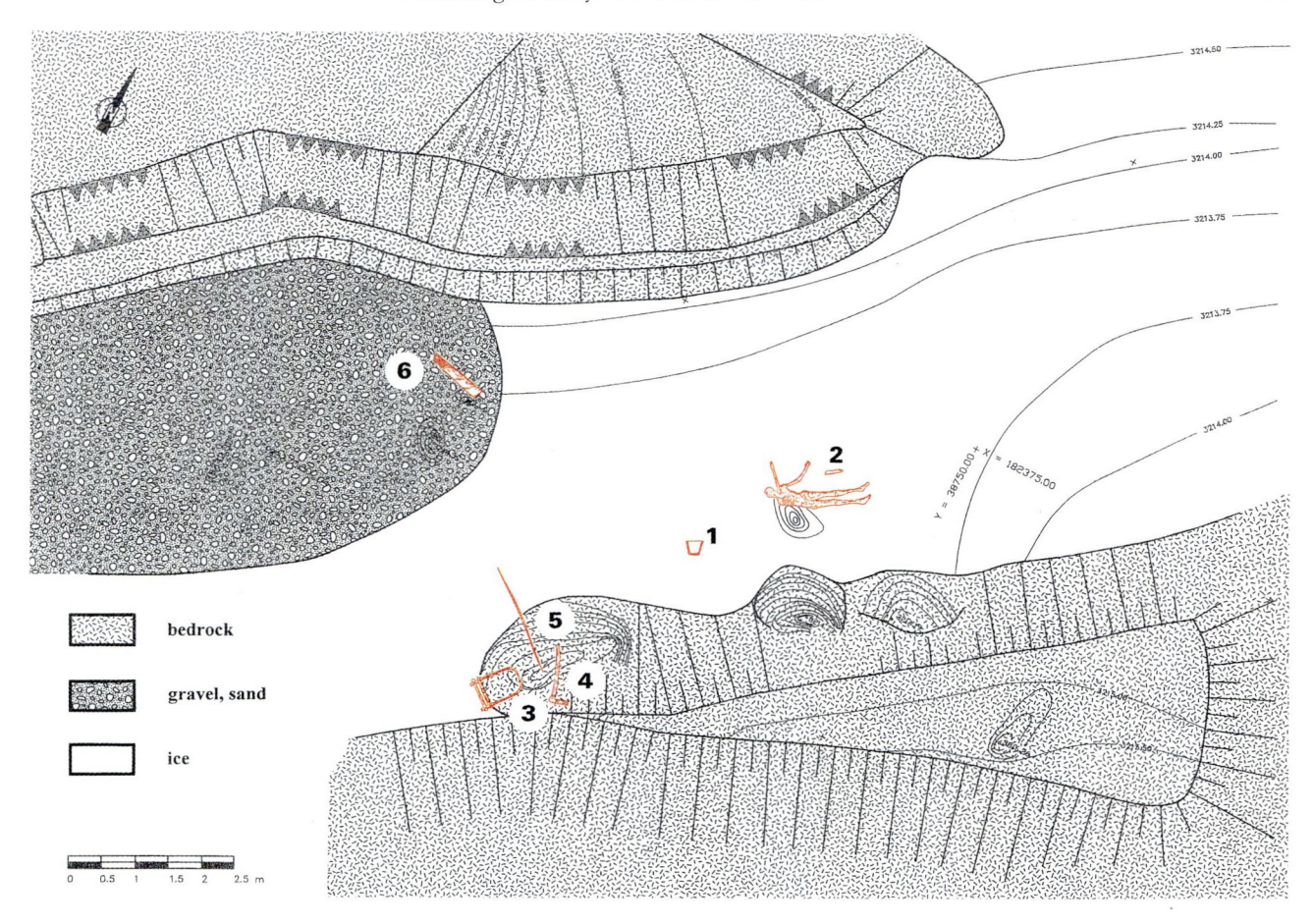

Fig. 2. The topography of the find site and the positions of the various artefacts: **1** birch-bark container, **2** the flint dagger, **3** the pack-frame, **4** the axe, **5** the bow, **6** the quiver (modified from the sketch in Lippert, 1992b)

A full-scale excavation took place in August of the following year. Both the high altitude and the nature of the site presented the excavation team under the leadership of Prof. Andreas Lippert, University of Vienna, with a great challenge. The ice surface melted under the influence of the sun's rays once the snow layer in the depression had been removed. So that no vital information should be lost, this meltwater was diverted into a channel, at the end of which a series of sieves of different mesh-sizes were installed. These sieves collected a large quantity of material of botanical and zoological interest. Larger-scale archaeological complexes, which could be discerned glimmering in the ice, were freed by using a jet of steam. The recovery of the broken-off piece of bow from the ice in the southern part of the gully was made in this way, as well as the samples of wood no. 92/275 in the vicinity of the quiver, pieces of twigs no. 92/48, 92/106, 92/116, 92/131 and 92/217, in the disturbed layer within the southern part of the rocky depression. Sample no. 92/285 was recovered from the rock surface, immediately northwest of the spot on which the corpse was lain as the ice melted. Yet another lath (92/400) was found between rock crevices in quadrant 2 of the excavation. In the meantime, the finds of the man's equipment and the fur remains were washed and cleaned, ready for preservation measures at the Römisch-Germanisches Zentralmuseum in Mainz. The water used for washing these finds was likewise filtered through various sieves and the resi-dues passed on to the Botanical Institute of Innsbruck University for further investigation. These samples contained wood fragments, for the majority of which the exact sites were not known. These samples were assigned samples no. 91/... etc.

Both the corpse and the equipment of the Iceman were in an excellent state of preservation. The special conditions pertaining to the find site had ensured, that ever since the deposition of the corpse, it had been subjected to only minor decay and minimal mechanical stress. Firstly, the site lies at an altitude of 3214 m, above the present snow limit, which in the Ötztaler Alps is considered to lie at a maximal altitude of 3100 m (Gross et al., 1976). In winter times the find site is covered by up to 7 m of snow. The mean annual temperature at the find site altitude is −5.6 °C (mean for the period 1982–1991). The mean monthly temperature only rises above 0 °C during the months July and August (values taken from Patzelt cited in Lippert, 1992a), i.e., only a superficial snow melt takes place. Even then at this altitude air temperatures above 4 °C are seldom of any long duration, wherefore microbial activity remains low and any deterioration and disintegration takes place very slowly. Secondly, permafrost conditions prevail at the find site, so that the ice in the depression remains frozen hard onto the underlying substrate most of the time. On this account, in view of the gentle slope involved, only a modest displacement, if any at all, of the Iceman from his point of deposition by glacier movement could be expected. Furthermore, the topography of the find site, a chamber-like depression, provided such protection, that the corpse will never have been exposed to glacier movement. These site conditions and the fact that both the body and the equipment were relatively quickly covered by snow, and later also by ice, ensured that the conditions for preservation were optimal and that the organic remains have been preserved virtually intact (Lippert, 1992a).

This dendrological analyses of the wood remains associated with the Iceman should help to provide answers to a variety of questions raised by the find. The immediate aim was to determine the plant species involved and their degree of preservation. The relationship between the species of wood used for the various equipment and their derivation from the different part of the trunk or branch is important since this affects their suitability for the particular usage involved, as well as their external form and their suitability for making the particular piece of wooden equipment. The answers given to such questions, as well as the analyses of the working traces left by the tools used in their construction, provide valuable pointers to the technology used in the making of the equipment. The assumption that has been made is that the Iceman will have used woods for his equipment for daily usage, that grew in the immediate vicinity of his settlement, where by conclusions can be drawn about the environment in which he lived from an evaluation of the regional floristics of the woody plant species that he utilized.

2. Results

2.1. The species determinations

The initial microscopic samples were taken from the wooden pieces of equipment in the laboratory of the Institut für Gerichtliche Medizin of Innsbruck University, October, 2nd and 3rd, 1991. There the corpse and its associated equipment were transported straight away after their excavation. The wood samples that were obtained during the excavation season from October, 3rd–5th, 1991, were assigned the prefix "B". Subsequent samples (prefixed "RGZM") were taken at the Römisch-Germanisches Zentralmuseum in Mainz on October 22nd and December 20th, where the artefacts had been stored prior to restoration and conservation. All the samples were washed and cleaned before conservation and restoration processes were commenced. The water used for washing was filtered through sieves and the residues afterwards examined for the presence of wood remains and charcoal particles. The wood samples were in an excellent state of preservation and exhibit, if at all, only minor traces of disintegration, wherefore, apart from a few exceptions, the species determinations were possible. The woody species used for the individual artefacts are listed in Table 1. (Detailed descriptions of the wood remains are given in the Appendix).

2.2. Charcoal analyses

A sample of leaves and grass (RGZM-91/139), all rolled up together with a volume of 20 cm, were recovered during the initial excavation in the vicinity of the northwestern extremity of the rock on which the corpse was found. The leaves were all of the Norwegian Maple (*Acer platanoides*). By dint of carefully unrolling the sample, the petioles of all the leaves were found to be missing, only the laminae were present. This is an unusual feature since in the normal course of events a callus forms between the stalk and the twig before the leaves are shed at the end of the vegetative period. In this way the leaves are shed with the petioles intact. The absence of the petioles is therefore unnatural and to be regarded as artefactual, i.e., the petioles must have been carefully removed. During the careful unrolling of the leaves, further plant remains and charcoal fragments were revealed inside. The charcoal particles measured 0.5–5 mm in size. Altogether six different species of plants were determinable in this sample (Table 2): probably Spruce (*Picea/Larix*-type), Pine (*Pinus mugo*-type), Green Alder (*Alnus viridis*), some Pomoideae, probably Juneberry (cf. *Amelanchier ovalis*), Dwarf Willow (*Salix reticulata*-type) and Elm (*Ulmus* sp.). Because the leaf sample was not found in association with other remains, it was at first uncertain, whether this sample was at all related to the find of the Iceman, or whether the leaves were younger. However, the finds of charcoal fragments in the sieve residues from the washings of the other artefacts, particularly those from sample

Table 1. The woody species identified from uncarbonized wood finds made during the excavations in 1991 and 1992

Sample no.	Specification	Woody plant species	Nos. found
B-91/15	Bow	Yew, *Taxus baccata* L.	1
	Axe shaft	Yew, *Taxus baccata* L.	1
	Arrow shafts	Wayfaring Tree, *Viburnum lantana* L.	14
		Dogwood, *Cornus* sp. L.	1
	Quiver stiffener	Hazel, *Corylus avellana* L.	1
	Dagger handle	cf. Common Ash, *Fraxinus* cf. *excelsior* L.	1
B-91/22	Wood found on hide from Q10	Larch, *Larix decidua* MILL.	1
B-91/23	Wood, stray find	Green Alder, *Alnus viridis* CHAIX. DC	1
B-91/27	Laths from the pack-frame	Larch, *Larix decidua* MILL.	1
B-91/28	The U-shaped spar of the packframe	Hazel, *Corylus avellana* L.	1
B-91/29	The retouching tool	Lime, *Tilia* sp. L.	1
RGZM-91/95	Leather, cords, hide	Yew, *Taxus baccata* L.	2
		cf. Willow, cf. *Salix* sp. L.	1
RGZM-91/96	Hair, fibres, cords	Hazel, *Corylus avellana* L.	1
RGZM-91/106	Sewn leather	Larch, *Larix decidua* MILL.	3
		Birch, *Betula* sp. L.	1
RGZM-91/138	Hair remains	Probably Spruce, *Picea/Larix*-type	4
		Larch, *Larix decidua* MILL.	2
RGZM-91/139	Grass and leaf remains	Larch, *Larix decidua* MILL.	3
		Hazel, *Corylus avellana* L.	1
RGZM-92/48	Wood	Wayfaring Tree, *Viburnum lantana* L.	1
RGZM-92/65	Wood and hide	indeterminate	1
RGZM-92/99	Hair and wood	Birch, *Betula* sp. L.	2
RGZM-92/106	Wood	Wayfaring Tree, *Viburnum lantana* L.	1
RGZM-92/116	Piece of the bow	Yew, *Taxus baccata* L.	1
RGZM-92/131	Wood	Wayfaring Tree, *Viburnum lantana* L.	1
RGZM-92/217	Wood	Larch, *Larix decidua* MILL.	1
RGZM-92/275	Wood	Green Alder, *Alnus viridis* (CHAIX.) DC	2
RGZM-92/285	Grass, hair, wood	Larch, *Larix decidua* MILL.	1
RGZM-92/292	Wood, grass	Pine, *Pinus* sp. L.	1
RGZM-92/400	Wood and hairs	Larch, *Larix decidua* MILL.	1
RGZM-93/117	Residues from washing the fur cap	Arolla pine, *Pinus cembra* L.	1

RGZM-91/132, allow the assumption that the charcoal fragments are of the same age as the Iceman. The species composition of sample RGZM-91/132 show a close correlation with that of the fragments obtained from the leaf sample RGZM-91/139. Sample RGZM-91/132, although primarily composed of birch bark fragments, was found to include fragments from a Maple tree (*Acer* sp.), Pine (*Pinus mugo*-type) probably Spruce (*Picea/Larix*-type), Green Alder (*Alnus viridis*), Birch (*Betula* sp.) and Elm (*Ulmus* sp.) The dominant taxa in both these samples are Spruce (*Picea/Larix*-type), Pine (*Pinus mugo*-type), Green Alder (*Alnus viridis*) and Elm (*Ulmus* sp.). This similarity makes it likely that the leaves (RGZM-91/139) had been closely associated with the birch bark fragments. Accordingly sample RGZM-91/132 are probably the remains of one birch bark container, which had held the Norwegion Maple (*Acer platanoides*) leaves (RGZM-91/139) with the charcoal fragments included in. When found, one birch bark container had already lost its base. Therefore the loss of leaves out of the container is readily understandable. The birch bark container probably served as a storage vessel for cold embers (Egg et al., 1993) which were rolled up inside the Norwegian Maple (*Acer platanoides*) leaves. In many cases the charcoal fragments show a marked curvature of the tree rings, an indication that they derive from twigs or branches. The mixture of montane, subalpine and alpine species probably arose from the practice of filling the storage container with the embers of several fires. By collecting the glow from different fireplaces a mixture of charcoals from different woody species would probably occur.

Taken as a whole, the Spruce (*Picea/Larix*-type) charcoal was predominant, especially in view of its

Table 2. Results of the charcoal fragments analyses: species composition of the samples

Sample no.	Specification	Woody plant species	Nos. found	Type of preservation
RGZM-91/102	Cords	probably Spruce (*Picea/Larix*-type)	9	charred
		probably Spruce (*Picea/Larix*-type)	3	uncharred
RGZM-91/103	Grasses, cords	Pine (*Pinus mugo*-type)	2	charred
		probably Spruce (*Picea/Larix*-type)	10	charred
		Elm (*Ulmus* sp.)	1	charred
RGZM-91/106	Leather	Maple (*Acer* sp.)	1	charred
		probably Spruce (*Picea/Latrix*-type)	47	charred
		Bones (Spongiosa)	1	charred
		Birch (*Betula* sp.)	2	uncharred
		Larch (*Larix decidua*)	3	uncharred
RGZM-91/132	Birch-bark fragments	probably Spruce (*Picea/Larix*-type)	42	charred
		Pine (*Pinus mugo*-type)	7	charred
		Maple (*Acer* sp.)	1	charred
		Green Alder (*Alnus viridis*)	2	charred
		Birch (*Betula* sp.)	1	charred
		Elm (*Ulmus* sp.)	4	charred
		Bark indeterminate	1	charred
RGZM-91/137	Leather, grass	probably Spruce (*Picea/Larix*-type)	1	charred
RGZM-91/138	Hair remains	probably Spruce (*Picea/Larix*-type)	9	charred
		probably Spruce (*Picea/Larix*-type)	4	uncharred
		Larch (*Larix decidua*)	1	uncharred
RGZM-91/139	Grass, leaf remains	probably Spruce (*Picea/Larix*-type)	71	charred
		Pine (*Pinus mugo*-type)	13	charred
		Green Alder (*Alnus viridis*)	5	charred
		Promoideae: probably Juneberry (cf. *Amelanchier ovalis*)	4	charred
		Dwarf willow (*Salix reticulata*-type)	1	charred
		Elm (*Ulmus* sp.)	16	charred
		Coniferous indeterminate	12	charred
		Bark indeterminate	9	charred
RGZM-93/117	Wooden laths	probably Spruce (*Picea/Larix*-type)	3	charred
		Arolla pine (*Pinus cembra*)	1	uncharred

constancy in all the samples (Table 2). The percentual composition of all the charcoal finds is shown in Fig. 3.

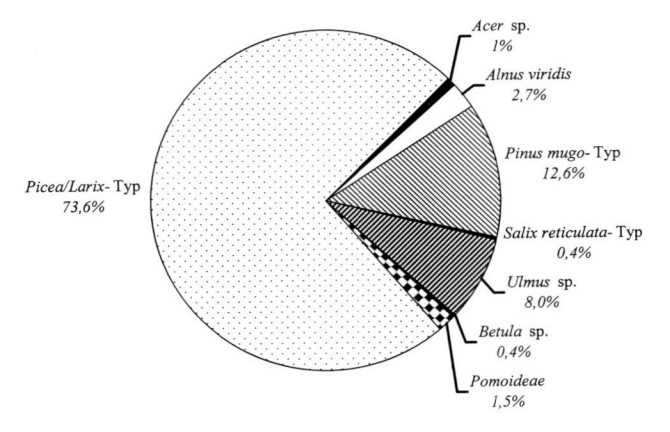

Fig. 3. The percent species composition of the charcoal fragments found

Picea/Larix-Typ 73,6%
Acer sp. 1%
Alnus viridis 2,7%
Pinus mugo-Typ 12,6%
Salix reticulata-Typ 0,4%
Ulmus sp. 8,0%
Betula sp. 0,4%
Pomoideae 1,5%

2.3. Technological-botanical analyses

Together with stone, wood was already in Palaeolithic times an important material for the manufacture of tools and utensils that were in daily use. A precise knowledge of the properties of the raw material would be essential for the manufacture and appropriate utilisation of such things, since the fitness of each tool would be to a high degree determined by the properties of the raw material in question. Functional analyses of wooden objects from the Neolithic period demonstrate that delibrate choices of particular woods for specific functions were made (Schweingruber, 1976). This relationship between the specific wood selected and the particular object to be manufactured also proves that a knowledge of the physical properties of the native wood species had already been gleaned empirically by that time (Beckhoff, 1965). The Iceman himself had made deliberate choices of the woody species used for the different pieces of his equipment, in direct relation to the purposes to which they would be put. The following analyses of the forms taken by his different pieces of equipment provide an insight into his technological know-how in their construction.

The bow: European Yew (*Taxus baccata*) was the wood chosen to make the bow. Under the microscope there was no indication of any decay and the wood still retained the reddish colouration characteristic of yew heartwood. This stain was also discernible on the wood surface when the bow was recovered, though after a while it changed to a dark violet shade, a phenomenon that is commonly found for yew wood that has lain in water for a long time (Hegi, 1935). The shape of the bow is perfectly clear, although unfortunately one limb was broken off. The find is a straight-shafted bow with a total length of 183.5 cm. The longer of the two pieces measures 142 cm, the shorter 41.5 cm in length. The maximum diameter of the bow in the middle section is 3.6 cm, the maximum height is 3.17 cm. The elongated bow shaft shows no sign of any special feature for giving a better grip and is in general rather coarsely worked. The whole of the upper surface is covered with obvious whittling marks (Plate 1). They run slightly rotating to the right from the tapering ends to the middle part of the bow. It is obvious that the bow was fashioned by cutting on alternate sides, because the whittling cuts change direction at about 50 cm – or measured from the other side – at 130 cm away from the ends. On the flattened inside of the bow shaft the cut marks change at about 80 cm. This side was fashioned from the middle part towards the limbs. The cut surfaces all along the length of the bow wood are smooth and show a deep initial entry of the tool used (Plate 1). The cuts at each end of the bow are appreciably longer than the rest. It is difficult to fashion the tough and high elastic wood of the European Yew (*Taxus baccata*) in a direction parallel to the fibres, because this readily produces unwanted grooving (Plate 1). Nevertheless such grooving is seldom visible on the bow shaft, which indicates that a sharp cutting edge must have been used for its construction.

Although the bow is still imperfectly finished, when examined, the profile in transverse section is still discernible. It shows the commonest shape for yew bows, which is plano-convex – D-shaped with a raised back and a flat belly side. The profile tapers off equally towards both ends. This reduces the weight on the limbs and thereby the overall dead-weight, wherefore the loss of power when shooting remains low (Beckhoff, 1963). The tree rings diminish on both conical ends and are partially cut through. These represent weak points on the bow, which would have needed a backing for its proper use (cf., Ballweg in Egg et al., 1993). Notches, or other forms of attachment for the bow string are absent from either end of the bow (Plate 1). The edges of the bow in these regions are entirely free from any signs of wear or tear from the attachment of a string, as was also obvious from examination under the microscope. This and the missing grip supports the view that this bow was unfinished and had never been used.

Following observations on the course of the tree rings and associated computer-tomographic examinations, the mode of construction and the original site of the piece of wood in the tree could be established. The curvature of the tree rings encompassed barely a quarter of a full circle. Side branches are visible and according this ramification the bow was made from a Yew trunk and not from a branch. The Yew trunk had a minimum diameter of 8 cm (Oberhuber and Knapp, this volume), and most probably even more, since the lighter-coloured sapwood was removed (cf., Ballweg in Egg et al., 1993). The profile is trimmed precisely to the direction of the tree rings in the body of the wood, and lay in such a way in relation to the trunk that the curvature of the D-shaped profile runs in parallel with the course of the tree rings (Oberhuber and Knapp, this volume). The above mentioned orientation of the wood used for the bow shaft makes the construction easier and ensures that the outer fibres would have been protected. However, this is not the optimal solution. It is a feature of coniferous woods in general that the maximal flexibility is achieved when the tree rings run diagonally at an angle of 45° through the profile, whereas for deciduous woods the maximum flexibility is achieved when the annual rings run radially, i.e., the course of the annual rings corresponds with the curvature of the profile (Beckhoff, 1968). However, with insufficient technological methods available and for a hard and tough wood like that of the Yew, it is probable more difficult to work on a piece with the tree rings running at an angle.

The efficiency of a bow shaft with a semi-circular or D-shaped profile is largely dependent on the drawing stance. Comparable finds of yew bows from the Neolithic period with a D-shaped profile reach their maximum power, when the curved side faces outwards and the flat side faces the archer. Normally, the notches for the attachment of the string provide an indication of the aiming position, and similarly a calculation of the power of a bow allows certain conclusions to be drawn concerning its use and effectiveness. Neither of these possibilities exist, since there are no notches and the profile is only roughly finished off. Therefore, in the present case the use can be indirectly ascertained by deficiencies in the wood and the manner and way in which the bow shaft was finished off. It is essential for an archer that the bow shaft does not fracture when in use. The danger of this occurring can be largely diminished already by the choice of the raw material. In

Plate 1. The bow (European Yew, *Taxus baccata*): **1** the broken bow, **2** enlargement showing the broken end of the bow shaft: the inner surface of the bow with a branch scar, **3** enlargement showing the tool marks on the bow surface, **4** the broken end of the bow, **5** end of the bow showing no features for inserting a string: **a)** ventral side, **b)** dorsal side (Photos nos. 1, 3 and 4 by the Römisch-Germanisches Zentralmuseum Mainz)

particular, attention must be paid to an absence of growth irregularities which might provide potential points of fracture. Readily recognisable potential breakage points lie in the knot-whole region where a side branch emerges and causes a disruption of the smooth course of the fibres. Even though the yew wood is the only material for bows which tolerates branching, the wood piece used should be as far as possible free of side branches. If this is unavoidable, the branch must run at a right-angle to the axis of a bow shaft. Furthermore, a branching point should never occur on the side subject to the bending stress, but only on the side subject to compression, i.e., the side facing the archer when he pulls his bow (Beckhoff, 1963). It is quite difficult to find a piece of wood suitable for a bow with a length of 1.8 m, that is free of branch knots. Even in the bow wood carried by the Iceman, small branch knots are present, but they lie exclusively on the flattened side (Plate 1). Therefore the bow was used in such a way that the rounded side of the D-shaped profile was the back of the bow and faced away from the archer, whereas the flat side was orientated towards him. Thereby the optimal power of this bow will have been attained.

In terms of pliability, great elasticity and toughness, the Iceman selected for his bow the best suited wood of all native tree species. Before being used to construct a bow, the wood needs to be well seasoned, so that it does not become warped during or after usage. The usually accepted length of time for maturation after felling is 1–1.5 months per tree ring (Beckhoff, 1968). To ensure a sufficient firmness of the wood, only yew stems with at least 8–10 tree rings per cm were used for construction. A wood with more than 14 tree rings per cm is reckoned to be of the highest quality (Beckhoff, 1968). Individual splinters, that had formed when the bow was broken, allow an investigation of the tree ring width (Table 4). On four of these splinters twenty tree rings were measured. The width ranged from 0.6 to 0.18 mm (0.45 mm average), which means a mean of 16 tree rings per cm. Therefore the Iceman's bow can be classed with top quality.

The quiver with arrows: It is quite exceptional that as well as the bow a quiver containing arrows could be found in such good condition, dating from the Neolithic period. The tubular container was made from the hide of a member of the Capridae and has a lateral seam running from top to bottom (Egg et al., 1993). The quiver has a stiffener made from Hazel (*Corylus avellana*). A deep groove has been cut in the stick to accommodate the margins of the container, where they were sewn together. This groove has holes bored in it at regular intervals, ca. 4.5 cm apart. Leather thongs were threaded through these holes and used to secure the quiver to the stiffener. The terminal ring is still recognizable on the Hazel (*Corylus avellana*) and allows to ascertain the felling date. Since the latest tree ring is completely formed, the Hazel (*Corylus avellana*) must have been cut outside of the growing period.

The arrows from the quiver provide further information about the bow, because the projectile and the firing weapon must perform perfectly together to guarantee a high degree of efficiency and accuracy. The elasticity of the arrow shaft is of vital importance in this respect. It affects the vibration property of the shaft, which must be exactly tuned to the particular bow used. In general, the more elastic the wood used for the shaft, the lower shaft-diameter is needed. This reduces the weight of the projectile, which is an advantage for both hunting and fighting. Furthermore the shaft must be completely straight and free from any warping to guarantee a good shot. This implies that only well-seasoned, dry wood with a high degree of stability – wood with the least degree of radial and tangential shrinkage – needs to be used (Beckhoff, 1965). In this respect too, the Iceman shows that he possessed the elementary knowledge about manufacturing arrow shafts. Already in his choice of wood, he had selected the most suitable species of his native woods for the arrow shafts. All the 14 arrows are made from sprouts of the Wayfaring Tree (*Viburnum lantana*). The wood of this shrub is distinguished by a high degree of elasticity and excellent wood-working qualities (Beckhoff, 1963). The shafts were made from complete stems (roundwood), that had been carefully prepared to different degrees. In most cases the periderm was removed completely. The wood surface, however, has not been worked on any further, apart from some coarse cuts intended to reduce the length to the desired even diameter. The side branches have been roughly removed. The anterior end is split to accommodate an arrowhead. On some of the arrows this notch is very carefully and cleanly cut out, though in the majority it is confined to a simple radial split. The shafts of the latter arrows also appear to have been only half-finished and were not capable of being shot. The shaft lengths vary from 84–87.5 cm (Table 3a, b). Arrow no. 13 is broken and incomplete, measuring only 69 cm. The diameter of either ends of the shafts are 8 mm and 13 mm, respectively. In general, prehistoric arrow shafts that have been made from native woods and finished off to a circular cross section, have a diameter of 8–9.5 mm (Beckhoff, 1965). This is true except for nos. 2 and 3 of the Iceman's arrow shafts, a fact which provides a further indication that the arrows need further work doing on them. Only two of the arrows, nos. 12 and 14, had been equipped with a flint point, notched on the posterior end, and fitted with feathers. Whether

Table 3a. Dimensions of the arrow. *The broken arrow no. 13, measurements made on the breakage point

Arrow no.	Shaft length	Diameter measured at the arrow head	Diameter measured at the shaft end
1	84.20 cm	0.70 cm	0.86 cm
2	87.40 cm	0.77 cm	1.20 cm
3	87.50 cm	0.75 cm	1.30 cm
4	84.00 cm	0.79 cm	0.83 cm
5	87.20 cm	0.68 cm	0.88 cm
6	87.10 cm	0.78 cm	0.88 cm
7	87.00 cm	0.73 cm	0.93 cm
8	87.00 cm	0.76 cm	0.93 cm
9	86.90 cm	0.69 cm	0.94 cm
10	86.70 cm	0.70 cm	0.92 cm
11	87.10 cm	0.70 cm	0.93 cm
12	87.20 cm	0.80 cm	0.89 cm
*13	69.00 cm	0.63 cm*	0.80 cm
14	81.30 cm	0.61 cm	0.95 cm

Table 3b. Measurements of the anterior parts of the arrow shafts nos. 12 and 14

Arrow no.	Length measured at the anterior part of the arrow shaft	Diameter of the thickened regions	
		at the arrow head	at the posterior end
12	10.60 cm	0.80 cm	1.50 cm
14	9.50 cm	0.61 cm	1.00 cm

they were capable of being shot is questionable, since both these arrows were also in a broken state when found (Egg et al., 1993). In addition, they both exhibit certain further peculiarities (Plate 2). Arrow no. 12 consists of two parts, which lay apart in the quiver. The posterior part of the shaft, like the other arrows, was made from wood of the Wayfaring Tree (*Viburnum lanata*), whereas the anterior part was made from Dogwood (*Cornus* sp.). The latter wood is also well suited for making arrow shafts and was found in excavations of the lake dwelling sites in Mondsee in appreciable numbers and uses (Franz and Weninger, 1927). The anterior part of the arrow no. 12 is pegged in a slot made in the posterior part and for safety bound and glued fast. The difference in the specific weight of these two woods is scarcely so great that the combination would have had any noticeable effect in altering the centre of gravity in the region of the shaft. It is more likely that this is a repair of a shaft. The posterior part comes from a broken arrow shaft, which is then joined onto a new anterior portion made of Dogwood (*Cornus* sp.). Shaft combinations of that kind are known from the Mesolithic period onwards (Rust, 1943; Beckhoff, 1966). Such fraction between the centre of gravity and the arrowhead can arise when an arrow hits a hard target at an angle, or, when fired the shaft axis becomes deflected either by a strong cross-wind, or by an intervening obstacle, by more than 2.5° from the ballistic curve.

Arrow no. 14 is similarly distinguished by its particular method of construction. Its shaft is carved in its entirety from a sappy shoot of the Wayfaring Tree (*Viburnum lantana*). Its peculiarity is a 9.5 cm long thickening in the region of the arrowhead. The tapering-off below, although distinct, is not marked in a way that it would have been intended as a deliberate breakage point. Even an intentional shift in the centre of gravity will have been only slight and would have been better achieved via the arrowhead. The intention underlying this thickening must therefore have been related to some other physical characteristics of the arrow shaft. As mentioned previously, the shaft length needs to be exactly harmonized with the bow in use, to achieve the optimal efficiency of the shot. The position adopted by prehistoric archers when firing arrows can only be conjectured. Nevertheless, to ensure an unerring aim, the arrow must always be positioned on the bow in the same way and shot off at a constant acceleration. Whichever the stance adopted during the draw – upright or crouched – the optimal utilisation of the bow's power is achieved when the bow is bent with the arm outstretched, and when the bowstring is pulled to the side of the archer's face, the nearer to the sight line the better for aiming (Tucker, pers. comm.). Aiming at the desired target is achieved by lining up on the direction of the arrowhead. From the above criteria the minimal length of an arrow shaft, in relation to the draw-length of the bow and to the thickness of the bow in the grip-region, can be calculated (Beckhoff, 1972). These considerations exclude any excessive projection of the arrowhead beyond the body of the bow, but sighting along the arrow is more positive with the arrow protruding. Hunting scenes depicted in rock paintings nonetheless confirm that prehistoric archers quite obviously allowed the arrow to project by a considerable

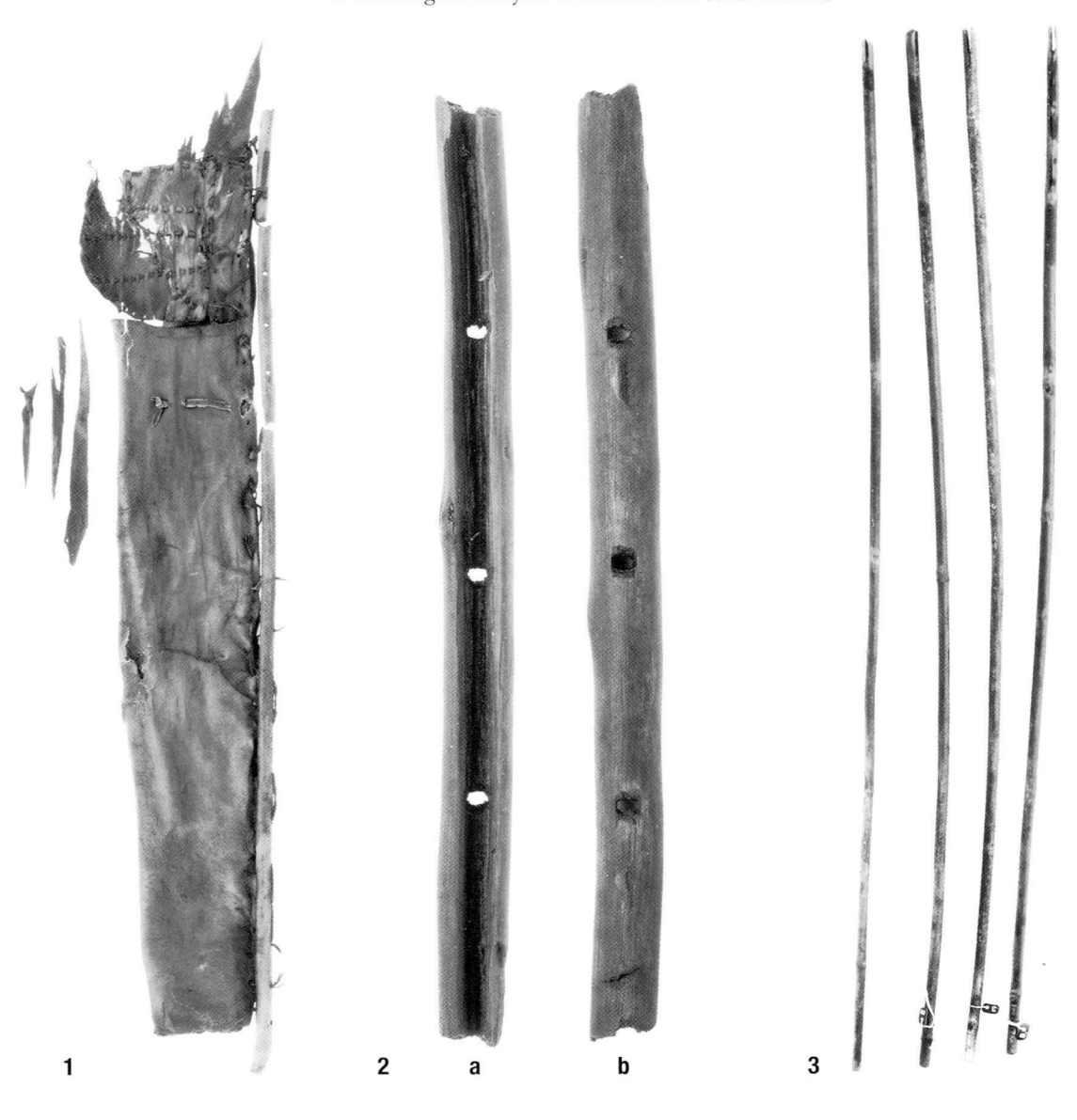

Plate 2. The quiver: **1** the quiver, **2** the broken piece of the stiffener made of Hazel, *Corylus avellana*: **a)** inner side **b)** outer side, **3** incompletely finished arrows made of Wayfaring Tree, *Viburnum lantana*, **4** detail of the arrows nos. 12 and 14: showing the thickenings and the repair made on the anterior ends. (Photos nos. 1, 2a, 2b, 3 and 4 by the Römisch-Germanisches Zentralmuseum Mainz)

amount (Clark, 1963; Beckhoff, 1963, 1966), but it is always dead weight in flight adding to gravity's pull. On the other hand this projection has a deleterious effect on the ballistic of the projectile, because arrow shafts are highly elastic objects, which when shot are subject to their free oscillation. Such an excessive projection has a bad effect on this natural oscillation of the shaft and calls for a strengthening of the shaft to avoid any deviation of the intended line of fire (Beckhoff, 1972). The height of the Iceman (1.60 m; according to Seidler et al., 1992) yields a theoretical minimum arrow shaft length of 75 cm, which taken together with the 9.5 cm thickening of the anterior shaft of arrow no. 14 matches up to the Iceman's shaft length.

In most cases the terminal ring is still recognisable on the arrow shafts, although only in the case of one arrow (no. 14) can an estimate be made of the time of the year at which the stem was cut. The last tree ring is almost fully formed, i.e., it has been cut sometime in the summer, just prior to the end of the growing period.

The axe: The axe handle, being made of Yew wood (*Taxus baccata*), is another surprising discovery. In contrast to the bow, the axe shaft has been most carefully finished off. Only the posterior end of the handle is coarsely trimmed, compared to the fine handwork of the rest of the shaft (Plate 3). Excepting that part, the surface is carefully smoothed off and scarcely any cut marks are visible. In this case too, the wood has the typical reddish tinge of yew heartwood. A remarkable feature is that the lower end of the handle is without a knob or another shape for ease in holding it. It has the form of an angled shaft. The raw material was so orientated in relation to the stem from which it was cut, that the handle piece was part of the stem and the fixing point of the axe blade is formed by a split branch. The angle formed between the branch and the stem segment is 77°. The entire shaft is carved from heartwood, the lighter coloured sapwood having been removed. The tree rings of the handle section are less than 1 mm thick, which ensures that it possesses the necessary solidity. In the Neolithic period yew was seldom used for axe shafts, or at least only a few samples have been preserved. Finds such as those made at Robenhausen (Switzerland), where in addition to clubs made of Yew wood (*Taxus baccata*) an axe shaft was discovered (Messikommer, 1913), are therefore an exception. Oak (*Quercus* sp.) was the preferred wood in the Neolithic period in the majority of cases (Schweingruber, 1976).

The pack-frame: The remains of a pack-frame were found just beside the axe on the rock rib (Fig. 2). It consisted of a roundwood bent into a U-shape, together with two coarsely-worked laths lying loose. The roundwood is 1.99 m in length and was made from a 3 cm thick branch of Hazel (*Corylus avellana*). The branch is straight and even for two-thirds of its length, wherefore, at the point where a side branch had diverged, growth had been slightly off vertical (Plate 4). All the side branches had been removed by coarse cutting. The periderm is absent though the terminal ring is obvious and undamaged. An analysis of the last tree ring has revealed that it was largely fully-formed but not finally completed. The presence of criss-cross impressions on the surface, in the region of the bow-shaped curvature, indicates that this part had been covered with a wrapping. These impressions are even more visible from a slight staining on the wood surface.

The roundwood is notched at both ends (Plate 4). A 6 cm long notch had been made on the bottom end of the thicker shank, to serve for the attachment of the frame. A 1 cm broad and 0.3 cm deep impression is clearly discernible in the lower corner of this notch, into which the narrow side of a lath fits. The thinner end of the roundwood has two, v-shaped, coarsely cut notches, one above the other and 6 cm apart. These probably served to hold the laths in place.

The two laths are 38 and 40.5 cm long, respectively, and 5.9 and 4.3 cm wide. Both are equally coarsely finished off. Pegs had been shaped out of the ends of the laths with rough cuts. The longer of the two laths had an 9.4 mm deep semi-circular hollowing 27.9 mm wide. The minimum curvature of the tree rings leads to the conclusion that the laths had been split off tangentially from a massive trunk of Larch (*Larix decidua*). Such pieces are readily obtained from trees that have been blown down by gales or struck by lightning. Ten and seven tree rings are visible on the radial breakage surface of the laths. The tree rings have an average width of 1 mm, indicating that the tree had grown on a relatively favourable site below the tree limit.

A further Larch lath (*Larix decidua*) was freed from the ice during the excavations in 1992 (Plate 4). This one, rhomboidal and wedge-shaped in cross-section, was appreciably shorter in length (16.5 cm) than the other two. The weak curvature of the tree rings indicated that this one had also been derived from a tangential split from a trunk. Eleven tree rings, with an average thickness of 4.96 mm, were counted on the radial breakage surface (Table 4). Like the other two longer laths, the surface of this one is rough and unsmoothed. A semi-circular hollow, 1 cm deep, had been cut into one end by making four coarse cuts. The diameter of this notch was 27 mm. This short lath could not be fitted together with the two longer ones. It was probably an additional lath that had been tied onto

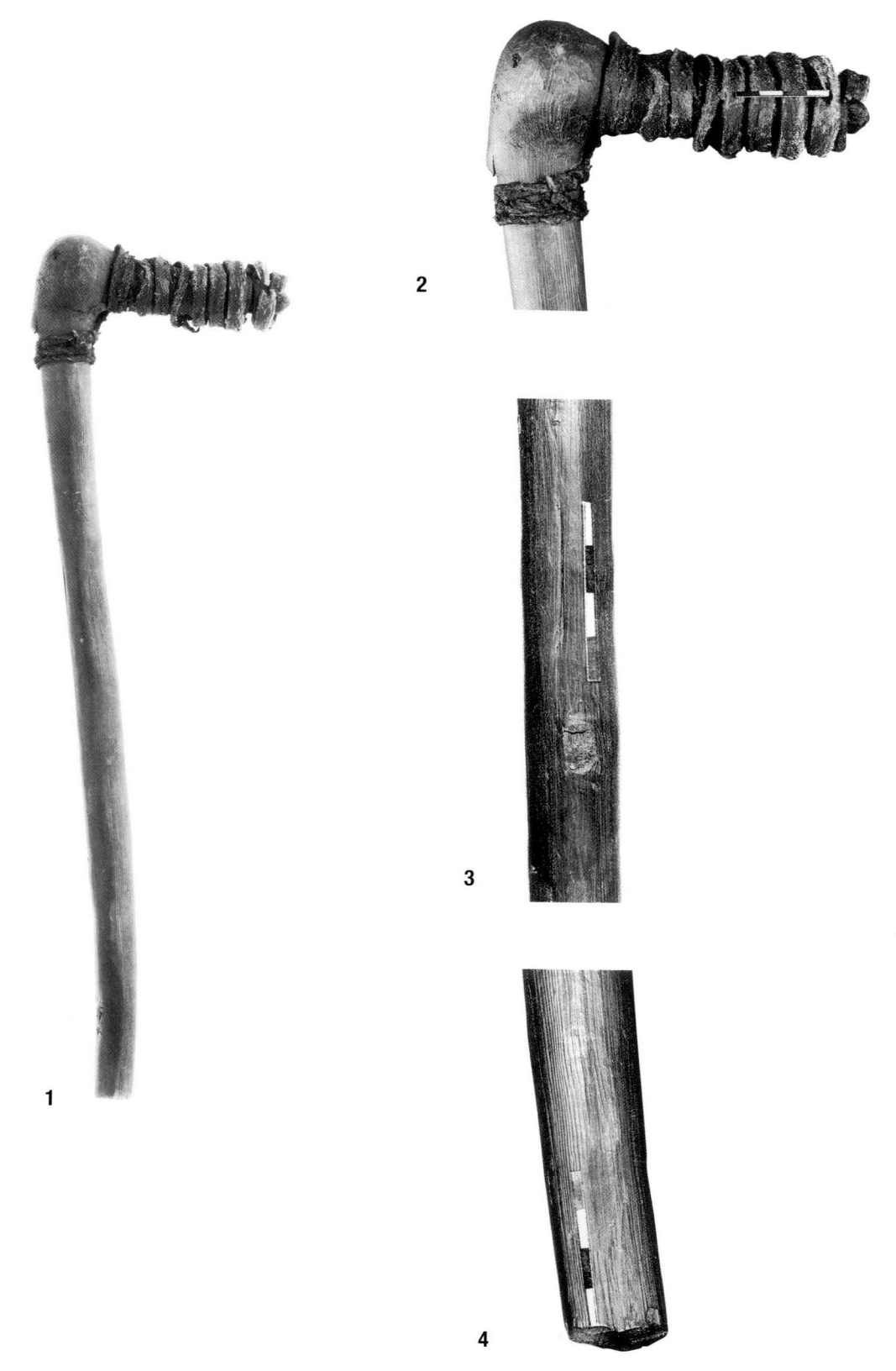

Plate 3. The axe made of European Yew, *Taxus baccata*: **1** the axe handle, **2** the axe head with the attachment points for the blade, **3** middle part of the handle showing a branch scar, **4** enlargement showing the surface of the lower end of the axe handle. (Photo no. 1 by the Römisch-Germanisches Zentralmuseum Mainz)

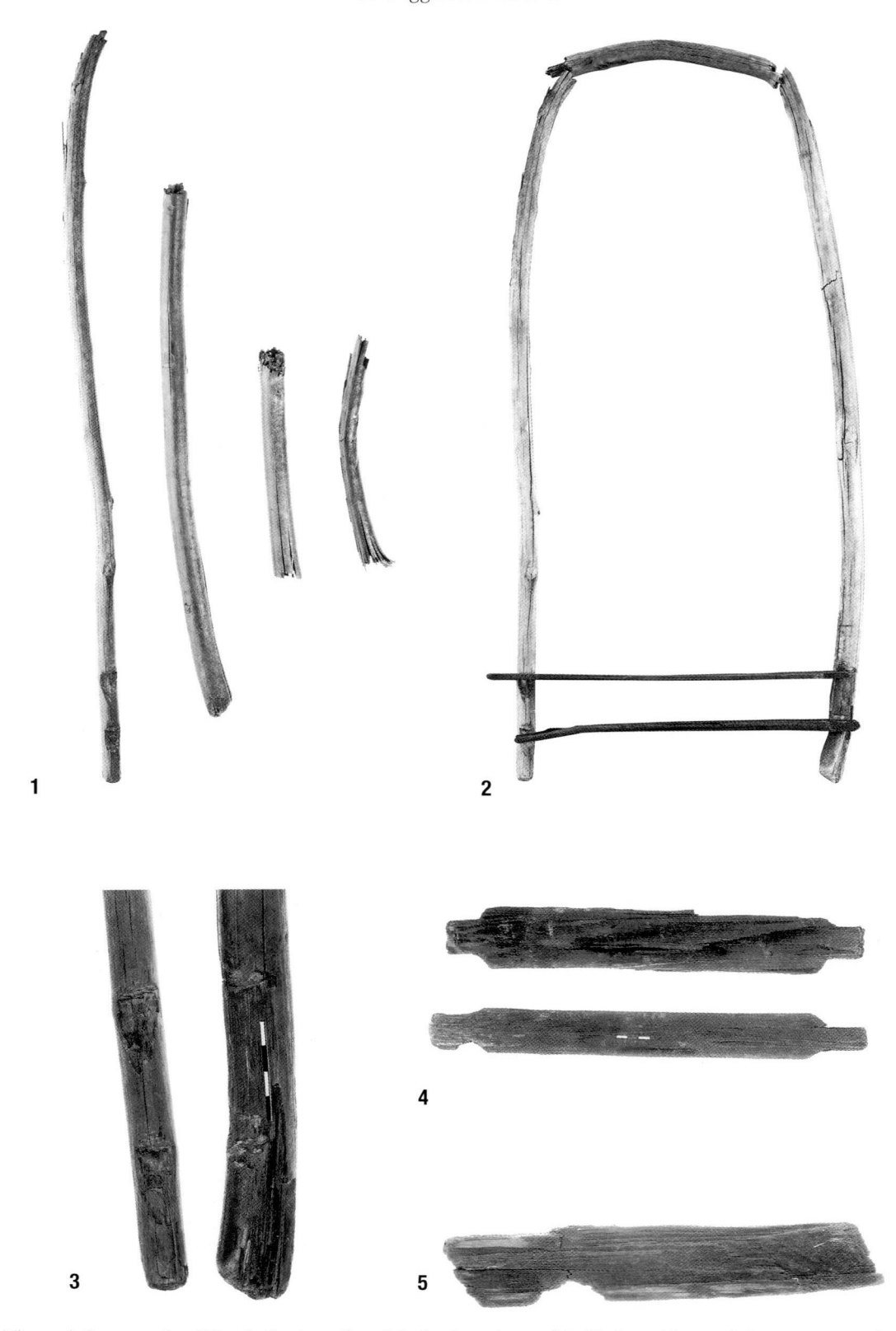

Plate 4. The pack-frame made of Hazel, *Corylus avellana*: **1** the broken pieces of the U-shaped frame, **2** the reconstructed pack-frame, **3** detail showing the ends of the U-frame roundwood with the notches for holding the laths in place, **4** the laths made of Larch, *Larix decidua*, **5** additional lath (sample no. RGZM-92/400) made of Larch, *Larix decidua*; (Photos nos. 1 and 2 by the Römisch-Germanisches Zentralmuseum Mainz)

Table 4. Tree-ring width measured on wood slivers taken from the bow (European Yew, *Taxus baccata*) and from one of the laths of the pack-frame (Larch, *Larix decidua*)

Yew, *Taxus baccata* ring width (mm)	Larch, *Larix decidua* ring width (mm)
0,48	1,2
0,6	0,9
0,54	0,4
0,6	0,5
0,57	0,9
0,54	1,0
0,36	1.1
0,51	1,0
0,51	1,0
0,54	0,9
0,54	0,7
0,57	–
0,50	–
0,24	–
0,18	–
0,18	–
0,16	–
0,48	–
0,40	–
0,38	–

the pack-frame. Such primitive pack-frames are still in use at present in Siberia.

The retouching tool: There were initial difficulties in the interpretation of this implement. It is a piece of a branch of Lime (*Tilia* sp.), 10.5 cm in length and 2.6 cm thick (Plate 5). The lower end is cut straight across. About 1 cm above the end, there is a circular groove which though rather shallowly made, allowed attachment of a cord. The branch piece is completely debarked, but the microstructure makes plain that the terminal ring remains intact over the entire surface. The latest annual ring is fully formed, showing that the branch was cut after the end of the growing period, i.e., between the late autumn and spring. The upper end is trimmed to a conical shape. A peg is inserted into the pith at this pointed end, which has resulted in the stub becoming split along one side. Only after an x-ray investigation, it was possible to conclude that the peg is one of the points from a deer antler (Egg et al., 1993), as well as providing a clue for its use as a tool for retouching flint blades.

The flint dagger: The handle of the dagger (Plate 5) was made of Common ash (*Fraxinus excelsior*). It has a length of 8.9 cm and a breadth of 2 cm, thinning to 1.1 cm at top. The surface is only roughly smoothed off. Two opposing grooves had been made near the lower end, in one of which a cord wrapped around the handle is held fast. The tree rings have only a slight curvature and run parallel to the greatest diameter. The dagger handle is made from a piece of lath that had been split off tangentially from a tree trunk. The blade is inserted in parallel with the course of the tree rings, this could have led to a tangential split arising in the handle during its use. Nevertheless, there is no particular preference known so far for shafting the blade related to the orientation to the tree rings of the handle in the Neolithic period (Schweingruber, 1976).

2.4. Results of the analyses to determine the felling date

Microscopic examination of the latest formed tree ring of a piece of wood yields information about the time of year at which the tree was felled. Different cell types are formed within the tree rings at different times during the growing period, which anatomical examination reveals. The precondition is that the terminal ring is clearly discernible in the wood piece under investigation. In ideal circumstances this means that the bark should still be present (Schweingruber, 1976). For many of the wood samples in the present investigation, this was not the case. Furthermore, because of the unique circumstances of these finds, only selected samples were available for examination. Various cutting dates were found for the individual artefacts (Table 5). This is not surprising, since the quality of an implement is directly related to the moisture content of the wood used during its construction. If an arrow shaft were to be made from fresh green wood, it would undoubtedly warp by shrinkage, both tangentially and radially, whereby no accuracy in flight could be guaranteed. On this account, the twigs of the Wayfaring tree (*Viburnum lantana*) used for the shaft of arrow no. 14 was cut at the end of the growing period and the remains of another shaft (RZGM-92/106) were cut outside the growing period, and then stored to season. At present, a minimum of 8 weeks storage is recommended (Beckhoff, 1965). Nor would excessive shrinkage be desirable in the case of the wood used for stiffening the quiver, therefore it was cut outside the growing period, too. The wood of the pack-frame, on the other hand, was cut during the early summer, because actively growing wood can more easily be bent to shape. The felling dates are thus related to the practical utilisation of the implement or object. The time of year at which the Iceman set off on his final trek into the mountains cannot be ascertained from such evidence.

2.5. An attempt to evaluate the finds floristically and geographically

Traces of prehistoric people in the Alps are not unusual. Numerous finds of Palaeolithic and Mesolithic

Plate 5. **1** birch-bark container, **2** the dagger made of Ash, *Fraxinus* sp., **3** the retouching tool made of Lime, *Tilia* sp., (Photos nos. 1, 2 and 3 by the Römisch-Germanisches Zentralmuseum Mainz)

Table 5. Results of the analyses made to determine the felling date

Artefact	Felling date
Pack-frame	Early summer, during the growing period
Arrow shaft no. 14	Summer or autumn, still during the growing period
Arrow shaft in sample no 92/106	Autumn or winter, before or after the growing period
Retouching tool	Autumn or winter, before or after the growing period
Quiver stiffener	Autumn or winter, before or after the growing period

remains have been made in the central regions of the Alps, at altitudes of 1900–2500 m (Bagolini and Pedrotti, 1992; Broglio, 1992; Dalmeri and Pedrotti, 1992). The frequency of finds declines in the subalpine and alpine zones at the end of the Mesolithic period, only increasing again in the late Neolithic, when human beings once more showed an interest in visiting the alpine regions (Broglio and Lanzinger, 1996). The total of Neolithic finds made in the alpine zone is nonetheless appreciably lower than for the Mesolithic period. Despite the number of Mesolithic finds made, no artefacts have been found in the Eastern Alps above an altitude of 2500 m. On this account, the find of the Iceman represents a special case and provides many enigmas. Ever since he was found in the nival zone in the main range of the Eastern Alps, the question has been raised as to whence he came and where his settlement was situated. This floristic and geographical assessment of the dendrological finds is an attempt to answer these questions.

Wood was an important raw material in the Neolithic period. A basic knowledge of the physical properties of different woods and their fashioning for different purposes, enabled the Neolithic man to make optimal use of the indigenous woody plant species. A ready and sufficient availability is equally important for the construction of implements. This is not to deny the importance of trade in the Neolithic period by any means, though the assumption nevertheless is, that the woodland resources in the immediate vicinity of the settlement will be the first to be utilized. It should be possible, therefore, to delineate the potential settlement area of the Iceman by reference to the spectrum of woods he used.

All the woody species so far identified have a circumpolar or alpine distribution, which, on a larger scale, do not permit any conclusions about their origin to be drawn. The smaller-scale distributional spectra of the woody species, however, do reveal certain differences, since the individual species extend to the inner alpine region to different degrees. The present-day distributional spectra are admittedly untrustworthy, because of the major anthropogenic influences and the introduction of non-native species of ornamental trees in modern times. Floristic accounts from the preceding century (Hausmann, 1851; Dalla Torre and Sarnthein, 1900) provide a good picture of the natural distribution of woody species in the inner-alpine region.

Were it not for the influence of human economic activities, the whole of the Ötztal and the Vinschgau, together with their side valleys, would be a wooded landscape, apart from some small-scale exceptional habitats and the alpine grass-sedge heaths. The climatic and edaphic conditions in the inner-alpine region favour coniferous forests, predominantly of Spruce (*Picea abies*) and Scots pine (*Pinus sylvestris*). The distribution of the more demanding deciduous forests (*Querco-Fagetea, Quercetea robori-petreae*) in the Inntal and Ötztal is hindered by the occurrence of late frosts, in particular. On this account, deciduous forest communities remain restricted to small patches in the upper Inntal and at the mouth of the Ötztal. In the southern Vinschgau and Schnalstal, on the other hand, warmth-demanding deciduous communities (*Quercetalia pubescentis*) are widespread in the valleys, from the valley-floor up to an altitude of 1100 m. However, the characteristic forest community in the inner-alpine region from the montane zone up to the subalpine zone, is Spruce forest (*Piceetum montanum, P. subalpinum*). This is in its turn replaced higher up by a Larch-Arolla pine forest (*Larici-Pinetum cembrae*) or pure Arolla pine forest (*Pinetum cembrae*). In a few places, such as the Niedertal, there is a transitional belt of Dwarf Mountain Pine (*Pinetum mugi*) between the latter and the alpine grass-sedge heaths.

Only two woody species from the alpine zone, the Green Alder (*Alnus viridis*) and a dwarf willow (*Salix reticulata*-type) have been represented among the finds associated with the Iceman, both in low numbers (Tables 1, 2, 6). The Green Alder (*Alnus viridis*) was found both as carbonized and uncarbonised fragments, the Willow (*Salix* sp.) only as charcoal fragments. The Green Alder (*Alnus viridis*) is widely distributed in the Central Alps on damp hillsides and in avalanche channels in the subalpine and lower alpine zones. It also occurs – in wet ravines and on mud-slides – even on valley bottoms in places. The lowest habitat known in the Vinschgau lies at 850 m.

Four species of dwarf willows occur in the investigation area: the Mountain Willow (*Salix herbacea*), the Alpine Willow (*Salix reticulata*), the Obtused-leaved Willow (*Salix retusa*) and the Thyme-leaved Willow (*Salix*

Table 6. Assessment of the palaeo-ecological and floristic comparative distributional data for the woody species so far found (distributional types based on Walter and Straka, 1970)

Woody plant species	Distributional area	Altitudinal zone	Vinschgau	Ötztal
European Yew, *Taxus baccata*	central European	montane	−	−
Norway Maple, *Acer platanoides*	central European	montane	+	−
Common Ash, *Fraxinus excelsior*	central European, submediterranean	montane	+	+? introduced
Juneberry, *Amelanchier ovalis*	central European, Alps	montane	+	+
Dogwood, *Cornus* sp.	central European, submediterranean	montane	+	+
Hazel, *Corylus avellana*	central European	montane	+	+
Lime, *Tilia* sp.	central European	montane	+	+
Elm, *Ulmus* sp.	central European	montane	+	+
Wayfaring Tree, *Viburnum lantana*	submediterranean	montane	+	+
Scotts Pine, *Pinus sylvestris*	subboreal	montane	+	+
Larch, *Larix decidua*	boreal	(montane), subalpine	+	+
Norway Spruce, *Picea abies*	boreal	montane - subalpine	+	+
Arolla Pine, *Pinus cembra*	boreal	subalpine	+	+
Dwarf Mountain Pine, *Pinus mugo*	central European-alpine	subalpine, alpine	+	+
Birch, *Betula* sect. *albae*	boreal	montane - subalpine	+	+
Green Alder, *Alnus viridis*	boreal-subalpine	subalpine	+	+
Dwarf Willow, *Salix reticulata*-Type	arctic-alpine	alpine	+	+

serpyllifolia). The Mountain Willow (*Salix herbacea*) colonizes damp humic soils in snowbed hollows and on moraines. It is one of those vascular plant species which are found at the highest altitudes in the Eastern Alps, in the investigated area up to 3300 m (Reisigl and Pitschmann, 1959). The Alpine Willow (*Salix reticulata*) is a basiphile species commonly found on damp, stony slopes, screes and humus-rich grassland; in the Niedertal it reaches 2500 m altitude. Leaves of both the above mentioned species are regularly to be found in the firn snow of the Niederjoch glacier just below the site where the Iceman was found. The other two species of willow (*Salix retusa* and *S. serpyllifolia*) occur more seldom in the area; both reach an altitude of 2500 m in the Niedertal (Dalla-Torre and Sarnthein, 1900).

Apart from the above, most of the wood finds belong to species that form forests in the montane zone (Table 6), sporadically penetrating as high as the subalpine zone. Among these the Spruce (*Picea abies*) is found in greatest number (Tables 1, 2, 6), a fact in agreement with the overall picture presented by the forests of the inner-alpine region (Mayer, 1974). It forms extensive forests from the montane to the subalpine zones, giving way to deciduous species only in the lowest part of the montane zone. In the subalpine zone it is replaced by Arolla Pine (*Pinus cembra*) and the Larch (*Larix decidua*). Some admixture of the Larch (*Larix decidua*) in the Spruce forests of the montane zone is also found. In the Ötztal, the montane Larch-Spruce forest (*Larici- Piceetum montanum*) reaches an altitude of 1500 m, thereafter changing at the entrance to the Vent valley into subalpine Larch-Spruce forest (*Larici-Piceetum subalpinum*) (Pitschmann et al., 1980). Spruce forests (*Piceetum montanum*) containing a mosaic of patches of Spruce-Fir forests (*Abietetum*) are found on the shaded north-facing slopes of the montane zone in the Vinschgau. Variants containing more Larch (*Larix decidua*) and Pine (*Pinus sylvestris*) are found in the montane Spruce forests growing on the sunnier, south-facing slopes (Karner et al., 1973). In the subalpine zone, the proportion of Larch (*Larix decidua*) in the forests increases on the south-facing slopes and that of Arolla Pine (*Pinus cembra*) on the north-facing slopes. The highest-lying occurrences of Spruce (*Picea abies*) in the area are at 1837 m in the Schnalstal and at 2014 m in the Ötztal (Dalla-Torre and Sarnthein, 1900).

Finds of Pine (*Pinus* sp.) wood are the second most common (Tables 1, 2, 6). There are three species of the genus *Pinus* present in the inner-alpine region. The Dwarf Mountain Pine (*Pinus mugo*) forms extensive spreads of scrub above the forest limit, above all on calcareous strata. On siliceous strata in the Central Alps the stands of the Dwarf Mountain Pine (*Pinus mugo*) are less extensive and are present at the transition to the subalpine Larch-Arolla Pine forest (*Larici-Pinetum cembrae*). Closed stands, at 2050–2370 m altitude, as in the Niedertal near Vent, the valley that leads directly to the Iceman's find site, are nowadays exceptional. The Dwarf Mountain Pine (*Pinus mugo*) growing on the south side of the watershed, in the Pfossental, a side valley of the Schnalstal, only reaches an altitude of 2200 m. However, studies of the vegetation history of the Central Alps have shown that, during the Atlantic period, Dwarf Mountain Pine scrub (*Pinetum mugi*) extended higher up (Oeggl and Wahlmüller, 1994). In general, it appears that in the Central Alps the stands of Dwarf Mountain Pine (*Pinus mugo*) had already been decimated at an early stage by clearance for pasturage (Kerner in Dalla Torre and Sarnthein, 1900; Mayer, 1974).

In the subalpine zone of the Eastern Alps, the Dwarf Mountain Pine (*Pinus mugo*) girdle of scrub forms a transition to a Larch-Cembran Pine forest (*Larici-Pinetum cembrae*) with varying proportions of Larch (*Larix decidua*) and Arolla Pine (*Pinus cembra*). This type of forest has formed the climax forest in the subalpine zone ever since the Preboreal period (Mayer, 1974; Kral, 1989, 1992; Oeggl and Wahlmüller, 1994). These stands have been much reduced over time in the course of forest thinning for pasturage and by now almost all the Arolla Pine (*Pinus cembra*) has been removed (Mayer, 1974; Kral, 1992; Zoller and Brombacher, 1984). Only a few pure stands of the Arolla Pine (*Pinus cembra*) now remain in the inner part of the Ötztal (Pitschmann et al., 1980); near Obergurgl and Vent they reach an altitude of 2330 m. The lowest occurrences in the Vinschgau have been recorded at 1432 m and in the Ötztal at 1090 m (Dalla Torre and Sarnthein, 1900).

Although there is no positive evidence for the use of Scots Pine (*Pinus sylvestris*) wood for any of the artefacts, because identification to species level was impossible due to the poor state of preservation of the wood samples, it should not be ruled out. It occurs in the region, both as pine forests and frequently as an admixture in other forests types. The Scots Pine (*Pinus sylvestris*) is primarily found on poor soils in the montane zone, but it also occurs in the subalpine zone. It forms extensive Pine forests (*Erico-Pinetum sylvestris*) in the upper Inntal on extreme calcareous sites. At Zwieselstein, near Vent in the Ötztal, it reaches an altitude of over 2050 m (Pitschmann et al., 1980). In the Vinschgau, with its low precipitation and continental climate, it forms the inner-alpine steppe-heath-pine forest association (*Astragalo-Pinetum*) at lower altitudes on south-facing slopes with shallow soils; at higher altitudes, in the transitional zone to the Downy Oak forest (*Quercetum pubescentis*) it forms a Downy Oak-Scots Pine forest (*Antherico-liliaginis-Pinetum*) (Mayer, 1974; Karner et al., 1973; Peer, 1981).

In the absence of human influence, the Larch (*Larix decidua*) would cover a greater area than it does today. Those areas of the subalpine zone that were cleared to increase pasturage, in particular, represent its potential habitats, where formerly it formed a mixed forest with Arolla Pine (*Pinus cembra*), the Larch-Arolla Pine forest (*Larici-Pinetum cembrae*), or with the Norway Spruce (*Picea abies*), the subalpine Spruce forest (*Larici-Piceetum subalpinum*) on siliceous rocks. The Larch (*Larix decidua*) frequently forms an admixture in the Spruce forest (*Piceetum montanum*) in the montane zone below.

Table 7. A taxonomic-uniformitarian approach of habitat recognition, in which the Iceman lived, based on the ecological demands of woody species found (① woodlands which do not occur in the Ötztal, ◆ woodlands and species distributed on both sides of the main Alpine range, ② species which occur only in the valley bottoms south of the main Alpine range)

	Downy Oak forests	Oak forests	Pine forests	Spruce forests	Gorges	Alluvial forests	Hedges, forest fringes	Timber-line	Krumm-holz	Snow bed communities
Ötztal < 100 km	①	◆	◆	◆	①	◆	◆	◆	◆	◆
Vinschgau < 25 km	◆	◆	◆	◆	◆	◆	◆	◆	◆	◆
Taxus baccata	②	–	–	–	②	–	–	–	–	–
Picea abies	–	–	–	◆	–	◆	–	–	–	–
Larix decidua	–	–	–	◆	–	–	–	◆	–	–
Pinus cembra	–	–	–	–	–	–	–	◆	–	–
Pinus mugo/sylvestris	◆	◆	◆	◆	–	–	–	–	◆	–
Salix reticulata	–	–	–	–	–	–	–	–	–	◆
Alnus viridis	–	–	–	–	◆	–	–	–	◆	–
Betula sect. albae	–	◆	◆	◆	–	–	–	–	–	–
Corylus avellana	◆	–	–	–	◆	◆	◆	–	–	–
Ulmus	◆	–	–	–	◆	◆	◆	–	–	–
Amelanchier ovalis	◆	◆	◆	–	–	–	–	–	–	–
Acer platanoides	–	–	–	–	②	②	–	–	–	–
Cornus	◆	◆	–	–	–	◆	◆	–	–	–
Fraxinus excelsior	◆	–	–	–	◆	◆	–	–	–	–
Viburnum lantana	◆	–	◆	–	–	–	◆	–	–	–
Tilia	◆	◆	–	–	◆	◆	–	–	–	–

48   K. Oeggl and W. Schoch

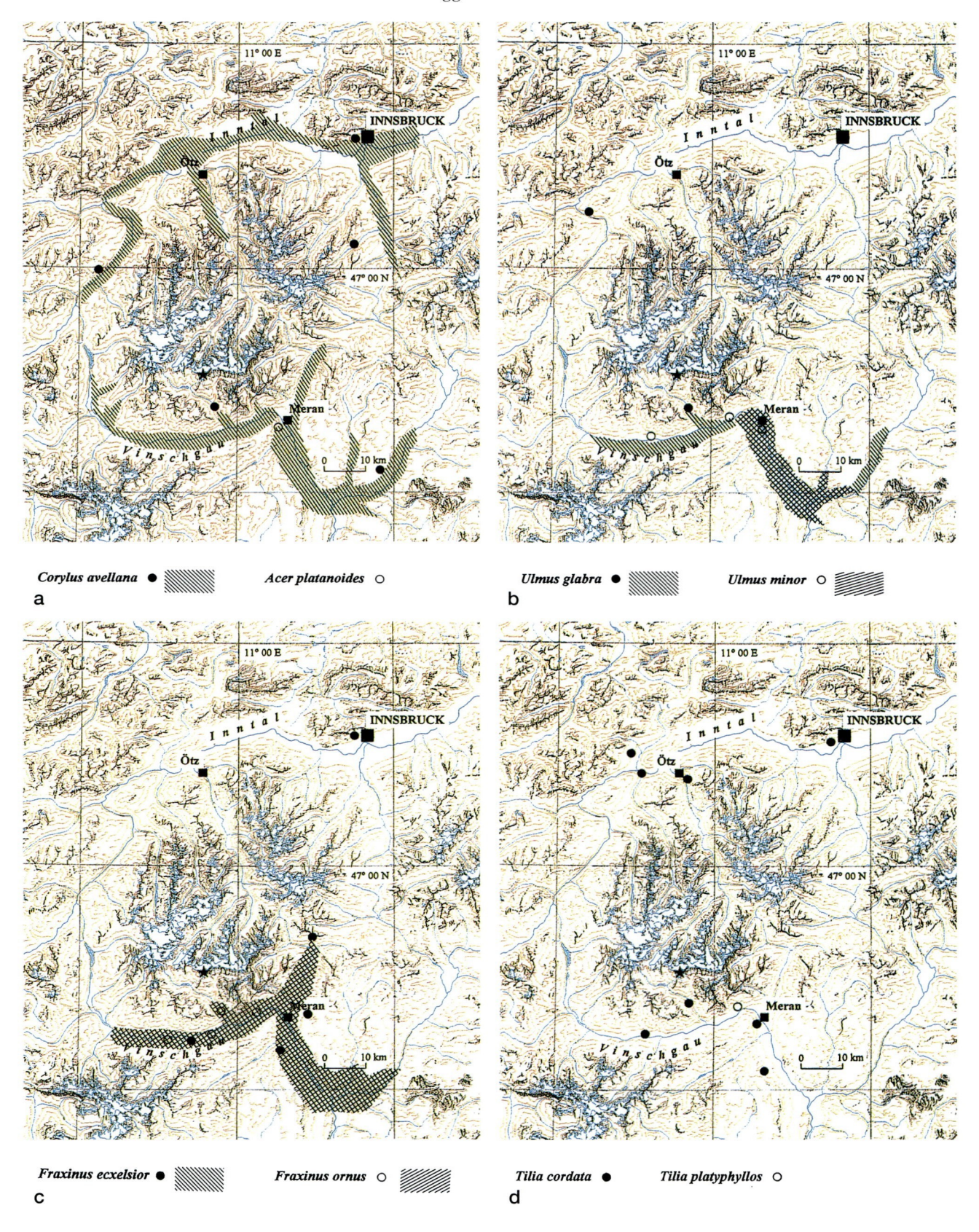

Corylus avellana ●   *Acer platanoides* ○

a

Ulmus glabra ●   *Ulmus minor* ○

b

Fraxinus ecxelsior ●   *Fraxinus ornus* ○

c

Tilia cordata ●   *Tilia platyphyllos* ○

d

Due to preferential felling of the evergreen conifers in these mixed forests, pure stands of Larch (*Larix decidua*) occur in both the montane and subalpine zone in places that are used as hay meadows or as pasturage. Nevertheless, according to Pitschmann et al. (1980) in the Schnalstal most of the Larch (*Larix decidua*) stands are natural. This view is supported by the very obvious limitation of the Larch (*Larix decidua*) by soil aridity, its distribution in all altitudinal zones, and the poor suitability of these habitats for farming use. The highest-lying sites for the Larch (*Larix decidua*) are at 2247 m in the Schnalstal and at 2100 m near Vent in the Ötztal (Dalla-Torre and Sarnthein, 1900).

The Birch (*Betula* sect. *albae*) has a wide ecological amplitude and is found in both the Ötztal and the Schnalstal from the valley floor up to the forest limit. As a pioneer tree species, it prefers unshaded habitats and grows throughout the region on poorer soils and on boulder-strewn hillslopes. In the Ötztal it ascends to an altitude of 2050 m. In the lower part of the Schnalstal it is abundant in the Gray Alder (*Alnus incana*) grooves of the riverine woodland (*Alnetum incanae*), at times forming pure stands there (Peer, 1981).

All the remaining finds of woods of deciduous tree species are of those restricted to the montane zone. The species concerned are those that are predominantly present in the richer deciduous forests and scrub. Because of the edaphic and climatic conditions in the upper part of the Inntal, deciduous forests of any extent are scarce on the valley floor, away from areas subject to regular flooding, these patches of forest have been further decimated by anthropogenic influences and converted into arable land and meadows. At the mouth of the Ötztal, about 100 km away from the find site of the Iceman, there is a solitary occurrence of a Pedunculate Oak-Small-leaved Lime forest (*Quercetum roboris*) (Pitschmann et al., 1980). The richer deciduous forest communities of the Vinschgau penetrate up the valley closer to the find site. Downy Oak forests (*Quercetum pubescentis*) and Manna Ash-Hophornbeam forests (*Orneto-Ostryetum*) still thrive in the lower part of the Schnalstal. These scrubby woodlands reach an altitude of 1100 m in the lower part of the Vinschgau, above which there is a transition to Downy Oak-Scots Pine forest (*Antherico-liliaginis-Pinetum sylvestris*) (Peer, 1981). Higher up still, there is montane Spruce forests (*Piceetum montanum*), although on the sun-exposed south side of the Vinschgau, because of human influence only fragments remain intact. In most cases variants of this community, rich in Larch (*Larix decidua*) and Scots Pine (*Pinus sylvestris*), are encountered (Karner et al., 1973).

In these deciduous forests, the species represented by finds of their woods show different frequencies to the north and the south of the main alpine chain (Table 7). The most numerous are those species that show a preference for the well-drained, mineral-rich soils of damp habitats. The Lime (*Tilia* sp.) is represented in this area by two species, the Small-leaved Lime (*Tilia cordata*) and the Large-leaved Lime (*Tilia platyphyllos*). The former grows on damp, well-drained to relatively dry soils in a climate characterized by summer-warmth. It is found in both the North Tyrol and the South Tyrol, and penetrates further up into the inner-alpine valleys than the Large-leaved Lime (*Tilia platyphyllos*) does (Fig. 5). The highest-lying site of the Small-leaved Lime (*Tilia cordata*) and the nearest to the find site of the Iceman is Ratteis in the Schnalstal at an altitude of 1000 m, in a damp, wooded ravine. The Large-leaved Lime (*Tilia platyphyllos*), on the other hand is seldom found in the North Tyrol, and in most cases they are planted trees, since it requires a mild winter climate (Gams, 1933). It is also absent from the Vinschgau and in the Etschtal only becomes at all common below Meran (Dalla-Torre and Sarnthein, 1900). The Norway Maple (*Acer platanoides*) has similar ecological requirements to those of Large-leaved Lime (*Tilia platyphyllos*). At the present time, the Norway Maple (*Acer platanoides*) does not even penetrate into the Vinschgau, its distributional limit lying just short of that valley, on the Marlinger Berg near Meran (Dalla-Torre and Sarnthein, 1900). In the North Tyrol, and especially in the upper Inntal, its indigenous status has been called into question, since it is seldom seen in natural habitats and frequently present as planted trees (Gams, 1933). The Common Ash (*Fraxinus excelsior*) also, only occurs as a planted tree in the upper Inntal and does not spread there by natural regeneration (Fig. 4). These floristic observations made by Dalla Torre and Sarnthein (1900) are confirmed by the results of recent pollen analytical investigations made in this area (Oeggl, unpubl. data). The Common Ash (*Fraxinus excelsior*) is commonly encountered in the Vinschgau on roadsides and near farmhouses, where it has been planted as a source of leaf-fodder. It occurs naturally in the lower Vinschgau near Laas (Dalla-Torre and Sarnthein, 1900). The second indigenous ash species, the Manna Ash (*Fraxinus ornus*), is a submediterranian element and only distributed in South Tyrol (Fig. 4).

Fig. 4. Distribution map of the woody species in the investigation area. The places of the highest occurrence in the investigation area is marked with a black dot: **a** Hazel (*Corylus avellana*) and Norway Maple (*Acer platanoides*), **b** Elms (*Ulmus* sp.): the Wych Elm (*Ulmus glabra*) and the Smooth-leaved Elm (*Ulmus minor*), **c** Ashes: the Common Ash (*Fraxinus excelsior*) and the Manna Ash (*Fraxinus ornus*), **d** Limes (*Tilia* sp.): the Small-leaved Lime (*Tilia cordata*) and the Large-leaved Lime (*Tilia platyphyllos*)

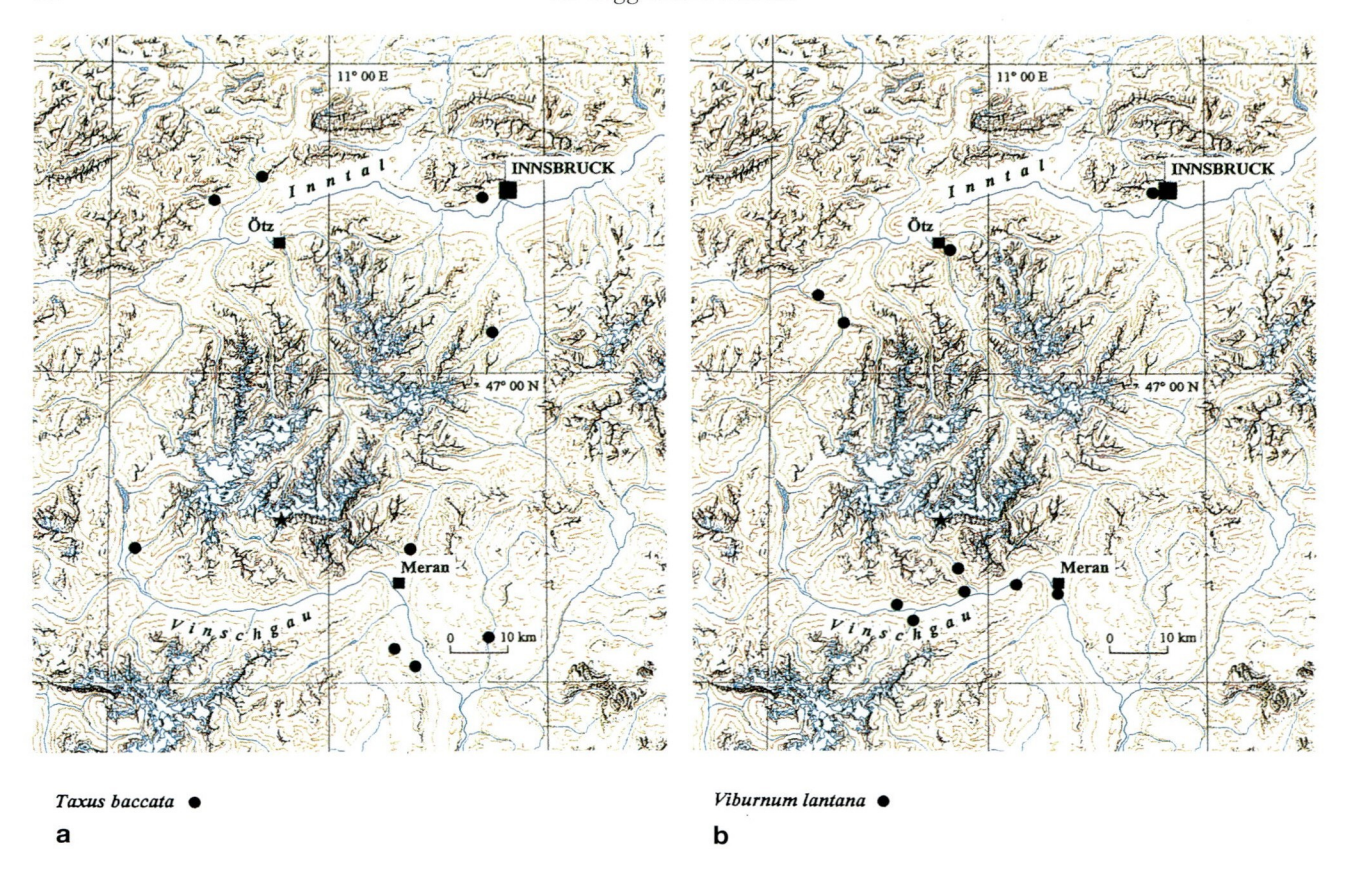

Taxus baccata ●

a

Viburnum lantana ●

b

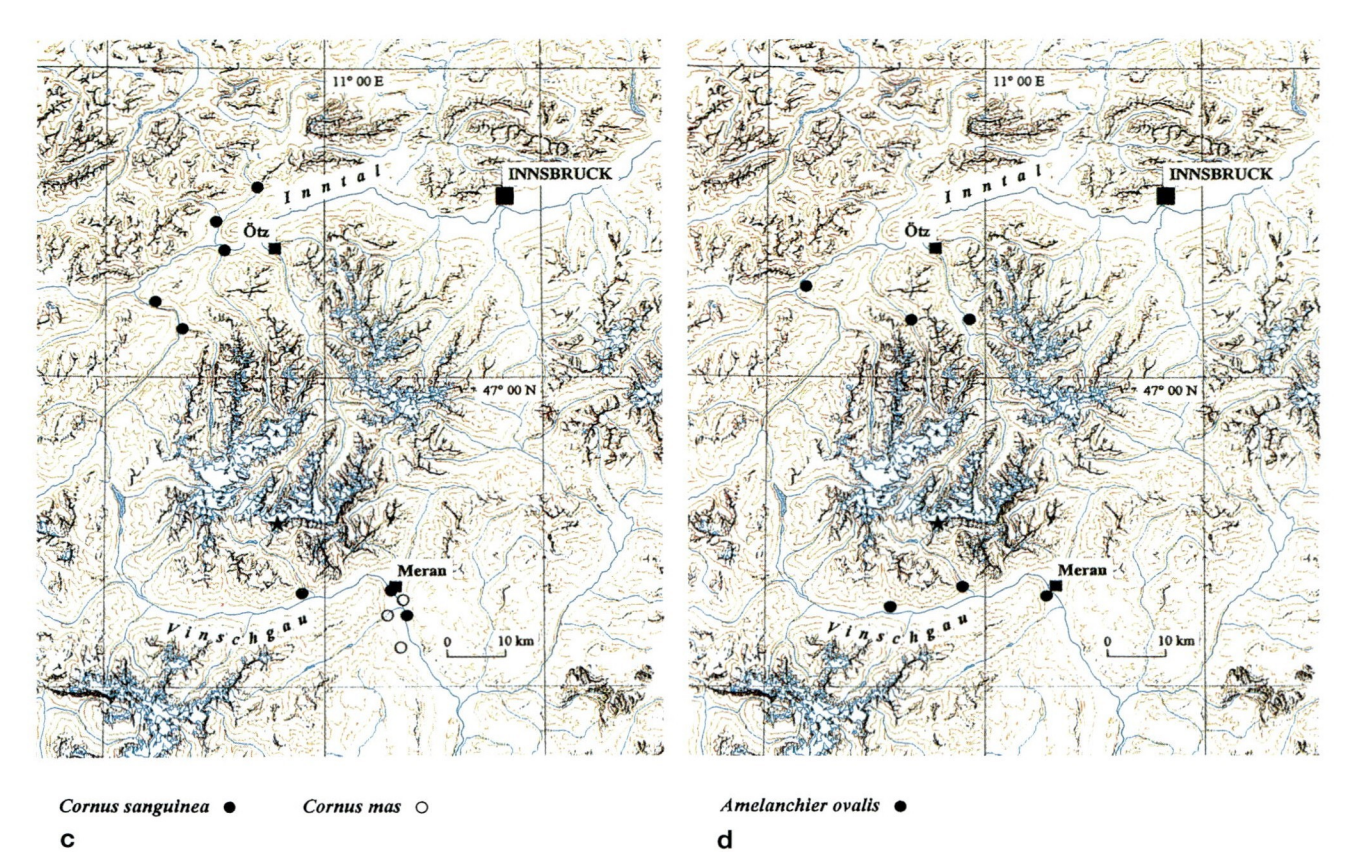

Cornus sanguinea ●   *Cornus mas* ○

c

Amelanchier ovalis ●

d

The Wych Elm (*Ulmus montana*) is generally distributed in wooded ravines and in shady hillside forests in the beech and Spruce region, up as far as the forest limit (Fig. 4). The Smooth-leaved Elm (*Ulmus minor*), on the other hand, is not present in either the North or the South Tyrol as a naturally occurring tree (Gams, 1933). The Wych Elm (*Ulmus montana*) occurs in the lower part of the Schnalstal as far up as Karthaus (1330 m). The Hazel (*Corylus avellana*) is commonly present as an undershrub in fairly open woodlands, on woodland margins, on ridges between fields and on waysides up to an altitude of 1300 m (Fig. 4). The nearest site to that of the Iceman is at Karthaus in the Schnalstal (Dalla-Torre and Sarnthein, 1900).

The European Yew (*Taxus baccata*) also prefers humid habitats with a mild winter climate. Its main distribution lies in the montane zone on the alpine margins. Although it has a mainly oceanic climatic distribution, it is resistant to drought and to late frosts and in habitats with a favourable microclimate can also grow in regions with a continental to subcontinental climate, as shown by isolated occurrences in the Central Alps in both North Tyrol (Gschnitztal, Wattental, Schwaz, Mayrhofen) and Eastern Tyrol (Lienz). Low winter temperatures are the controlling factor in regard to a continental climate. Frost damage is only observed below −23°, but on account of its evergreen habit it becomes so weakened by frost-drought, that the stands die off (Leuthold, 1980). This may be one reason why the European Yew (*Taxus baccata*) is absent from lower-lying parts of valleys, where in winter temperature inversions regularly occur. Its altitudinal distribution is decisively related to the accumulated warmth recorded during the growing period. Apart from valleys subject to warm wind influences, where it exceptionally reaches 1600 m, it is seldom found in the Alps above altitudes of 1200–1400 m (Hegi, 1935). Its ecological requirements are similar to those of the Beech (*Fagus sylvatica*), although, in the inner-alpine region, the European Yew (*Taxus baccata*) extends beyond the climatic spectrum of that tree. There the European Yew (*Taxus baccata*) also thrives in such low-rainfall transitional regions together with Sessile Oak (*Quercus petraea*), Downy Oak (*Quercus pubescens*) and the Scots Pine (*Pinus sylvestris*), as well as in mixed Lime forest (Leuthold, 1980). On the margins of its distribution, the European Yew (*Taxus baccata*) is restricted to sites with higher humidity (gullies and ravines). At the present-day the European Yew (*Taxus baccata*), north of the main alpine chain, occurs in the mixed beech forests (*Abieti-Fagetum*) growing in the northern limestone Alps. It is absent from the Ötztal and the Vinschgau, probably on account of the continentality of the climate (Dalla-Torre and Sarnthein, 1900). The nearest localities for European Yew (*Taxus baccata*) to the find site of the Iceman are near the mouth of the Ötztal (Imst) to northward, and in the Etschtal to southward (Fig. 5). In the latter area it occurs in Tisens in a montane mixed forest (*Abieti-Fagetum*) on the mid-altitude terrace above the valley, and at Andrian the European Yew (*Taxus baccata*) is present in a Hophornbeam-Manna Ash forest (*Orneto-Ostryetum*) (Stuefer, 1992). A dendrological dating from bore samples taken from the latter trees showed that these stands are relatively young, 130 years at most. Nevertheless, this yew-rich Hophornbeam-Manna Ash forest (*Orneto-Ostryetum taxetosum*) at Andrian has a special importance in the region since it proves that the European Yew (*Taxus baccata*) can thrive on the margin of its distributional limit at the transition from a submediterranian to a central European type of vegetation. The former presence of Yew (*Taxus baccata*) in Hophornbeam-Manna Ash forests (*Orneto-Ostryetum*) also in the lower Vinschgau would appear to be a reasonable assumption, especially in shaded localities with a high humidity, such as wooded ravines. Since given a lower degree of forest clearance in the Neolithic period compared to today, the humidity in the Vinschgau would have been greater.

The remaining species among the wood finds such as Juneberry (*Amelanchier* cf. *ovalis*), the Dogwood (*Cornus* sp.) and the Wayfaring Tree (*Viburnum lantana*) are predominately found in the dry, mixed oak forests (*Quercetalia pubescentis*) that thrive in the lower Vinschgau as far up as Schlanders and in the Schnalstal up to the altitude of the Karthaus village (Fig. 5). They thrive best in sunny, warm and dry habitats up to the middle montane zone, where they also occur in wayside associations (*Prunetalia*) and on forest margins. The Juneberry (*Amelanchier ovalis*), the Common Dogwood (*Cornus sanguinea*) and the Wayfaring Tree (*Viburnum lantana*) are found on or near rocky outcrops in warm localities in degraded Downy Oak forests (*Quercetalia pubescentis*) on the Sonnenberg in the Vinschgau and in the lower part of the Schnalstal. The above mentioned species are also to be found in coniferous forests where the thermophile oak forests (*Quercetum pubescentis*) dovetails into areas of montane Spruce forest (*Piceetum montanum*). These shrubs do not penetrate far up into the Ötztal northern side of the main alpine chain. The Common Dogwood (*Cornus sanguinea*), which is the only naturally occurring *Cornus* species in the North Tyrol, can be

Fig. 5. Distributions as in Fig. 4: **a** European Yew (*Taxus baccata*), **b** Wayfaring Tree (*Viburnum lantana*), **c** Dogwood and Cornelian Cherry (*Cornus sanguinea* and *C. mas*), **d** Juneberry (*Amelanchier ovalis*) (map production: Atlas of the Tyrol, design by Ernest Troger, Institut für Landeskunde der Universität Innsbruck)

found on waysides and in the scrub communities on the valley bottoms nearby the mouth of the Ötztal (Imst and Nasserreith). The Wayfaring Tree (*Viburnum lantana*) grows at the entrance to the Ötztal on limestone screes formed by rockfalls just north of Ötz (Dalla-Torre and Sarnthein, 1900). The Juneberry (*Amelanchier ovalis*) goes furthest up into the Ötztal, with its distributional limit on the Engelswand at Umhausen.

Taken as a whole, the evidence from the distribution of these woody species represented in the wood finds associated with the Iceman points to a settlement site south of the main alpine chain.

3. Conclusions

The dendrological analyses have yielded a rather complex picture, which makes a generalized assessment difficult. This complexity is related to both the woody species identified and their geographical distributions, as well as to the different techniques utilized in constructing the artefacts. From comparison of the individual artefacts, it immediately becomes obvious that some have been carefully constructed whereas others have been roughly improvised. Twelve of the arrows are unfinished and the bow is quite crudely finished off. Therefore, these pieces of equipment could not have been used by the Iceman and could not have served as weapons. Nor could he have shot, with any accuracy, from the bow he possessed, even the two fully-finished arrows. The degree of workmanship shown by the other completed objects — such as the dagger shaft, the retouching tool and the pack-frame — is essentially makeshift as regards the finish of the wood used and was essentially expedient. Care in the construction can only be shown for the axe handle, the wood stiffener of the quiver and the two completed arrows.

In his choice of woods for a particular purpose the Iceman has shown that he possessed full knowledge of the physical properties of his native woody species. Judging from the way in which he took his raw material from stems and trunks, the Iceman knew about the heterogeneity of the woods and their different stress capacities in relation to the orientation of the fibres. The axe shaft is thickest in the vicinity of the axe head, the point subject to the greatest strain when in use. With regard to the Iceman's workmanship in finishing-off his wooden artefacts, however, usage of primitive techniques is obvious. Fabrication of the bow from stem wood in which the tree rings ran in parallel to the shaft, can be considered primitive. The use of roundwood of the Wayfaring Tree (*Viburnum lantana*) for the arrow shafts and their excessive length indicates use of primitive technique too (Beckhoff, 1963; 1968). Attachment of the blade of the flint dagger in parallel to the course of the tree rings is not an ideal solution; although no particular preference in regard to shafting of knives in the Neolithic period has been noted (Schweingruber, 1976).

Some pieces of equipment were already damaged when they were discovered. The U-shaped pack-frame had been broken by torsion strain at one point at least on the thicker shank of the bent section. Similarly, the two arrows that were in a firing condition lay, broken in two pieces, separately in the quiver (Egg et al., 1993). Even a piece of the stiffening rod of the quiver was broken off. This bit was discovered in fact before the quiver had been found (Zissernigg, 1992), wherefore breakage in the course of the excavations can be ruled out. In each of the above cases, since the damage to the artefacts had occurred before their discovery, they must have been exposed to considerably kinetic energy at some earlier period. Because these finds, after becoming embedded in the ice, were subject to little, if any, subsequently movement (Lippert and Spindler, 1991; Kuhn, in Seidler et al., 1992), this exposure to force must have occurred at the time of their deposition.

The phytogeographic evaluation of the woods found in association with the Iceman initially yields a heterogeneous picture, too. Woody species from all the different vegetational zones are represented, although those from the subalpine and alpine zones are in minority. These alpine species (*Salix reticulata*-type) are typical of snow-bed vegetation and therefore belong to the local vegetation of the find site. The subalpine species, Dwarf Mountain Pine (*Pinus mugo*), Arolla Pine (*Pinus cembra*) and Green Alder (*Alnus viridis*), as well as the montane species, will necessarily have been brought there by the Iceman. The obvious majority of the wood species belong to species growing in the montane zone (Table 6). In particular, the species from which his pieces of equipment were made, all have a montane distribution. Their ecological demands indicate a fairly humid habitat, with warm summers and mild winters. They are mainly species of the thermophile oak forests (*Quercetalia pubescenti-petraea*) and the deciduous high forests (*Fagetalia*), as well as the shrubs of the forest margins and waysides (*Prunetalia*). In the Inntal and at the entrance of the Ötztal, these forest communities are only rudimentarily represented and cover only small areas, whereas in the Vinschgau they cover the valley floors and the lower part of the Schnalstal. The Norwegian Maple (*Acer platanoides*), the European Yew (*Taxus baccata*), the Ash (*Fraxinus* sp.), the Lime (*Tilia* sp.) and the Elm (*Ulmus* sp.) point to forests growing on steeper slopes and in ravines, with a humid local climate and well-drained to seasonally wet soils, since truly thermophile species like Downy Oak (*Quercus pubescens*) or the Manna

Ash (*Fraxinus ornus*) are not represented in the species list. Species that grow in dry habitats, such as Juneberry (*Amelanchier ovalis*), the Wayfaring Tree (*Vibrunum lantana*), the Dogwood (*Cornus* sp.) and possibly those such as Small-leaved Lime (*Tilia cordata*), together with the Norway Spruce (*Picea abies*), the Larch (*Larix decidua*) and the Pine (*Pinus* sp.) point in the direction of the ecotone between the thermophile oak forest (*Quercetalia pubescentis*) and the montane Spruce forest (*Piceetum montanum*). In the lower part of the Vinschgau, these forest types are found at the entrances to the north-south orientated side valleys or ravines, as also at altitudes above 1100 m in the main valley itself. In view of these considerations, the potential site of the Iceman's home settlement probably is to be sought somewhere in the vicinity of the lower Vinschgau and the lower Schnalstal, or at the entrance of the latter valley (Table 7).

Abstract

The results of the analyses of the wood and charcoal finds associated with the Iceman are presented. Altogether 17 different woody plant species were provable: 9 of these had been used by him to construct various pieces of equipment. These plant species, together with 8 more, were also represented in the charcoal fragments that were picked out of the samples. All of the species occur naturally in the wider extent of the investigation area. An attempt has been made to reconstruct the Iceman's environment and to localize his settlement by floristic and phytogeographic means. The majority of the species found belong to the montane region, although some subalpine and alpine species were also represented. The ecological requirements of the woody plant species point to the transition zone between thermophile mixed-oak forest communities (*Quercetalia pubescenti-petreae*) and the montane Spruce forest (*Piceetum montanum*). Norwegian Maple (*Acer platanoides*), European Yew (*Taxus baccata*), Ash (*Fraxinus sp.*), Lime (*Tilia* sp.) and Elm (*Ulmus* sp.) allows us to infer a humid habitat with a mineral rich, free-draining soil and a mild winter climate, similar to the present-day conditions in the woodlands found on the slopes and in gorges in the lower Schnalstal and Vinschgau in South Tirol.

The technological and botanical analyses of his equipment show that the Iceman already knew from experience the different physical properties of his native woods. Each woody species was chosen by the Iceman precisely for the particular requirements of each piece of equipment. Nevertheless, the techniques utilized to make these pieces were still relatively primitive. The fact that certain finds had already become broken before the body had been embedded in the ice and excavated, indicates that they had been subjected to stronger stresses at the time of their deposition than at anytime afterwards.

Zusammenfassung

Die Analyseergebnisse der Hölzer und Holzkohlen, die beim Eismann gefunden wurden, werden vorgestellt. Insgesamt konnten 17 verschiedene Holzarten nachgewiesen werden. Neun davon hat er zur Herstellung seiner Artefakte verwendet. Diese Holzarten, zusammen mit acht weiteren, und die Holzkohlenfragmente, die aus den Proben ausgelesen wurden, werden beschrieben. Basierend auf einer floristisch-phytogeographischen Auswertung der Funde wird versucht die Umwelt des Eismannes und die Lage seiner Siedlung zu rekonstruieren. All die gefundenen Gehölzarten kommen im Untersuchungsgebiet vor. Der Großteil der nachgewiesenen Arten stammt aus der montanen Stufe, obwohl auch einige subalpine und alpine Arten vorkommen. Die ökologischen Ansprüche der Gehölze weisen auf die Übergangszone zwischen thermophilen Eichenwäldern (*Quercetalia-pubescenti-petreae*) und den montanen Fichtenwäldern (*Piceetum montanum*) hin. Spitzahorn (*Acer platanoides*), Eibe (*Taxus baccata*), Esche (*Fraxinus*), Linde (*Tilia*) und Ulme (*Ulmus*) zeigen ein humides Habitat mit mineralreichen frischen Böden und wintermildem Klima, ähnlich wie die heutigen Steilhang- und Schluchtwälder am Eingang des Schnalstales bzw. des Vinschgaus in Südtirol, an.

Die technisch-botanischen Analysen seiner Ausrüstungsgegenstände bestätigen, daß der Eismann bereits eine empirische Kenntnis der physikalischen Eigenschaften der einheimischen Hölzer hatte. Jede Holzart wurde präzise für den jeweiligen Verwendungszweck ausgewählt; dagegen ist die Fertigungstechnik noch relativ primitiv. Aus der Tatsache, daß einzelne Gegenstände schon vor ihrer Einbettung im Eis zerbrochen sind, ist eine starke mechanische Belastung auf die Hölzer bei der Deposition abzuleiten.

Riassunto

Vengono presentati i risultati degli esami di relitti di legno e carbone di legna individuati sulla mummia. Si sono riscontrati complessivamente 17 tipi di legno di cui 9 utilizzati per la fabbricazione dei suoi artefatti. Tali tipi di legno vengono descritti insieme ad altri 8 tipi e frammenti di carbone di legna. Sulla base di una valutazione floristico-fitogeografica si intenta una ricostruzione ambientale ed insediativa. Tutte le specie arboree ed arbustive utilizzate sono presenti nell'area di riferimento. La maggiore parte delle specie rilevate proviene dal livello montano, alcune da quello subalpino ed alpino. Le esigenze ecologiche sono indice di una zona di transizione tra querceti termofili (*Quercetalia-pubescenti-petreae*) e pinete montane (*Piceetum montanum*). *Acer platanoides, Taxus baccata, Fraxinus, Tilia, Ulmus* sono indici di un habitat umido dal suolo fresco e ricco di minerali e dal clima invernale mite, simile a quello delle foreste sui ripidi pendii e nelle forre all'inizio delle valli Senales e Venosta.

Le analisi tecnico-botaniche dell'equipaggiamento dell'uomo dello Hauslabjoch confermano una conoscenza empirica delle caratteristiche fisiche delle specie arboreo-arbustive locali dell'uomo del ghiacciaio. Ogni specie veniva prescelta accuratamente a seconda della rispettiva utilizzazione, mentre la fabbricazione stessa era relativamente grossolana. Dal fatto che alcuni oggetti fossero già spezzati prima del loro ricovero nel ghiaccio fa pensare ad un forte carico meccanico sui legni esercitato nel depositarli.

Résumé

L'article présente le résultat des analyses des bois et du charbon de bois découverts auprès de l'Homme des glaces. Au total, 17 espèces de bois ont été identifiées, dont neuf lui avaient servi à fabriquer son équipement. Ces espèces et huit autres ainsi que les fragments de charbon de bois dégagés des vestiges, sont décrits par la suite. Partant de l'exploitation floristico-phytogéographique des objets façonnés, les auteurs essaient de reconstituer l'environnement de l'Homme des glaces et de localiser son habitat. La totalité des essences découvertes sont présentes dans la région étudiée. La majorité des espèces identifiées proviennent de l'étage montagnard, complétées par quelques espèces subalpines et alpines. Les exigences écologiques des essences retrouvées font penser à la zone de transition entre chênaies thermophiles (*Quercetalia-pubescenti-perteae*) et pessières montagnardes (*Piceetum montanum*). L'érable plane (*Acer platanoides*), l'if (*Taxus baccata*), le frêne (*Fraxinus*), le tilleul (*Tilia*) et l'orme (*Ulmus*) indiquent un habitat humide comportant des sols frais et riches en minéraux et bénéficiant d'un climat doux en hiver, similaire à celui des forêts de pente et de gorges à l'entrée du Val di Senales et du Val Venosta au Tyrol italien.

Les analyses botanico-techniques de l'équipement de l'Homme des glaces ont confirmé que celui-ci avait une connaissance empirique des caractéristiques physiques des essences endogènes. Chaque espèce avait été choisie soigneusement en vue de son usage: la technique de fabrication des objets par contre est assez rudimentaire. Le fait que certains objets se soient cassés avant même d'être ensevelis dans la glace, laisse supposer une forte sollicitation mécanique des bois au moment du dépôt.

Acknowledgements

The authors acknowledge the Austrian Science Fund for supporting this research (Project No 10151-SOZ). We thank the following persons for their help and for placing valuable literary sources at our disposal: PD Dr. Markus Egg of the Römisch-Germanisches Zentralmuseum in Mainz for the measurements of the arrows; Dr. Walter Oberhuber of the Institut für Botanik der Universität Innsbruck for the dendrological evaluation of the yew stands at Andrian; Douglas Elmy and W. E. Tucker of the Society of Archer-Antiquaries, GB, for valuable information about finds of Neolithic bows, OR Mag. Dr. Wilfried Keller of the Institut für Geographie of Innsbruck University for supplying cartographic material and Dr. Philip A. Tallantire, Penrith for the English translation of this manuscript.

References

Bagolini B. und Pedrotti A. (1992) Vorgeschichtliche Höhenfunde in Trentino-Südtirol und im Dolomitenraum vom Spätpaläolithikum bis zu den Anfängen der Metallurgie. In: Höpfel F., Platzer W. und Spindler, K. (eds.): Der Mann im Eis, Bd. 1. Veröffentlichungen der Universität Innsbruck,187: pp 463.

Beckhoff K. (1963) Die Eibenholz-Bogen vom Ochsenmoor am Dümmer. Die Kunde, Neue Folge 14: 63–81.

Beckhoff K. (1964) Der Eibenbogen von Vrees. Die Kunde, Neue Folge 15: 113–125.

Beckhoff K. (1965) Eignung und Verwendung einheimischer Holzarten für prähistorische Pfeilschäfte. Die Kunde, Neue Folge 16: 51–61.

Beckhoff K. (1966) Zur Morphogenese der steinzeitlichen Pfeilspitze. Die Kunde, Neue Folge 17: 34–65.

Beckhoff K. (1968) Eignung und Verwendung einheimischer Holzarten für prähistorische Pfeilbogen. Die Kunde, Neue Folge 19: 85–101.

Beckhoff K. (1972) Über die Größenbeziehung zwischen prähistorischen Bogenschützen und seiner Waffe. Die Kunde, Neue Folge 23: 49–61.

Broglio A. (1992) Mountain sites in the context of the north-eastern Italian Upper Palaeolithic and Mesolithic. Preistoria Alpina 28(1): 293–310.

Broglio A. and Lanzinger M. (1996) The human population of the southern slopes of the Eastern Alps in the Würm Late Glacial and Early Postglacial. Il Quaternario 9: 499–508.

Clark J. G. D. (1963) Neolithic Bows from Somerset, England. Prehistory of Archery in N.W. Europe. Proceedings of the Prehistoric Society 29: 50–98.

Dalla-Torre K. W. von und Sarnthein L. von (1900) Flora der gefürsteten Grafschaft von Tirol, des Landes Vorarlberg und des Fürstenthumes Lichtenstein. 6 Bände. Innsbruck.

Dalmeri G. und Pedrotti A. (1992) Distributione topographica dei siti del Palaeolithico Superiore finale e Mesolithico in Trentino Alto-Adige e nelle Dolomiti Venete (Italia). Preistoria Alpina 28/2: 247–267.

Egg M., Goedecker-Ciolek R., Groenman-Van-Wateringe W. und Spindler K. (1993) Die Gletschermumie vom Ende der Steinzeit aus den Ötztaler Alpen. JbRGZ 39: pp 128.

Egg M. (1992) Zur Ausrüstung des Toten vom Hauslabjoch, Gem. Schnals (Südtirol). In: Höpfel F., Platzer W. and Spindler K. (eds.): Der Mann im Eis, Bd. 1. Veröffentlichungen der Universität Innsbruck 187: pp 463.

Franz L. und Weninger J. (1927) Die Funde aus den prähistorischen Pfahlbauten am Mondsee. Mat. 1927: pp. 87.

Gams H. (1933) Die Pflanzenwelt Tirols. In: Tirol – Land, Natur, Völker und Geschichte. DÖAV, München: 92–108.

Gross G., Kerschner H. und Patzelt G. (1976) Methodische Untersuchungen über die Schneegrenze in alpinen Gletschergebieten. Zeitschrift für Gletscherkunde und Glazialgeologie 12: 223–252.

Hausmann F. Freih. von (1851) Flora von Tirol. Ein Verzeichnis der in Tirol und Vorarlberg wild wachsenden und häufiger gebauten Gefäßpflanzen. 1. - 3. Band. Wagner, Innsbruck.

Hegi G. (1935) Illustrierte Flora von Mitteleuropa. München.

Karner A., Kral F. und Mayer H. (1973) Das inneralpine Vorkommen der Tanne im Vinschgau.–Centralblatt für das gesamte Forstwesen 90: 129–163.

Köllemann C. (1979) Der Flaumeichenbuschwald im unteren Vinschgau (Vegetationskundliche, bodenkundliche und ökologische Untersuchungen). Diss. Universität Innsbruck: pp 222.

Kral F. (1989) Spät- und postglaziale Waldentwicklung in den italienischen Alpen. Bot. Jahrb. System. 111: 213–229.

Kral F. (1992) Die postglaziale Entwicklung der natürlichen Vegetation Mitteleuropas und ihre Beeinflussung durch den Menschen. Österr. Akad. Wiss., Veröffentlichungen der Kommission für Humanökologie 3: 7–36.

Leuthold C. (1980) Die ökologische und pflanzensoziologische Stellung der Eibe (*Taxus baccata* L.) in der Schweiz. Veröffentlichungen des Geobotanischen Institutes der Eidg. Techn. Hochschule, Stiftung Rübel, in Zürich 67: pp 217.

Lippert A. und Spindler K. (1991) Die Auffindung einer frühbronzezeitlichen Gletschermumie am Hauslabjoch in den Ötztaler Alpen (Gem. Schnals). Archäologie Österreichs 2/2: 11–17.

Lippert A. (1992a) Topographie der Fundstelle und die weiteren Forschungen. In: Payrleitner A. (ed.): Der Zeuge aus

dem Gletscher. Das Rätsel der frühen Alpen-Europäer. Wien: 35–51.
Lippert A. (1992b) Die erste archäologische Nachuntersuchung am Tisenjoch. In: Höpfel F., Platzer W. und Spindler K. (eds.): Der Mann im Eis, Bd. 1. Veröffentlichungen der Universität Innsbruck 187: pp 463.
Mayer H. (1974) Die Wälder des Ostalpenraumes. Stuttgart: pp 344.
Messikommer H. (1913) Die Pfahlbauten von Robenhausen. Zürich: pp 132.
Oeggl K. und Wahlmüller N. (1994) Holozäne Vegetationsentwicklung an der Waldgrenze der Ostalpen: die Plancklacke 2150 m, Sankt Jakob im Defreggental (Osttirol). In: Lotter A. und Ammann B. (eds.): Festschrift Lang. Diss. Botanicae 234: 389–411.
Peer T. (1981) Die aktuellen Vegetationsverhältnisse Südtirols am Beispiel der Vegetationskarte 1:200.000. Angewandte Pflanzensoziologie 26: 151–168.
Pitschmann H., Reisigl H., Schiechtl H. M. und Stern R. (1980) Karte der aktuellen Vegetation von Tirol 1/100 000. VII. Teil: Blatt 10, Ötztaler Alpen–Meran. Documents de Cartographique 23: 47–68.
Reisigl H. und Pitschmann H. (1959) Obere Grenzen von Flora und Vegetation in der Nivalstufe der zentralen Ötztaler Alpen (Tirol). Vegetatio 8: 93–129.
Rust A. (1943) Die alt- und mittelsteinzeitlichen Funde von Stellmoor. Neumünster: pp 240.
Schweingruber F. (1976) Prähistorisches Holz. Die Bedeutung von Holzfunden aus Mitteleuropa für die Lösung archäologischer und vegetationskundlicher Probleme. Academica Helvetica: pp 106.
Stuefer J. (1992) Eibenreiche Busch- und Waldbestände bei Andrian. Amt für Naturparke, Naturschutz und Landschaftspflege, Bozen: pp 67.
Seidler H., Teschler-Nicola M., Wilfling H., Weber G., Traindl-Prohaska M., Platzer W., Zur Nedden D. und Henn R. (1992) Zur Anthropologie des Mannes vom Hauslabjoch: Morphologische und metrische Aspekte. In: Höpfel F., Platzer W. und Spindler K. (eds.): Der Mann im Eis, Bd. 1. Veröffentlichungen der Universität Innsbruck 187: pp 463.
Walter H. und Straka H. (1970) Arealkunde. Floristisch-historische Geobotanik. Stuttgart. pp 478.
Zissernig E. (1992) Der Mann vom Hauslabjoch: Von der Entdeckung bis zur Bergung. In: Höpfel F., Platzer W. und Spindler K. (eds.): Der Mann im Eis, Bd. 1. Veröffentlichungen der Universität Innsbruck 187: pp 463.
Zoller H. und Brombacher Ch. (1984) Das Pollenprofil "Chavalus" bei St. Moritz – ein Beitrag zur Wald- und Landwirtschaftsgeschichte im Oberengadin. Diss. Botanicae 72 (Festschrift Welten): 377–398.

Appendix

Catalogue of the wood finds

Explanation of the various types and derivations of wood fragments described in the finds catalogue (according to Schweingruber, 1976):

radial fragment = the tree rings run at right angles to the longest axis of the fragment;
tangential fragment = the annual rings run parallel to the longest axis;
oblique fragment = the tree rings form an angle of 45° transversely to the longest axis;
fragments sensu lato = no definite relationship exists between the direction of the annual rings and the orientation of the fragment from the trunk, branch or tree stems.

Sample no. B-91/22

Description: wood found on the "chamois fur" in quadrant Q8;
Form: elongated wood fibres;
Dimensions: Length = 0.97 mm, Diameter = 0.02 mm;
Working traces: none;
Species: probably Larch (*Larix/Picea*-Type); clustered biserate bordered pits in the longitudinal tracheids of the late wood clearly visible.

Sample no. B-91/23

Description: stray find of wood;
Comment: may have been derived from the packframe;
Form: an elongated radial sliver from a branch, periderm lost, split down 3/4 of its length; the sliver was twisted around its own axis;
Dimensions: Length = 6.3 cm, breadth = 0.15 cm; radial thickness = 0.31 cm;
Working traces: none visible;
Species: Green Alder (*Alnus viridis* (CHAIX) Dc.);
Source: presumably as binding material.

Sample no.: RGZM-91/95

Description: leather, cords, fur;
Comment: 3 slivers of wood recovered during sample washing;
Sliver no. 1: Species: Yew (*Taxus baccata* L.); conspicuous spiral thickenings of the tracheids;
Form: elongated slivers, frayed at both ends, with a trapezoid cross section, the ends thin and sharp edged, reddish-brown coloration, nine annual rings recognisable on the cross section seen on the radial breakage point;
Dimensions: Length = 46 mm;   breadth = 7 mm; thickness = 3.7 mm;
Source: a sliver detached from the bow;
Sliver no. 2: Species: Yew (*Taxus baccata* L.);
Form: tangential sliver, elongated quadrate in shape, three annual rings visible on the radial breakage point;
Dimensions: Length = 23.4 mm, breadth = 4.2 mm; thickness = 2.8 mm;
Source: a sliver detached from the bow;
Sliver no. 3. Species: probably Willow (cf. *Salix* sp. L.);
Dimensions: Length = 3.6 mm; diameter = 0.3 mm;
Source: unknown, a twig fragment.

Sample no.: RGZM-91/96
Description: Hairs, fibres, cords;
Comment: wood sliver recovered during sample washing;
Species: Hazel (*Corylus avellana* L.);
Form: elongated radial sliver; spathulate, with a branch knot-hole at the upper end;
Dimensions: Length = 21.3 mm; breadth = 6.4 mm; thickness = 2.1 mm;
Source: a sliver broken off from the U-shaped spar of the pack-frame.

Sample no. RGZM-91/102
Description: cords and grass;
Comment: wood slivers recovered during sample washing;
Species: probably Spruce (*Picea/Larix*-type);
Form: radial slivers, maximum 1 mm in length;
Number: 3;
Source: unknown.

Sample no. RGZM-91/106
Description: sewn leather;
Comment: wood slivers recovered during sample washing;
Sliver no 1: Species: Birch (*Betula* sp. L.);
Dimensions: < 2 mm;
Form: radial sliver;
Number: 1;
Source: unknown;
Sliver no 2: Species: Larch (*Larix decidua* MILL.); biseriate bordered pits distinctly visible, clustered in the latewood tracheids;
Form: radial sliver;
Dimensions: < 2 mm;
Number: 3;
Source: slivers from the laths of the pack-frame.

Sample no. RGZM-91/138
Description: Hair remains from quadrant Q12; 6 wood slivers recovered during sample washing;
Sliver no 1: Species: probably Spruce (*Picea/Larix*-type);
Form: radial sliver;
Dimensions: Length = 2 mm;
Number: 4;
Source: unknown;
Sliver no 2: Species: Larch (*Larix decidua* MILL.);
Form: radial sliver;
Dimensions: Length = 15.2 mm; breadth = 1 mm;
Number: 1;
Source: presumably from the transverse laths of the pack-frame;
Sliver no 3: Species: Larch (*Larix decidua* MILL); biseriate bordered pits clustered and distinctly visible;
Form: elongated sliver, conically pointed, triangular in cross-section, dark-brown coloration, 4 tree rings visible, average ring width 0.77 mm (0.6–0.84 mm);
Dimensions: Length = 62 mm; maximum breadth = 7.4 mm; thickness = 3.9 mm;
Number: a single one;
Source: a piece broken off from the laths of the pack-frame.

Sample no.: RGZM-91/139
Description: Grass and leaf remains recovered from the meltwater channel; with 4 wood slivers;
Sliver no 1: Species: Hazel (*Corylus avellana* L.);
Form: radial sliver of irregular shape;
Dimensions: maximum length = 11.5 mm;
Number: a single sliver;
Source: a piece broken off from the U-shaped roundwood of the pack-frame;
Sliver no 2: Species: probably Larch (*Larix/Picea*-type);
Form: radial slivers, triangular in cross section;
Dimensions: maximum length = 5 mm;
Number: 3;
Source: broken pieces from the laths of the pack-frame.

Sample no.: RGZM-91/141
Description: Leather and grass remains;
Comment: wood;
Form: rectangular wood sliver, frayed at both ends, externally charcoal-grey in colour, internally light brown;
Dimensions: Length = 1.99 cm; breadth = 0.26 cm; thickness = 0.48 cm;
Working traces: none;
Species: indeterminate; presumed to be recent wood;
Source: unknown.

Wood samples from the excavations made in August 1992 and recovered when the artefacts were being cleaned at the Römisch-Germanischen Zentralmuseum in Mainz.

Sample no.: RGZM-92/48
Description: wood;
Comment: found in the west side of the southern rockrib;
Form: piece split off a branch, ca. 1/3 of the original thickness; periderm lost, pith visible. This piece could

be joined up with the pieces of twig of samples nos. 92/106 and 92/131 to form a 2/3 of the original complete twig (Plate 6);
Dimensions: maximum length = 68.0 mm; breadth = 6.6 mm; thickness = 3.9 mm;
Working traces: none;
Species: Wayfaring Tree (*Viburnum lantana* L.);
Source: probably the remains of an arrow shaft.

Sample no.: RGZM-92/65

Description: wood, fur;
Comment: from the southern rock-rib;
Form: a triangular sliver in cross-section, blackened;
Dimensions: maximum length = 39.6 mm; breadth = 2.2 mm; thickness = 1.7 mm;
Working traces: none;
Species: a sliver of recent wood; indeterminate;
Source: unknown.

Sample no.: RGZM-92/99

Description: wood and hairs;
Comment: not in fact wood, but periderm;
Form: a rectangular flattened-out piece of stem bark;
Dimensions: Length = 19.7 mm, breadth = 10.8 mm;
Number: 2;
Species: Birch (*Betula* sp. L.); pieces of bark;
Source: probably remains of the birch-bark container.

Sample no.: RGZM-92/106

Description: wood (Plate 6);
Comment: none;
Form: an elongated sliver from a twig; broken into three pieces, the three bits fitted together perfectly, almost semicircular in cross-section, pith visible, periderm lost, terminal ring visible, only one tree ring present, though fully formed, twig cut outside the growing period. This piece could be joined up with the pieces of twig of samples nos. 92/48 und 92/131 to form 2/3 of the original complete twig (Plate 6);
Dimensions: maximum length = 114 mm; breadth = 8.0 mm; radius = 4.2 mm;
Working traces: none;
Species: Wayfaring Tree (*Viburnum lantana* L.);
Source: probably the remains of an arrow shaft.

Sample no.: RGZM-92/116

Description: piece of the bow;
Comment: found embedded in the ice;
Form: an elongated sliver, reddish-brown coloration;
Dimensions: maximum length = 21.5 mm; thickness = 0.8 mm;
Working traces: none;
Species: Yew (*Taxus baccata* L.);
Source: part of the bow.

Sample no.: RGZM-92/116

Description: wood;
Comment: found in the ice;
Form: elongated, polygonal bark scale;
Dimensions: Length = 35.0 mm; breadth = 8.5 mm; thickness = 2.4 mm;
Species: carbonised periderm from an indeterminate deciduous species;
Source: unknown.

Sample no.: RGZM-92/131

Description: Wood (Plate 6);
Comment: on the southern rock-rib;
Form: rectangular sliver of a twig, in cross-section forming an obtuse angle; pith visible, periderm absent, only one tree ring visible; late-wood fully formed (twig cut outside the growing period), the ends slightly frayed. This piece could be joined up with the pieces of twig of samples nos. 92/48 und 92/106 to form 2/3 of the original complete twig (Plate 8);
Dimensions: maximum length = 48 mm; breadth = 7.7 mm, radius = 3.3 mm;
Working traces: none;
Species: Wayfaring Tree (*Viburnum lantana* L.);
Source: probably the remains of an arrow shaft.

Sample no.: RGZM-92/217

Description: Wood (Plate 6);
Comment: recovered from the meltwater flowing to the glacier below;
Form: streamlined sliver, the bottom end with rounded edges, broken off at the top, wedge-shaped above with sharp edges, tree ring pattern discernible, probably a desquamation of a callus tissue overgrowing a stem wound;
Dimensions: maximum length = 56 mm, breadth = 21 mm; thickness = 3.1 mm;
Working traces: there were cut marks on the external surface as signs of deliberate working; the edges of the cut surfaces on the bottom end were rounded and broken off (Plate 6);
Species: probably Larch (*Larix/Picea*-type);
Source: unknown.

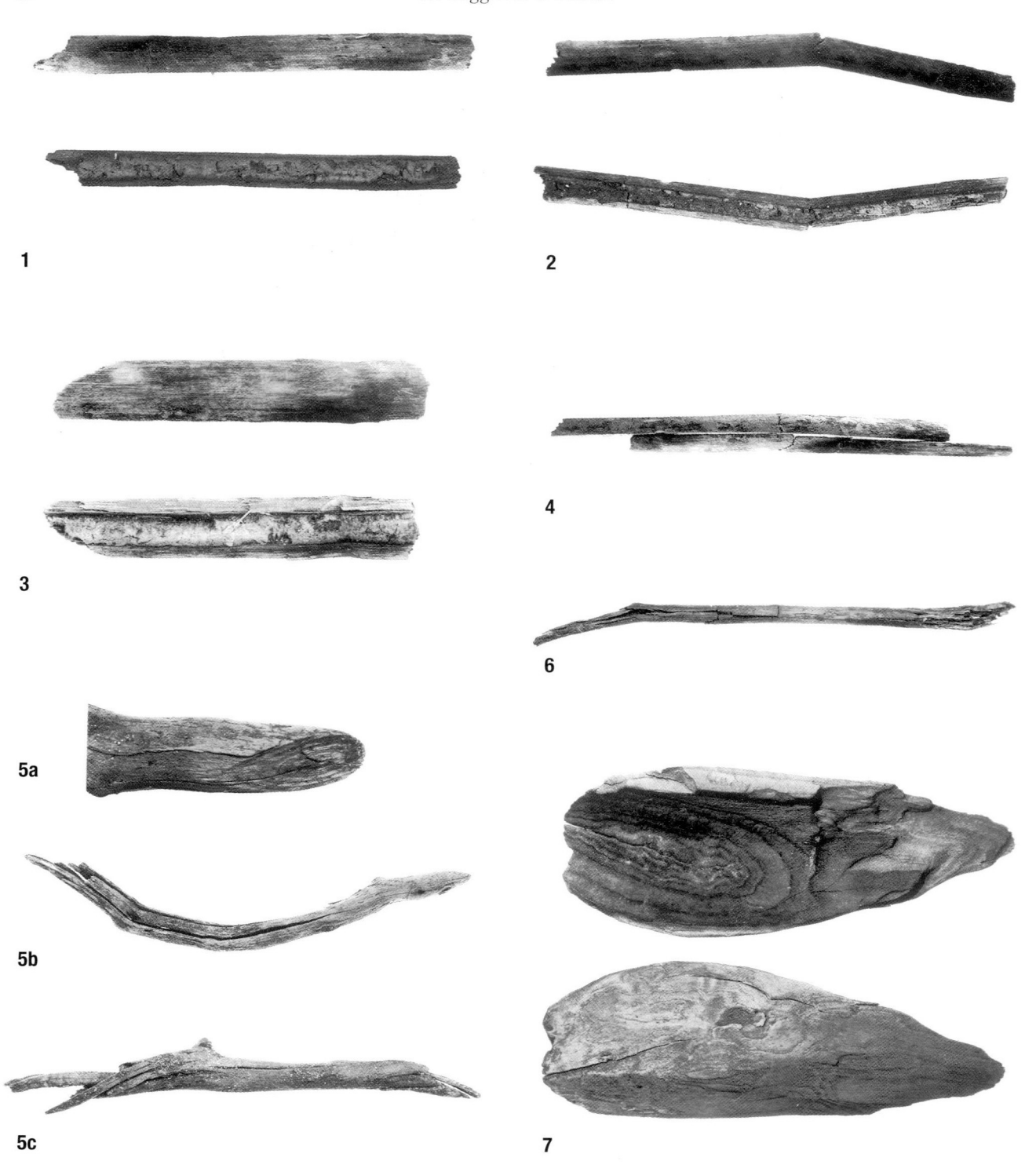

Plate 6. Wood remains recovered during the excavations in 1992: **1–3** fragments of an arrow shaft made from Wayfaring Tree (*Vibrunum lantana*): inner and outer surfaces: **1** wood fragment RGZM- 92/48, **2** wood fragment RGZM-92/106, **3** wood fragment RGZM-92/131, **4** fragments RGZM-92/48, RGZM-92/106 and RGZM-92/131 fit together to an arrow shaft, **5a** detail of the Green Alder (*Alnus viridis*) with precise cutting mark; **5b** piece of Green Alder (*Alnus viridis*) wood used as binding material (RGZM-92/275 dated to the Ironages!), **5c** twisted piece of Green Alder (*Alnus viridis*) wood used as binding material, **6** wood fragment RGZM-92/292: twig from a Pine (*Pinus* sp.), **7** streamlined Larch (*Larix decidua*) sliver (RGZM-92/217): inner and outer surfaces

Plate 7. Thin sections of the wooden artefacts found (from left to right cross-radial–tangential section): **1** Yew, *Taxus baccata*, **2** Larch, *Larix decidua*, **3** Hazel, *Corylus avellana*, **4** Ash, *Fraxinus* sp

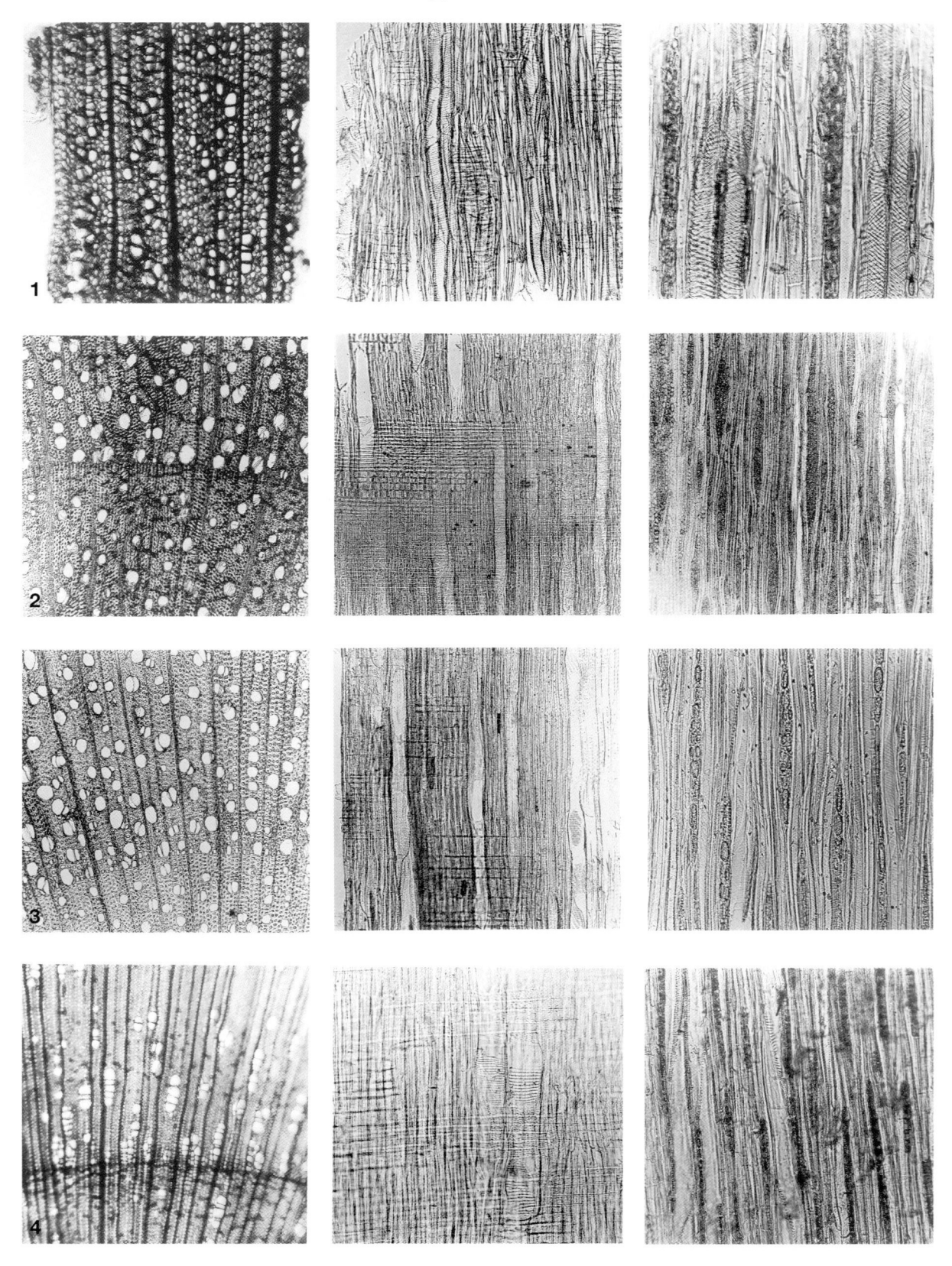

Plate 8. Thin sections of the wooden artefacts found (from left to right cross-radial–tangential section): **1** Lime, *Tilia* sp., **2** Wayfaring Tree, *Viburnum lantana*, **3** Dogwood, *Cornus* sp., **4** Green Alder, *Alnus viridis*

Sample no.: RGZM-92/275

Description: 2 pieces of wood (Plate 6);
Comment: found in the western part of the southern rock-rib. This piece of wood is younger than the Iceman and dated to the Ironage (see Kutschera et al., this volume);
Form: sliver no 1: elongate, round branch or twig, periderm absent, the thicker end cut off smoothly at an angle, six tree rings discernible on the cut surface, the thinner end split and frayed and twisted around;
Dimensions: maximum length = 170.0 mm; diameter of the thicker end = 10.4 mm;
Working traces: smoothly cut off at an angle at the thicker end, the thinner end split by blows from a blunt instrument, with subsequent distortion and twisting of the individual split pieces;
Sliver no 2: an elongated piece of a branch, periderm absent, split and the individual split pieces twisted together, the thinner end worn and rounded (Plate 8);
Dimensions: maximum length = 97 mm; diameter = 7.8 mm;
Working traces: the branch piece split into several bits by blows from a blunt instrument, with subsequent twisting together of the individual split pieces;
Species: Green Alder (*Alnus viridis* (CHAIX.) Dc.); uniseriate rays, vessels in radial groups, scalariform perforations;
Source: probably binding material.

Sample no. RGZM-92/285

Description: grass, hair and wood;
Comment: found just north-west of the rock on which the mummified body was lying (in the ice);
Form: an elongated, rectangular sliver, broken off tangentially from the trunk;
Dimensions: maximum length = 15.2 mm; breadth = 2.5 mm; thickness = 1.1 mm;
Working traces: none;
Species: Larch (*Larix/Picea*-type);
Source: probably from the larch laths of the pack-frame.

Sample no.: RGZM-92/292

Description: wood, grass;
Comment: found on the southern rock-rib. This piece of wood is older than the Iceman and dates to the early Neolithic (see Kutschera et al., this volume);
Form: a straight piece of a branch or twig, broken off at both ends, periderm completely absent;
Dimensions: maximum length = 140 mm; diameter = 5.75 mm;
Working traces: none;
Species: Pine (*Pinus* sp. L.); closer identification to species level was impossible, due to poor preservation of the vertical tracheids;
Source: unknown.

Sample no.: RGZM-92/400

Description: Wood and hair (Plate 4);
Comment: found in quadrant 2, between rock crevices;
Form: rhomboidal board with a semicircular hollow cut out at one end, wedge-shaped in cross-section, the lath broken off tangentially from the trunk, the surface rough and unsmoothed, 13 tree rings visible, their average breadth = 0.87 mm (0.4–1.2 mm);
Dimensions: maximum length = 165 mm, breadth = 36 mm, Thickness at base h_1 = 11.5 mm, thickness at top h_2 = 6.1 mm; the diameter of the semicircular hollow = 27 mm;
Working traces: the surface shows no working traces; the end with the semicircular hollow shows a slanting cut on the cut surface. The incision was made with four coarse cuts from opposite sides;
Species: Larch (*Larix decidua* MILL.);
Source: probably a piece of the pack-frame.

Sample no.: RGZM-93/117

Description: the fur hat;
Comment: wood sliver recovered after washing the hat;
Number: a single one;
Form: a radial sliver;
Dimensions: length = 2.5 mm; thickness = 0.8 mm;
Species: Arolla pine (*Pinus cembra* L.);
Source: unknown.

The bow of the Tyrolean Iceman: A dendrological investigation by computed tomography

W. Oberhuber[1] and **R. Knapp**[2]

[1] Institut für Botanik, Universität Innsbruck
[2] Abteilung für Röntgendiagnostik und Computertomographie, Universitäts-Kliniken Innsbruck

1. Introduction

Yearly tree-ring width sequences can be used for dendrochronological analysis, i.e., establishing the exact year in which each ring was formed (Stokes and Smiley, 1968; Baillie, 1982). This method has proved as a very useful dating tool in archaeological as well as ecological investigations (Creber, 1977). Measuring of tree-ring widths is traditionally carried out on cross-sections, which require some surface preparation before tree rings are readable (Schweingruber, 1988). For valuable wood samples (e.g., tools, vessels, woodcarvings from archaeological sites, art history) this method is not applicable.

On the other hand computed tomography (CT) offers the possibility of getting cross-section images within any part of the object investigated without the requirement of sample preparation. There are some reports on the use of CT X-ray scanning for dendrological investigations (Tout et al., 1979; Reimers and Riederer, 1986) showing that this method proves useful for analysis of wooden objects. Due to the simpler anatomical structure and sharp density boundaries between growthrings, softwoods like fir or spruce are generally better suited to CT X-ray analysis than hardwoods (Fig. 1). The only limitation with medical CTs is the inadequate spatial resolution, i.e., only tree rings broader than about 1 mm can be identified with certainty on the CT-pictures (see Material and Methods). Nonetheless, as shown in this report, CT X-ray scanning can be of great help in analyzing the structure of valuable wood samples.

2. Material and Methods

2.1. Wood samples

Cross-cuttings of spruce (*Picea abies* Karst.) and fir (*Abies alba* Mill.) in dry and wet condition were used to determine the most appropriate instrument settings. Then, the bow-fragment (about 30 cm long) of the prehistoric Tyrolean Iceman (Höpfel et al., 1992) carved out of yew (*Taxus baccata* L.) was analyzed using these settings.

2.2. Computed tomography

In a CT-scanner, X-rays are caught by a number of detectors (instead of a photographic film), which transform them to electrical impulses. These impulses are further processed by the computer to create a spatial picture. Our analyses were made with a high quality CT Somatom Plus (Siemens, Erlangen, Germany). The sample was entered with the longitudinal axis (stem-axis) oriented parallel to the z-axis of the CT. Cross-sections (2 mm thick) were made perpendicular to the stem-axis. Tube voltage and tube current were set at 137 kV and 145 mA, respectively. Exposure time was 6 seconds for every image.

Image reconstruction was made by means of an edge enhancing algorithm in ultra high resolution mode. A laser camera was used to create radiographs on conventional X-ray film (size 35 × 43 cm). Window settings for center and width were −217 and 1312 Hounsfield units, respectively. Since the absorption of X-rays for water is much higher than for air, a wood sample in wet condition was investigated, too. We found that the difference in density between water filled cell-lumina and the cell wall is still sufficient to get images rich in contrast. Sample size is confined to maximum area of field of view, which is 50 cm × 50 cm in this system.

2.3. Image resolution

With zoom set to 16, the minimum pixel area we can get with our equipment is $0.0037\ mm^2$, i.e., 0.061 mm edge length (field of view/[image matrix × zoom]). To test maximum image resolution a coniferous wood sample (fir – *Abies alba* Mill.) in dry condition, showing a continuous decrease in ring-widths without major

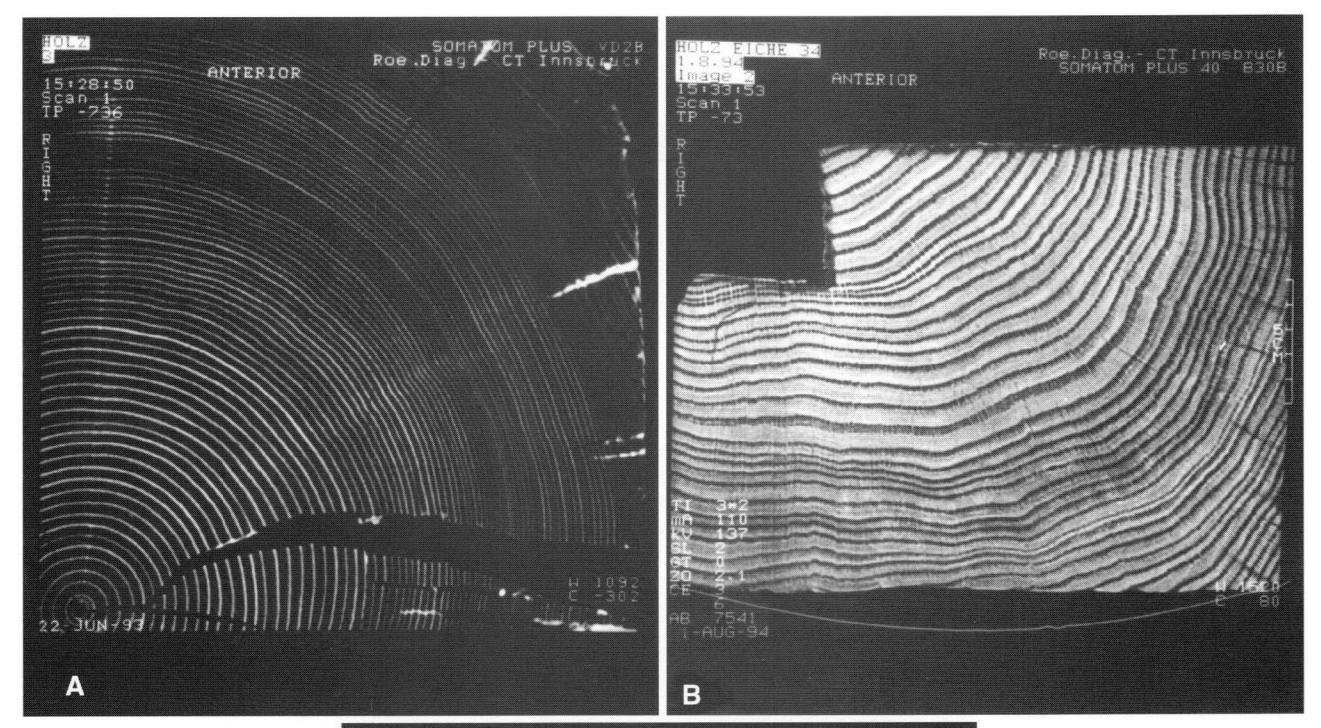

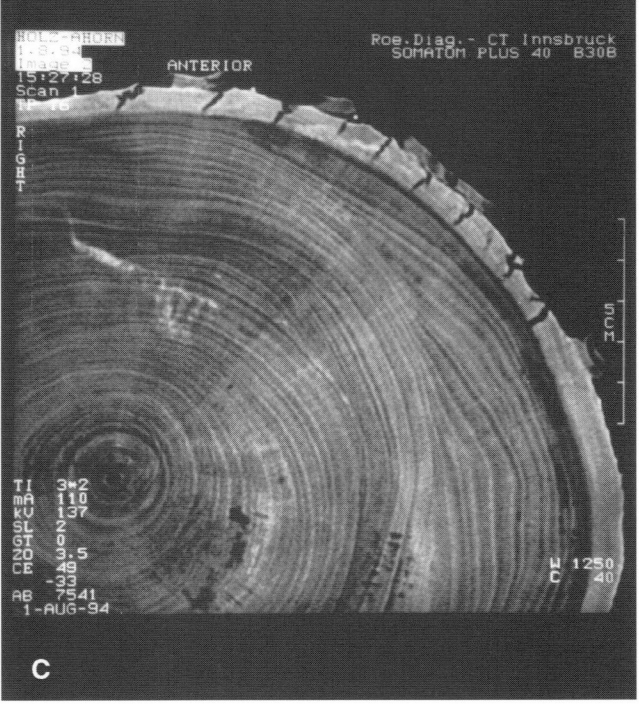

Fig. 1. Examples of CT-images: **A.** softwood fir (*Abies alba* Mill.); **B.** ring-porous hardwood oak (*Quercus* sp.); **C.** diffuse-porous hardwood maple (*Acer* sp.). Tree rings in diffuse-porous hardwoods can hardly be distinguished on radiographs, since there are neither variations in the size of early- and latewood pores nor does the pore density vary substantially

disturbances, was analyzed. Ring-widths on the CT-image and on the wood disk (cross-cutting) were measured in the same radial direction. Synchronization of ringwidth series is shown in Fig. 2. Furthermore, two dendrochronological parameters, which are used for judging the strength of agreement between ring-series, are mentioned. Since both the percentage of agreement of the year-to-year ring-width changes

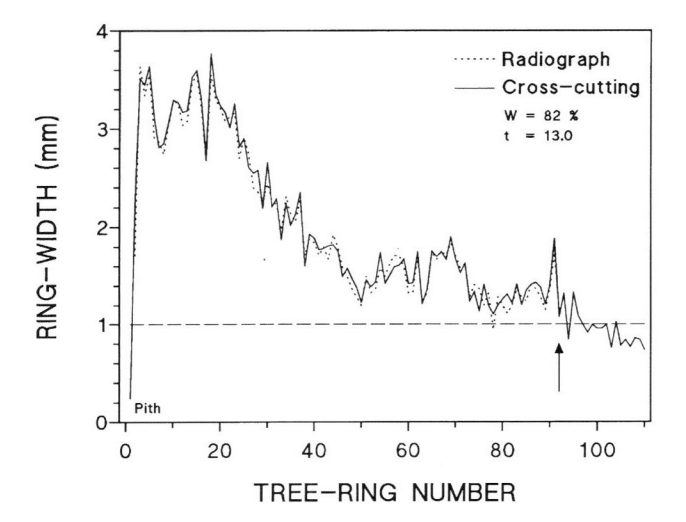

Fig. 2. Synchronization of ring-widths measured directly on the sample (thick line) and on the CT-image (dotted line). The arrow indicates the last clearly distinguishable growthring on the radiograph

(W-statistic; Eckstein and Bauch, 1969) and the correlation-coefficient (t-value; Baillie and Pilcher, 1973) are very high, CT-images can be a substitute in cases where cross-sections are not available. Clear image resolution, however, is restricted to a tree-ring width of about 1 mm, depending on the difference in radiologic density between latewood and earlywood cell walls (Fig. 2). Preuss et al. (1991) obtained similar results in their investigation with oak.

3. Results and Discussion

Even though scan of the bow-fragment gave a blurred image (Fig. 3a), the curvature of tree rings can largely be detected (Fig. 3b). Due to the limited resolution only growthrings wider than 1 mm are clearly distinguishable on radiographs (see Material and Methods). Therefore, neither number nor mean ring-width can be determined in this case. On the basis of the given curvature of ring segments, the minimum stem diameter can be determined mathematically. Without taking any removed sapwood into account we find a value of the order of 8 to 9 cm most likely. However, it has to be considered, that measurements of width and diameter in the middle of the bow are about 20% higher than at the scanned position (Egg and Spindler, 1992). As shown in Fig. 3b, the cross-section covers a sector of about 70°. If the shape is enlargened to allow for additional waste, the angle will increase to about 90°, which is $1/4^{th}$ of the total circular area.

The blurred image of the CT-picture might be caused in part due to heartwood substances, since X-ray absorption rates of these compounds differ from those of cell wall components. Unclear radiographs can also emerge from oblique fiber direction, i.e., earlywood and latewood zones are getting mixed.

On the basis of several radiographic scans within different parts of the sample, CT-images have already been used primarily by art historians to detect restoration processes or homogeneity of archaeological objects (Tout et al., 1979; Trenard and Perrault, 1990). However, due to the limited resolution of medical CT X-ray scanners dendrochronological dating by CT has not been very successful so far (Preuss et al., 1991).

In conclusion, computed tomography creates cross-section images within any part of the wooden sample without any treatment or damage. Therefore, this method is of great help in dendrological analysis of the inner structure of valuable objects.

Abstract

Computer-tomographical X-ray scanning was used to analyze growthrings in the prehistoric bow-fragment of the Tyrolean Ice-man (about 3300 years BC) made out of yew (*Taxus baccata* L.). This nondestructive investigation gave helpful information about the interior wood structure and the way of manufacture. We also found out that the minimum stem diameter must have been approximately 9 cm (without sapwood). Computed tomography can be highly recommended for dendrological investigations of valuable wood samples.

Zusammenfassung

Mit Hilfe von CT-Aufnahmen wurden die Wachstumsringe in dem aus Eibe (*Taxus baccata* L.) gefertigten Bogenfragment des prähistorischen Eismanns (ca. 3 300 v. Chr.) analysiert. Diese zerstörungsfreie Untersuchung lieferte wertvolle Hinweise auf die Holzstruktur und die Fertigungsmethode. So konnte auch festgestellt werden, daß der Mindeststamm-Durchmesser (ohne Splintholz) ca. 9 cm betrug. Generell kann man sagen, daß sich die Computertomographie als ideale Untersuchungsmethode für dendrologische Analysen wertvoller Holzproben anbietet.

Résumé

La scanographie a été utilisée pour analyser les cernes du fragment de l'arc en if (*Taxus baccata* L.) retrouvé auprès de l'Homme des glaces préhistorique (env. 3300 ans av. J.-C.). Cet examen non-destructif a fourni des informations utiles sur la structure du bois et le mode de fabrication de l'arc. L'examen a également révélé un diamètre minimum de 9 cm (sans le bois d'aubier) du tronc d'arbre utilisé. D'une manière générale, la scanographie est une méthode qui se prête parfaitement à l'analyse dendrologique d'échantillons de bois particulièrement précieux.

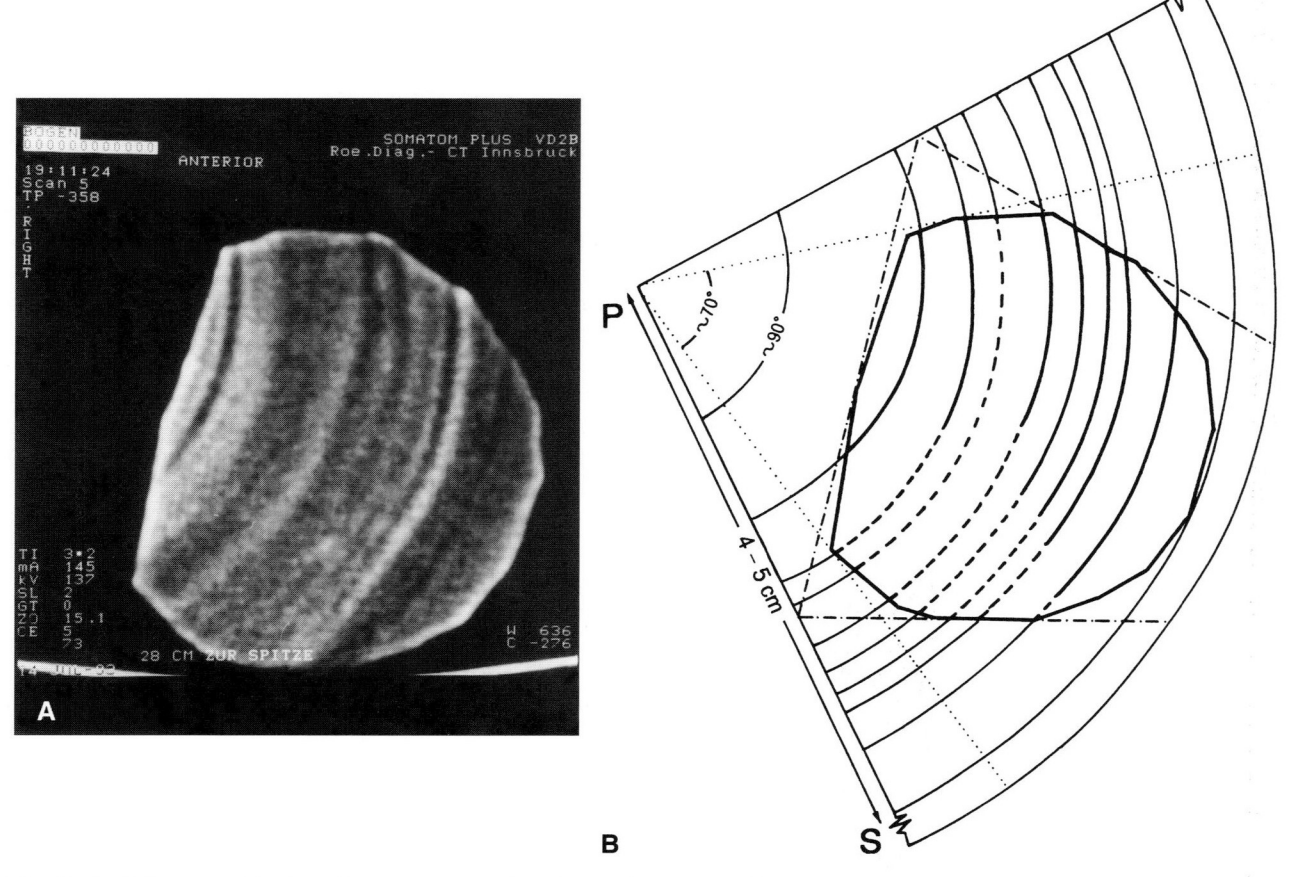

Fig. 3. A. CT-image of the bow-fragment scanned at 28 cm off the top (width: 3 cm, diameter: 2,65 cm); **B.** Curvature of identifiable tree rings taken from radiograph and supposed location in the stem. Note that the area close to the pith has not been used for manufacture (P = pith; S = sapwood)

Riassunto

Per analizzare gli anelli di accrescimento del frammento dell'arco dell'uomo del ghiacciaio, fabbricato in legno di *Taxus Baccata* L. ci si è avvalsi della tomografia assiale compiuterizzata. Tale esame non distruttivo ha fornito utili informazioni sull'interno della struttura lignea ed il metodo di fabbricazione.

Il diametro minimo del fusto è ritenuto fosse di 9 cm circa, senza alburno. In linea di massima la tomografia assiale computerizzata è ritenuta altamente raccomandabile per effettuare esami dendrologici di campioni di legno scientificamente prezioso.

References

Baillie M. G. L. (1982) Tree-ring dating and archaeology. London, Croom Helm.

Baillie M. G. L. and Pilcher J. R. (1973) A simple cross-dating program for tree-ring research. Tree-Ring Bulletin 38: 35–43

Creber G. T. (1977) Tree rings: a natural data-storage system. Biological Review of the Cambridge Philosophical Society 52: 349–383.

Eckstein D. und Bauch J. (1969) Beitrag zur Rationalisierung eines dendrochronologischen Verfahrens und zur Analyse seiner Aussagesicherheit. Forstwissenschafltiches Centralblatt 88: 230–250.

Egg M. und Spindler K. (1992) Die Gletschermumie vom Ende der Steinzeit aus den Ötztaler Alpen. JbRGZ 39, Mainz 1993.

Höpfel F., Platzer W. und Spindler K. (eds.) (1992) Der Mann im Eis, Bd. 1. Veröffentlichung der Universität Innsbruck 187.

Preuss P., Christensen K. and Peters K. (1991) The use of computer-tomographical X-ray scanning in dendrochronology. Norw. Arch. Rev. 24(2): 123–130.

Reimers P. and Riederer J. (1986) Dendrochronology by means of X-ray computed tomography (CT). International Symposium on Archaeometry, Greece 1986.

Schweingruber F. H. (1988) Tree rings. Basics and applications of dendrochronology. Kluwer, Dordrecht Boston London.

Stokes M. A. and Smiley T. L. (1968) An introduction to tree-ring dating. The University of Chicago Press, Chicago London.

Tout R. E., Gilboy W. B. and Clark A. J. (1979) The use of computerised X-ray tomography for the nondestructive examination of archaeological objects. Proceedings of the 18th International Symposium on Archaeometry and Archaelogical Prospection, Bonn, 14–17 March 1978. Archaeo-Physika 10: 608–616.

Trenard Y. und Perrault G. (1990) Apport de la tomographie à l'analyse d'objets en bois. In: La conservation du bois dans le patrimoine culturel. Journées d'études de la S.F.I.I.C. Besancon–Vesoul, 8–10 November 1990.

Analysis of the bast used by the Iceman as binding material

K. Pfeifer and K. Oeggl

Institut für Botanik, Universität Innsbruck

1. Introduction

The bark of dicotyledenous plants is made up of phloem (bast) and the hard bark (Fig. 1A). Phloem is the living bark. It is the term applied to the secondary tissue that is formed from the cambium layer in dicotyledenous plants. It is comprized of conducting elements, parenchyma and sclerenchyma cells, the bast or phloem fibres. Phloem fibres and parenchyma are formed alternately by the cambium. This explains the accumulation of phloem fibres in the tissue, known as hard phloem, which is followed by layers of conducting elements and parenchyma, the soft bark. The bundles of phloem fibres have a protective and supporting function under mechanical load and among other things are responsible for the ability of plants to return to the vertical after yielding to lateral pressure. These fibres are accordingly soft, elastic and have a very high breaking strength. That makes them ideal for making yarns and durable cords. These properties of phloem fibres or bast were known to the Neolithic man. Numerous plaited and textile specimens from the Neolithic lakeside settlements of the northern pre-Alps were made of tree bast (Körber-Grohne, 1987; Winiger, 1995; Vogt, 1938). The most common source of the bast was the lime tree (*Tilia*), although use was also made of willow (*Salix*), oak (*Quercus*) and to a lesser extent elm (*Ulmus*) (Körber-Grohne, 1987). In addition to tree bast, bast fibres from herbaceous plants were also used. Flax (*Linum usitatissimum*) was grown in the Neolithic period for oil and fibre production, and the technique of fibre processing was already known. From the Swiss pile-dwelling settlements of the Late Neolithic, exquisitely woven textiles made of flax (*Linum usitatissimum*) have survived (Vogt, 1938). The process used to extract the individual bast fibres from the bark is complicated and labour-intensive. For that reason whole strips of bast were also used at first to make rough cording material and wickerwork.

A large amount of binding material was also found with the Iceman and his equipment, i.e., cords and cord fragments and even pieces of plaited material. In the vicinity of the pannier, for example, several cords joined together to form a kind of net indicative of some plaited article were found (Lippert, 1992). The grass lining for the Iceman's shoes was also held in place by a kind of net knotted from cords, with a cord used to tie the shoes themselves round the ankle. In the cape the vertically arranged grasses are also knotted together with horizontal twines. The scabbard for the dagger was made of bast in a similar way. In this case bast warp threads are plaited into longitudinal strips of bast (Goedecker-Ciolek, 1992). The quiver was found to contain a coarse coiled string. Cords and threads were also used as binding materials for various other items of equipment. The sewing materials for the Iceman's fur garments are mainly carefully twisted threads, although lightly twisted bast strips were used for temporary repairs. The arrow points and fletching were also attached with threads, as was the blade of the dagger in the ash-wood handle. On the handle itself there is a torn loop of string. At the end of the shaft of the retoucheur, too, there is a groove which suggests that a loop of cord had been fastened to it (Egg, 1992). Numerous fragments of cording material were also found during the process of cleaning the various artefacts. In addition to leather and hairs, plant fibres were also used. The purpose of this study was to identify the plants from which the bast or bast fibres were harvested.

2. Material and method

The bark of native woody plants can be identified histologically through their characteristic arrangement of conducting elements, parenchyma, phloem fibres, and xylem and bast rays (Holdheide, 1951). In an archaeological context it is often impossible to employ sections for bast analysis, and identification must be performed on the basis of the tangential surface. The reliability of the analysis can be further reduced by the degree of histological variation within one and the same species. In addition, processes of selective decomposition often cause the loss of characteristic features, like crystals, phloem ray parenchyma etc., so that a collection of recent specimens is often required for purposes of comparison.

Our specimen collection was prepared on the basis of the guidelines and reports presented by Körber-Grohne (1977). First of all those native woody plants were selected that are a source of good-quality bast. On the basis of the ease of separation of the bast from the wood, we can distinguish the following four categories:

1. Bast easily removed in long strips:
   lime (*Tilia*), oak (*Quercus*), willow (*Salix*), aspen (*Populus tremula*), common maple (*Acer campestre*), wych elm (*Ulmus glabra*), ash (*Fraxinus excelsior*), spindle tree (*Euonymus europaeus*)
2. Bast easily removed in short strips:
   Norway maple (*Acer platanoides*), sycamore (*A. pseudoplatanus*), honeysuckle (*Lonicera xylosteum*), common elder (*Sambucus nigra*), bird cherry (*Prunus padus*), rowan (*Sorbus aucuparia*), hornbeam (*Carpinus betulus*), spruce (*Picea abies*), fir (*Abies alba*)
3. Bast cannot be separated:
   hazel (*Corylus avellana*), silver birch (*Betula pendula*), yew (*Taxus baccata*), juniper (*Juniperus communis*), dogwood (*Cornus sanguinea*), wayfaring tree (*Viburnum lantana*), guelder-rose (*Viburnum opulus*), hawthorn (*Crataegus monogyna*), alder buckthorn (*Frangula alnus*), whitebeam (*Sorbus aria*)
4. Bast without bast fibres, cannot be separated:
   beech (*Fagus sylvatica*), grey alder (*Alnus incana*)

Categories 3 and 4 are unsuitable for harvesting bast and thus for cord making and were therefore excluded from the analysis. A collection was built up of specimens from the woody plants included in categories 1 and 2. For this purpose several-year-old twigs were taken from the various species at the beginning of the vegetation period in May when sap flow through the phloem had begun, and 5 cm long strips of bark were removed. They were boiled for 15 minutes in 10% KOH. This chemical treatment replaces the retting process in which the bark is soaked in water for a long period of time (up to four months). As a result, decomposition processes set in. The middle lamella in particular is affected by slimy decomposition, and the phloem layers and fibres can be more easily separated form the woody tissue. Then the specimens were washed and the bark detached from the phloem. The strips of bast produced in this way were then embedded in glycerine to conserve the specimens for purposes of comparison with the subfossil bast.

The analysis was performed using selected specimens obtained from the follow-up excavations at Tisenjoch in 1991. In almost all these botanical specimens, binding material is present in varying quantities. In a first macroscopic inspection using a pair of binoculars at 10–20x magnification, the binding material was separated from the other materials, and materials of animal origin were also excluded. For the analysis, five specimens containing a lot of bast and three containing only little bast were selected on a random basis. Tissue was obtained from the following samples (see also Table 1): 91/8, 91/9 (birch bark), 91/102 (cords), 91/103 (grass, fibres, hair, cords), 91/106 (sewed leather), 91/132 (birch bark fragments), 91/133 (tree fungus with leather straps) and 91/139 (grass and leaf remains).

Specimens 91/8, 91/9 and 91/133 contained only few bast fragments, and they were analyzed in their entirety. In the other cases, fifty sub-samples were taken from the bast tissue on a random basis. A total of 295 pieces of bast, including both individual bast layers and fragments of string and cord, were taken for analysis. Tissue analysis was performed under the light-optical microscope at a magnification of 100–400x using both the bright field and polarised light. The latter is required to reveal the crystals that are important for correct diagnosis and also the characteristic microfeatures of the fibres (see Plates 3A–D and 4A–C).

3. Results

On the whole the subfossil bast specimens studied were well preserved. Some of the separated bast layers are twisted (Plate 4D); some are fragments of strings (Plate 4E) and some are the remains of flat bast strips.

Table 1. Results of the bast analysis

Sample no.	Specimen	No. of fragments analysed	Result
91/8		5 (all bast material)	lime, *Tilia*
91/9	birch bark	22 (all bast material)	lime, *Tilia*
91/102	cords	50	lime, *Tilia*
91/103	grass, fibres, hair, cords	50	lime, *Tilia*
91/106	sewed leather	50	lime, *Tilia*
91/132	birch bark fragments	50	lime, *Tilia*
91/133	tree fungus with leather straps	18 (all bast material)	lime, *Tilia*
91/139	grass and leaf remains	50	lime, *Tilia*

Both forms of bast were used as binding material. The antler fragments and the animal sinew found in the quiver, for example, were bound together with broad bast strips (Egg, 1992). Aerobic processes of decomposition were detected on some of the bast layers, which also show signs of fungal infection, in some cases severe. The signs of decomposition in the cell walls can be considered minor. The xylem ray parenchyma has mostly separated. Only few diagnostically relevant crystals were found in the subfossil material. Most of the specimens were whole bast layers, which were used as binding material in that form or twisted to make a yarn or cord. In only a few cases (specimen 91/106) were bast fibre bundles seen that had been carefully separated from the surrounding tissue, twisted into a yarn and used for sewing.

At a binocular magnification of 15x, the bast layers analyzed all have the same appearance. The surface is smooth with a longitudinal fibre arrangement interrupted at more or less regular intervals by single or multiple rows of xylem rays (Plates 2B and C). At a more powerful magnification (100x), the xylem ray cells – where present – can be seen to be isodiametric and rectangular, in some cases slightly rounded or, in the case of the wider xylem rays, with tangential broadening (Plate 3B). The parenchyma cells are wide and have extensively pitted longitudinal walls (Plates 3E and F). The cross walls of the parenchyma cells are thicker than the longitudinal walls. Two different types of parenchyma cells are to be found: highly elongated cells, and short cells with a length that is two or three times the width at most. The latter are primarily present in regions adjoining the xylem rays. In the recent specimens collected for purposes of comparison, these cells contain the characteristic monohydrates of calcium oxalate (Plate 3A). They can be clearly seen in polarised light, as can the bast fibres. The bast fibres are long, with pointed ends and oblique pitting on the surface (Plates 4A and B). This histology is characteristic of the bast of the lime (*Tilia*), and according to Körber-Grohne (1977) can only be confused microscopically with bast from the elder (*Sambucus*). Elder bast fibres, however, have no oblique pitting and the parenchyma cells have extremely thick walls.

From the macroscopic picture, consideration must also be given to oak (*Quercus*) and willow (*Salix*). Oak bast (Plates 3C and D), however, is compact and cannot be separated into thin annual layers like the bast of the lime (Plate 1C). In addition, the bast fibres of the oak (*Quercus*) have characteristic knobby areas of greater thickness. The bast of the willow (*Salix*) can also be harvested in thin layers, but has only single-row xylem rays. The wider longitudinal gaps (Plate 2A) are not xylem rays but dilatation cracks (Körber-Grohne, 1977).

4. Discussion

For his binding materials, the Iceman used the best raw material available in his environment. In the Neolithic – and probably in the Palaeolithic, too – the bast of the lime (*Tilia*) was the material of choice for producing binding materials in Central Europe. The fibre is of good quality, easily worked and produces a good yarn that can be used to make cords and textiles. It is also readily available from the native flora. In the histological section, lime phloem displays characteristic banding caused by the periodic release of bast fibre and parenchyma cell layers from the cambium. Young trees have between two and five such alternating tissue layers per growth ring. In old trees there is normally only one bast fibre and parenchyma layer per annual ring (Holdheide, 1951). This layered structure guarantees a good harvest. As an annual bast ring is usually covered with wide-lumen parenchyma cells, the individual annual rings are easily separated. This fact was known to Neolithic man in the Alpine region. In the contemporaneous Neolithic lakeside settlements of the pre-Alps, lime bast is the most common material used for making cord or plaited articles (Körber-Grohne, 1977).

As mentioned above, the analyzed binding material can be divided into two types: on the one hand the rough strips of bast layers that the Iceman used either in that form or, more frequently, twisted to make cords, and on the other hand pure bast fibres twisted into a yarn for use in sewing. In size, shape and features, the latter fibres are identical with the lime bast fibres taken from the bast layers. The yarn used for sewing consists of bast fibres that are either loose or still adhere to each other. Only isolated remains of parenchyma cells were detected, which is indicative of the quality of the yarn-making process. In order to obtain bast fibres of such quality, the bast must be subjected to a long process of soaking or retting. That was not done in the case of the bast strips, some of them multilayered, that the Iceman used. In that case the bast fibre bundles are still surrounded by parenchyma, or they comprise several annual layers of bast. In the case of the lime tree they are easily stripped from the wood at the beginning of the vegetation period. During the analysis work, some of the bast strips separated into their individual annual layers on being handled. This can also be the result of a brief retting process, but in the material analyzed it is to be interpreted as the first stage of decomposition, as the bast strips had remained in a damp environment or even underwater for a long period of time. This incipient decomposition can have developed either over the years on Tisenjoch or during the brief period that has elapsed since the Iceman's discovery. According

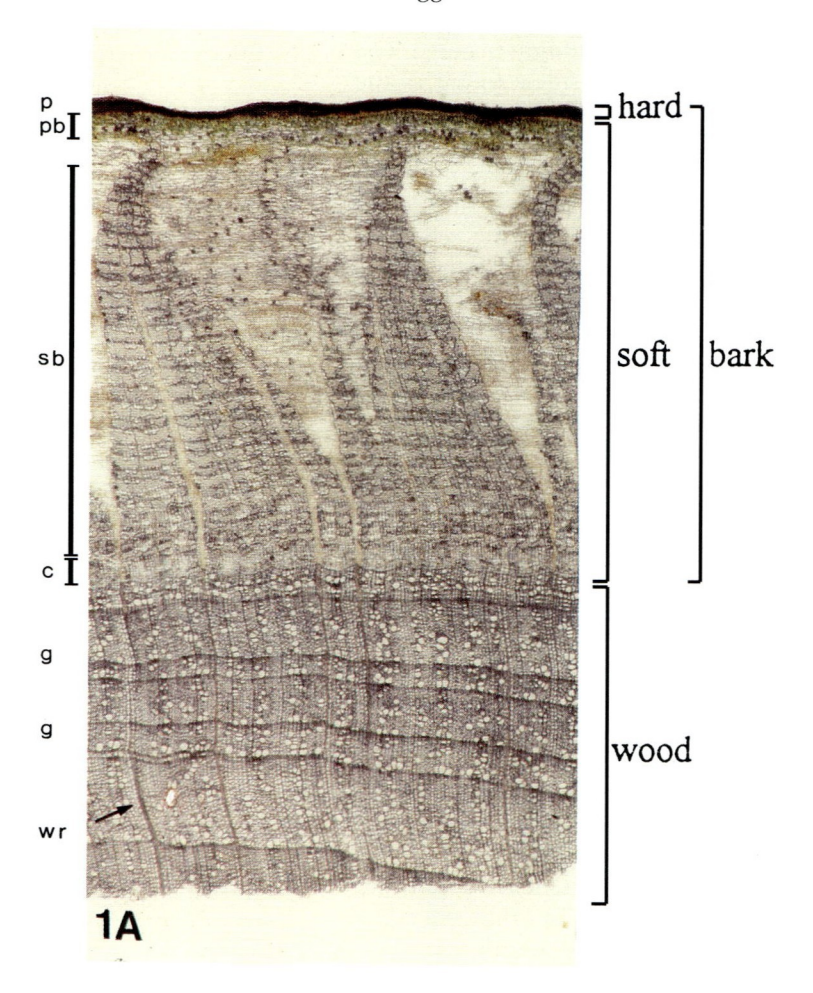

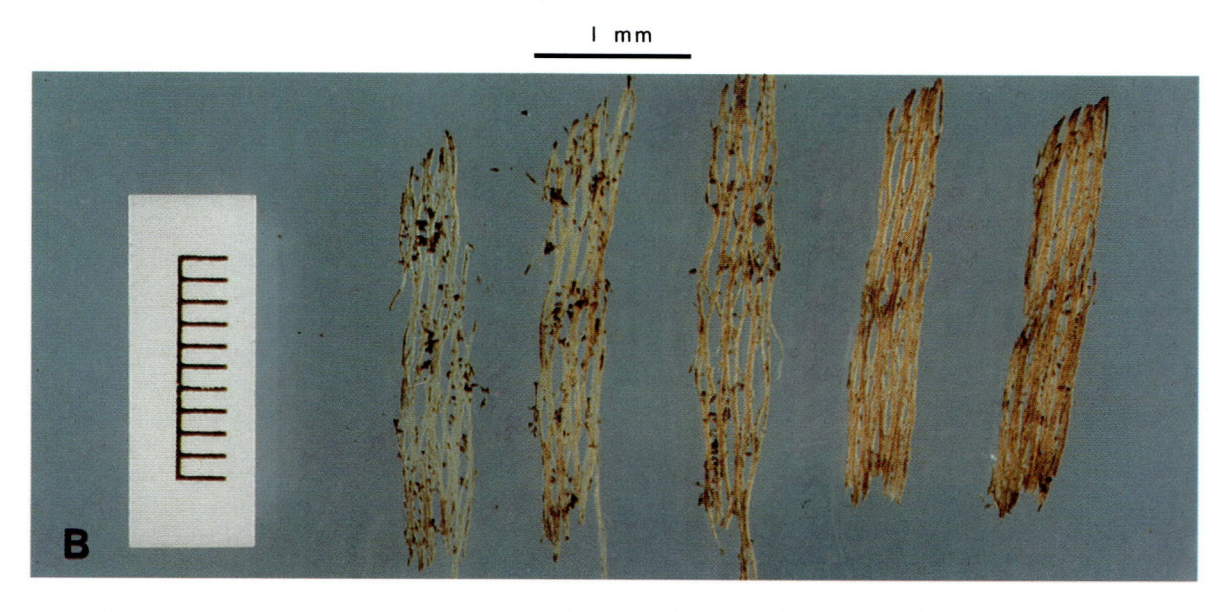

Plate 1. **A.** Microphotograph of a branch in cross section. Wood, soft (= bast) and hard bark of large-leaved lime (*Tilia platyphyllos*) and cell structure along a cross section of a young branch: g growth-ring boundary; c cambium; sb secondary bark (phloem); pb primary bark; p multilayered periderm; br wedge-shaped widened bast ray out of a primary multiseriate wood ray (wr). **B.** Fossil soft bark of lime (*Tilia*) divided into single annual bast layers (1 scale unit ≅ 1 mm)

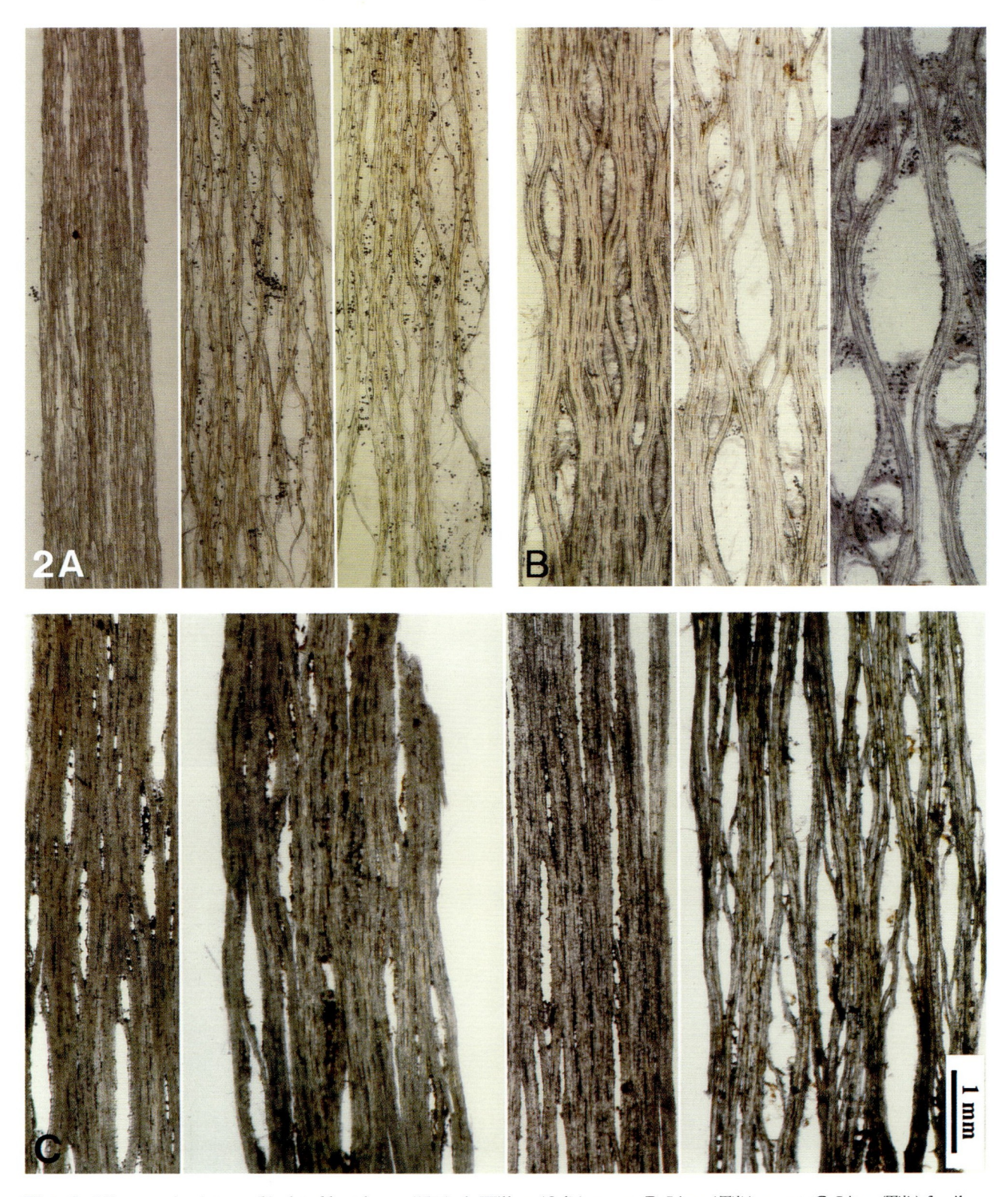

Plate 2. Microscopic picture of isolated bast layers (15×). **A.** Willow (*Salix*) recent; **B.** Lime (*Tilia*) recent; **C.** Lime (*Tilia*) fossil

Plate 3. Microscopic pictures of bast layers in polarized light (100×). **A, B.** Lime (*Tilia*) recent; **C, D.** Oak (*Quercus*) recent. Pitting of bast parenchyma in lime (250×). **E.** recent; **F.** fossil

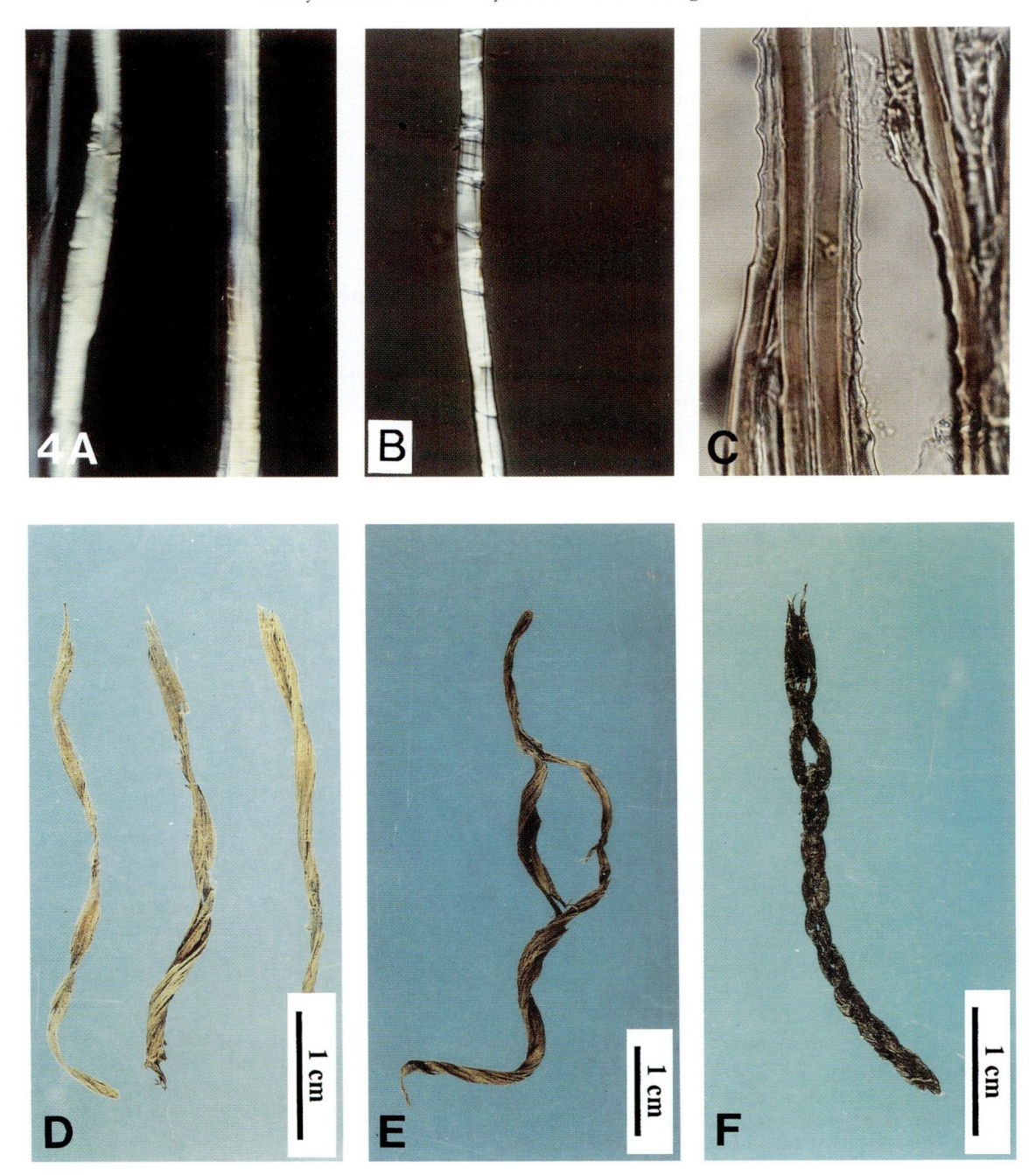

Plate 4. Highly enlarged view of isolated libriform fibres in polarized light (400×). **A.** Lime (*Tilia*) recent; **B.** Lime (*Tilia*) fossil; **C.** Oak (*Quercus*) recent. Fossil bast of lime (*Tilia*). **D.** Isolated twisted bast layers (shafts, "threads" respectively); **E.** Remnant of a multilayered twisted bast strip–partly untwisted; **F.** Fragment of a twined cord out of two shafts (sample 91/133)

to investigations into changes in the lipid composition of the Iceman's skin performed by Bereuter et al. (1996), the Iceman had lain in water for several months prior to desiccation. When lime bast is placed in water for three months, it starts to separate into its annual layers through a process of slimy decay and decomposition of the pectin lamellae of the wide-lumen parenchyma cells (Körber-Grohne, 1977). This incipient decomposition also explains the numerous isolated annual bast layers found among the specimens obtained from the vegetable residues in the water used to wash the Iceman's clothing and equipment.

Those bast strips that have survived in multilayered form were worked by the Iceman while still fresh. They were not subjected to retting, as the annual layers are still bonded to each other. In the simplest case he used the bast as plain strips, while others were twisted to form single threads (Plates 4D and E), and then sometimes doubled and twisted into cords (Plate 4F).

Abstract

The botanical remains found with the Iceman include a number of well preserved Neolithic cords. On macroscopic inspection it can be seen that bast (phloem) was one of the raw materials used for the binding materials. Microscopic analysis of 295 tissue samples taken from the cords on a random basis show that the Iceman used only bast from the lime tree (*Tilia*) for this purpose. The bast was harvested by the simple technique of stripping the fresh bast layers from the wood without soaking or retting. This can be seen from the multilayered bast strips and bast fibre bundles which were simply twisted together to make the cords.

Zusammenfassung

Unter den botanischen Resten, die beim Eismann gefunden wurden, befinden sich eine Reihe von Schnüren. Die Reste sind gut erhalten und schon bei makroskopischer Betrachtung läßt sich Bast als eines der Rohmaterialien für das neolithische Bindematerial erkennen. Die mikroskopische Analyse von 295 nach dem Zufallsprinzip gezogener Gewebeproben dieses Bindematerials hat ergeben, daß der Eismann ausschließlich Bast der Linde (*Tilia*) verwendet hat. Die Gewinnung des Bastes erfolgte einfach durch Abziehen der frischen Bastlagen vom Sproß ohne Wasser- bzw. Tauröste. Dafür sprechen die mehrschichtigen Bastlagen und die zusammenhängenden Bastfaserbündel, die in einfacher Verarbeitungstechnik zu Schnüren gedreht wurden.

Résumé

Parmi les vestiges végétaux, découverts auprès de l'Homme des glaces, il y a toute une série de cordons. Ces restes sont bien conservés, et l'analyse macroscopique permet déjà d'identifier le liber comme l'une des matières premières utilisées pour la confection des moyens d'attache néolithiques. L'analyse microscopique de 295 échantillons de tissu prélevés au hasard sur ces matériaux a mis en évidence que l'Homme des glaces avait utilisé exclusivement du liber de tilleul (*Tilia*). Ce matériau avait été récolté en détachant simplement les couches fraîches de liber des pousses, sans les exposer à l'écrouissage à l'eau ou à la rosée. C'est ce qu'indique leur structure multicouche et les faisceaux continus de fibres libériennes, tordus selon un procédé très simple pour obtenir les cordons.

Riassunto

Tra i resti botanici, ritrovati sulla mummia si trovano anche vari tipi di corde. Tali resti sono in on buono stato di conservazione ed al microscopio si è riscontrata la filaccia tra i materiali utilizzati nel neolitico per corde o cordicelle.

L'analisi microscopica di 295 campioni, scelti a caso, di tessuto di tale materiale hanno evidenziato che l'uomo del Similaun utilizzava esclusivamente filaccia di tiglio, ricavata semplicemente sbucciandi gli strati freschi di filaccia dal germoglio. Ciò è comprovato dalla presenza di più strati di filaccia riuniti in fasci coerenti a loro volta attorcigliati onde formare corde.

Acknowledgement

This research was funded by the Austrian Science Fund (project no. 10151-SOZ).

References

Bereuter T. L., Reiter C., Seidler H., and Platzer W. (1996) Postmortem alterations of human lipids - part II: lipid composition of a skin sample from the Iceman. In: Spindler K., Wilfling H., Rastbichler-Zissernig E., Zur Nedden D., Nothdurfter H. (eds.): Human Mummies. The Man in the Ice, vol. 3: 275–278.

Denffer D.v. (1983) Morphologie. In: Lehrbuch der Botanik für Hochschulen/begr. von E. Strasburger et al., 32. Aufl./ neubearb. von Dietrich von Denffer et al. Stuttgart New York: Gustav Fischer Verlag, pp. 1161.

Egg M. (1992) Die Ausrüstung des Toten. In: Egg M., Goedecker-Ciolek R., Groenman-van-Wateringe W. und Spindler K. (1993) Die Gletschermumie vom Ende der Steinzeit aus den Ötztaler Alpen. JbRGZ 39: pp. 128.

Goedecker-Ciolek R. (1992) Zur Herstellungstechnik von Kleidung und Ausrüstungsgegenständen. In: Egg M., Goedecker-Ciolek R., W. Groenman-van-Wateringe und Spindler K. (1993) Die Gletschermumie vom Ende der Steinzeit aus den Ötztaler Alpen. JbRGZ 39: pp. 128.

Holdheide W. (1951) Anatomie mitteleuropäischer Gehölzrinden (mit mikrophotographischen Atlas). In: Freund H. (ed): Handbuch der Mikroskopie in der Technik. Band V. Teil 1: 195–367.

Körber–Grohne U. (1977) Botanische Untersuchungen des Tauwerks der frühmittelalterichen Siedlung Haithabu und Hinweise zu Unterscheidung einheimischer Gehölzbaste. Berichte über die Ausgrabungen in Haithabu. Untersuchungen zur Anthropologie, Botanik und Dendrochronologie 11: 64–111.

Körber–Grohne U. (1987) Textiles, fishing nets, wickerwork and rope from the Neolithic sites of Hornstaad and Wangen on Lake Constance (Bodensee): Botanical Investigations. In: Textiles in Northern Archeology, NESAT III, Textile Symposium York, 11–20.

Lippert A. (1992) Die erste archäologische Nachuntersuchung am Tisenjoch. In: Höpfel F., Platzer W. und Spindler K. (eds.): Der Mann im Eis, Bd. 1. Veröffentlichungen der Universität Innsbruck 187: pp. 463.

Vogt E. (1938) Geflechte und Gewebe der Steinzeit. Monographien zur Ur- und Frühgeschichte der Schweiz. Bd.1. Basel.

Winiger J. (1995) Die Bekleidung des Eismannes und die Anfänge der Weberei nördlich der Alpen. In: Spindler K., Rastbichler–Zissernig E., Wilfling H., Zur Nedden D., Nothdurfter H. (eds.): Der Mann im Eis. Neue Funde und Ergebnisse. The Man in the Ice, vol. 2. Wien: 119–187.

Bryology and the Iceman: Chorology, Ecology and Ethnobotany of the Mosses *Neckera complanata* Hedw. and *N. crispa* Hedw.

James H. Dickson

Division of Environmental and Evolutionary Biology, University of Glasgow

1. Introduction

The first published paper on bryology and the Iceman is that by Dickson et al. (1996) in the Proceedings of the Royal Society of London, Series B. It listed all the bryophytes so far identified after removal from the Iceman's clothes as well as the bryophytes extracted from the coarse mineral sediments that had accumulated in the hollow where the corpse had lain frozen for over 5000 years. The main point of the paper was to discuss the Iceman's provenance but because of the rules of the journal the paper was short and could not give a detailed account of the crucially important species of *Neckera*. The second paper on bryology and the Iceman (Dickson, 1997) concerns the single, tiny leaf from a filiform shoot of *Neckera complanata* recovered from a small sample of the contents of the Iceman's colon.

Within the studied area which is a rectangle of 35 squares, each of 10 square km (Fig. 5), four species of the genus *Neckera* have been noted since bryological recording began more than 100 years ago: *N. besseri* (Lob.) Jur., *N. complanata*, *N. crispa* and *N. pumila* Hedw. Widespread but local and declining in central Europe, *N. pennata* Hedw. has not been recorded. None of the *Neckera* remains found with the Iceman belong to *N. besseri* which is very distinct in its small size and very obtuse leaves (He, 1997). Nor do any belong to *N. pumila*, again distinct in its small size but with undulate, long pointed leaves. Some of the Iceman material referred to *N. crispa* is small with somewhat pointed leaves but there is nothing strongly suggesting *N. pennata*.

This third paper gives a fully comprehensive account of the distributions, habitats, Late Quaternary and archaeological records and ethnobotany of the two species of *Neckera*. All the bryological samples discussed were washed from the Iceman's clothing or gear in 1991 and 1992 at the Römisch-Germanisches Zentralmuseum Mainz, Germany. There are not less than 30 bryophytes in these samples. The only bryophyte that occurs as a coherent mass is the moss *Neckera complanata* (in sample 1991/124). It was also found in two other samples (1991/123 and 137). *Neckera crispa* occurs in six samples (1991/95, 96, 104, 106 and 137 and 1992/36). Sample numbers and descriptions were sent to JHD by Dr. Markus Egg of Mainz. Details of the number of subfossil mosses and preservation follow Dickson (1973, p. 52–53): st = stem(s), lf = leaf or leaves, v = very, b = bad, g = good, t = tentative (species identification uncertain).

2. Samples with the *Neckera* spp

2.1. *Neckera complanata*

"91/123 Lederstücke (Leggings)" [Leggings]. 2st; 5 mm; vb.
"91/124 Gras, Leder-bzw. Fellteile (Oberkörperbekleidung und Leggings)" [Grass, pieces of leather or hide (upper body clothing and leggings]. Mass of st to 65mm; g. Many detached lf; g. See Figs. 1 and 2.
"91/137 Haare, Gras und Fellreste" [Hair, grass and remains of hide]. 1st; 7 mm; b.

2.2. *Neckera crispa*

"91/95 Leder, Fell (Oberkörperbekleidung) und gedrehte Riemen (Quaste)" [Leather, hide (upper body clothing) and twisted thongs (tassel)]. 2st; 30 mm; vb; t.
"91/96 Haare, Schnurreste, Fasern NG 91" [Hair, remains of cord, fibres NG 91]. 1st; 7 mm; g. See Fig. 3.
"91/104 Leder bzw. Fell (Oberkörperbekleidung), Haare und gedrehte Lederriemen (Quaste) NG91" [Leather or hide (upper body clothing) hair and twisted leather thongs (tassel)]. 1st; 9 mm; vb.
"91/106 Leder- bzw. Fellteile (Oberkörperbekleidung und leggings)" [Pieces of leather or hide (upper body clothing and leggings)]. 1st; 3 mm; b.
"91/137 Haare, Gras und Fellreste" [Hair, grass and remains of hide]. 2st; 18 mm; g.
"92/36 Fell- und Lederreste (Schurz, Leggings und Oberköperbekleidung) und gedrehte Lederriemen

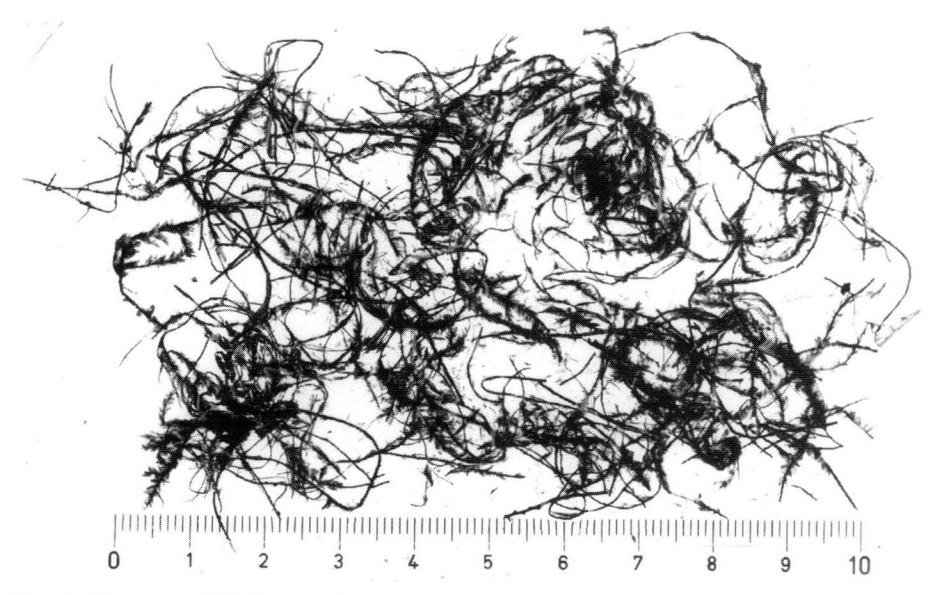

Fig. 1. The mass of *Neckera complanata* from sample 124

Fig. 2. A leafy stem with a leafless side branch of *Neckera complanata* from sample 124. The stem is about 18 mm long

Fig. 3. A small leafy stem of *Neckera crispa* about 7 mm long from sample 96. Well-preserved material of *Neckera* is readily recognized by leaf shape and areolation. The characteristic rugosity of the leaves of *N. crispa* can be seen

(Quaste)" [remains of hide and leather (apron, leggings and upper body clothing) and twisted leather strings (tassel)].

3. Archaeological and Holocene records of the *Neckera* spp

If asked to consider which mosses the Iceman might have taken on his last journey as part of his gear JHD might well have guessed these very *Neckera* species which have highly detailed Quaternary histories and have occurred many times in archaeological contexts in many places in Europe. Their archaeological occurrences are so numerous and their abundance sometimes so great in these contexts that two unconflicting hypotheses arise.

1. The *Neckera* spp were formerly very abundant and so were readily gathered.
2. The *Neckera* spp were sometimes selectively collected for particular purposes.

In the British Isles archaeological occurrences of *N. complanata* totalled nine (from the Neolithic period to Medieval times) as collated more than twenty years ago (Dickson, 1973). Since then there have been at least eleven other discoveries to add; this makes a total of not less than 20 archaeological sites. Perhaps the most noteworthy of the 20 is the use of the species as caulking for two Bronze Age sewn boats from the Humber Estuary (Fig. 4; Wright, 1990); these craft had virtually pure *N. complanata* packed into all the many seams (Dickson, 1973, p. 193). Why did the builders not simply use any moss or mosses that could have been easily found? It seems evident that the *Neckera* was part of their building technique and that they had a ready supply of that particular moss. The boat builders had also selected the tall, strong moss *Polytrichum commune* Hedw. to make into ropes, also used as caulking for a third boat.

Mosses including both *Neckera* spp, especially *N. crispa*, have been recorded many times from the central European prehistoric lake villages. As early as 1865 Heer wrote "Die Moose dienten ohne Zweifel zum Verstopfen der Löcher in den Hütten und als Polster." By 1905, Neuweiler had listed 11 archaeological discoveries of *N. crispa*: nine from Switzerland (5 Neolithic, 4 Bronze Age), two from Italy (both Bronze Age) and one from Austria (Neolithic). He thought that the uses were for plugging, for bedding and for cushions and stated that *N. crispa* was the most abundantly used of the many mosses he discussed. Neuweiler also listed four Swiss Neolithic finds of *N. complanata*.

Much more recently Ochsner (1975) discussed the use of *N. crispa* by prehistoric Swiss without adding much to Neuweiler's ideas but he speculated that the moss had been traded. From southern Germany, Rösch (1988) has listed nine recoveries of *N. crispa* and nine of *N. complanata* from prehistoric sites in the Bodensee area. Insulation of wattle walls and hygiene (toilet paper) were the purposes he discussed for the *Neckera* spp and the many other mosses he found. Dr. Ursula Maier has informed JHD of excavations of the Neolithic Hornstaad-Hörnle IA of Bodensee dating from c. 4000 BC. Abundant moss pads were found with *N. crispa* and *N. complanata* among the commonest species. The pads were mostly inside the houses where they may have been used for construction, bedding or floor coverings. Karg (1990) found both the *Neckera* spp from Neolithic layers near Kr. Konstanz, as did Dick (1989) from Mozartstrasse in Zürich and Bollinger (1994) from Egolzwil 3, Luzern. From the Neolithic village at Weier, Switzerland, Robinson and Rasmussen (1989) reported *N. complanata*. Knörzer (1995) found *Neckera complanata* to be the most abundant moss recovered from the Neolithic well at Kuckhoven, Germany. From the Neolithic site at Lake Dummer, Northwest Germany, Grosse-Brauckmann (1979) listed *N. crispa*, *N. complanata* and *N. pumila* Hedw.; he thought that only the first had been collected for use.

Jennifer Miller has investigated the botanically rich occupation layers from the Oakbank Crannog (a lake dwelling which was inhabited some 2500 years ago) in Loch Tay, a large lake in the Scottish Highlands. Eleven of the investigated eighteen samples of contained *N. complanata*.

Körber-Grohne (1979) listed *N. crispa* and *N. punctata* [sic; probably intended as *N. pennata* Hedw.] from a well in the Roman site at Welzheim near Stuttgart. Two other Roman wells, both in England yielded *N. complanata* (at Denton: Dalby, 1971 and at York: Kenward et al., 1986). Both *N. crispa* and *N. complanata* were listed from an Iron Age Dutch site at Enzige by Van Zeist (1974), both from a Roman site at Aachen-Burtscheid, Southwest Germany, by Knörzer (1980) and both from a Pictish (Dark Age) midden at Dundurn in the Scottish Highlands (Dickson et al., 1989).

In a shallow pit in Viking age Dublin, Ireland, were eleven species of woodland mosses mostly in large masses, including both *Neckera* spp, mixed with human excrement (Mitchell et al., 1987; Dickson, 1973; 1986); mosses were the toilet paper of those times, just as

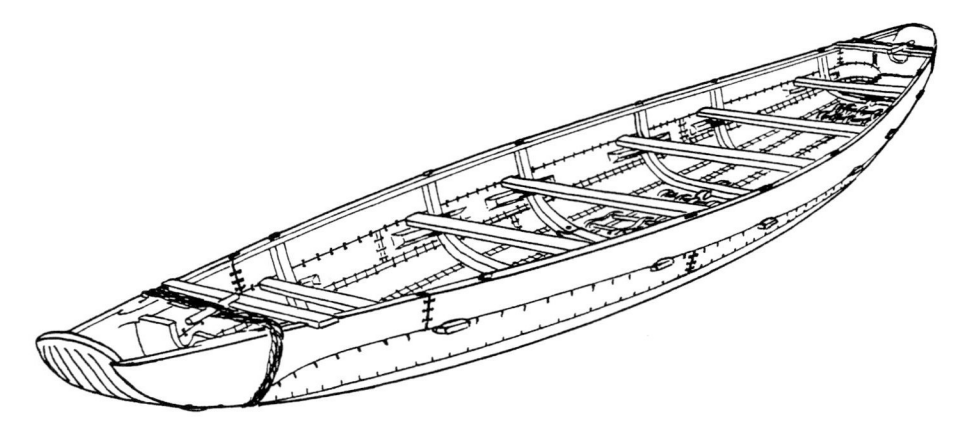

Fig. 4. The most recent reconstruction of the three boats of sewn *Quercus* planks made in the Bronze Age at North Ferriby, England. The boats were about 15.4 m long and 2.6 m wide. The seams between the planks were packed with *Neckera complanata* and in the case of boat 3 with ropes made of *Polytrichum commune*

they have been at other periods and places. Such may have been the use of *N. complanata* (in ten samples) and *N. crispa* (two samples) listed from Viking Haithabu, northern Germany by Behre (1969). Okland (1988) found *N. complanata* in a Medieval latrine in Oslo. Wiethold recovered it from Medieval and later wells and cesspits in Kiel, Germany. Cappers and van Zanten (1994) reported *N. complanata* from layers (c. 50 BC–c. 400 AD and c. 800–c. 1300 AD) in a dwelling mound at Heveskesklooster, the Netherlands, and Jörgensen (1980) reported it from a Medieval site at Svendborg, Denmark. Fraser (1980) recovered it from Medieval layers in the three Scottish towns of Aberdeen, Elgin and Perth and Stevenson (1986) recognized it from layers of Saxon-Medieval age in the English town of Ipswich. Van Zeist (1988) reported it from the Medieval site at Pesse and from Leeuwarden, another Medieval site in the Netherlands, he found *N. complanata* (in ten samples), *N. crispa* (six samples) and *N. pumila* (in three samples); he speculated that the first two may have been imports. Culikova (1995) listed *N. complanta*, *N. crispa*, *N. pumila* and *N. pennata* from four different samples of 13th and 14th century age within the town of Most in the Czech Republic. Earlier Duda and Opravil (1988) had listed *N. crispa* from seven Medieval sites and *N. complanata* from five in the former Czechoslovakia.

Another noteworthy archaeological discovery of the genus *Neckera* concerns *N. intermedia* Brid. which was recovered from the eviscerated abdomen of a "mummy" of a Guanche, a prehistoric inhabitant of the Canary Islands (Home et al., 1991).

No doubt there have been other published records, some in obscure journals, of the *Neckera* spp from archaeological layers. However, the many records listed above are more than enough to make the point that these mosses are very often found during excavations.

Recovery of the *Neckera* species from archaeological contexts need not indicate that these mosses had been specially gathered for a purpose. *N. complanata* was found under the enormous Neolithic mound, called Silbury Hill, in southern England (Williams, 1976) but it had not been selectively collected; it had been a minor component of the chalk grassland that had been cut into turves to make the mound. Another way by which *Neckera complanata* could have been brought to archaeological contexts unintentionally is shown by the observations of Bates (1993). He found epiphytic bryophytes in situ on bark left on structural timbers of *Quercus* in a farmhouse in Calvados, Normandy; among the ten species was *N. complanata* as a minor component. Such an incorporation into the archaeological layers at Heveskesklooster is contemplated by Cappers and van Zanten (1994) for *N. complanata* and other epiphytic mosses. Other workers such as Dick (1989) and Van Zeist (1988) have discussed this matter. See Ando and Matsuo (1979) concerning mosses as plugging for log cabins.

Nevertheless, most of these archaeological records make clear that species of *Neckera* have been used, either selectively or merely as part of general gatherings of mosses, over long periods of time in many parts of Europe and beyond.

4. Present distribution and habitats of the *Neckera* spp

N. complanata has an extensive geographical range at present in Europe as does *N. crispa* (map of the latter in Störmer, 1969). They are woodland mosses, only very rarely if ever reported from above the altitudinal treeline. Often growing together, these species, especially *N. crispa*, are strongly connected with basic or ultra basic, often very calcareous, geology throughout their European ranges including the Tyrol (Buchner et al., 1993; Düll, 1991 and 1994; Kuc, 1964; Marstaller, 1992; Preston, 1994; Venturi, 1899).

Across much of their ranges their habitats are both shady rocks (often more or less vertical rock faces) and the trunks and branches of trees. Only rarely do they occur on the ground. A few authors have noted a change with geography from epiphytism to epilithism, for example Sjögren (1964) referring to southern Sweden. Both species tend to form large patches that protrude from the substratum and are easily pulled away in often more or less flat handfuls without too much adhering soil or fragments of bark. They may well have been available in the greatest quantities in areas of base-rich geology. However, that neither species is dependent on substrata of alkaline reaction is well shown by their occurrences in Vinschgau, Southern Tyrol, as observed by JHD and co-workers in the summers of 1994 and 1995 and reported long ago by Venturi (1899). Here they can be found alone or together, often in large amount, on shady rock whether of marble or of non-calcareous schist. During field work in 1997 JHD was accompanied by his geological colleague Dr. J. G. MacDonald who noted the rock types that were the substrata for *N. complanta*. They were as follows: marble; volcanic ash flows, often calcareous; augen gneiss, a widespread rock in the studied area, which is not usually calcareous but is sometimes veined with coarse-grained calcite; amphibolite which often has small amounts of calcite. Other lithologies included quartz muscovite schist, biotite schist, paragneiss and granite, none of which is usually calcareous. In the Tyrol the *Neckera* spp often occur in the greatest abundances in gorges where the necessary shady, humid conditions prevail.

During the 1994, 1995 and 1997 field work, the *Neckera* spp were found at altitudes from about 500 m near Meran (the lowest at which bryological recording took place) to about 1450 m above Partschins (more than 600 m below the present tree-line). In small quantity on rock, *N. complanata* was found at c. 1700 m at Leiter Alm, above Algund.

For Ötztal [OT] and Pitztal [PT], Northern Tyrol, Düll (1991, p.278), states concerning *N. complanata* "Hohenstufe: - O (OT: −1500 m!; PT: 960! −1290 m!). Bis 1700 ml. ziemlich verbreitet in tieferen Lagen, aber nur steril" and considering *N. crispa* "Höhenstufe: - O - (S) (OT: 750! −>900m!, PT: −960 m!). Bis 1800! −2200 m. ziemlich verbreitet, nicht selten c. spg." O he defined as "obere montane bzw. oreale Stufe; bis fast 2000 m s.m." and S as "subalpine bzw. Krummholzstufe; ca. 2000–23000 m s.m."

5. Discussion

One of the great successes of pollen analysis over the last 50 years has been the revelation of how great in both space and time has been the extent of anthropogenic clearance of woodland all over Europe. In prehistoric times the woodlands were much more extensive than they are now. Given that woodlands with their shade and humidity are and were the main habitat of the *Neckera* spp the abundance of these mosses may very well have been much greater than now. The fossils from non archaeological contexts are consistent with that estimate. So impressed was JHD by the British fossil record of *N. complanata* that in 1973 (p.127) he wrote "If one divided the Pleistocene into moss periods the Flandrian [Holocene] would be the *Neckera complanata* period." Additional work since 1973 strengthens this statement. As an example, there is the work of Hamilton et al. (1983). From a plant-rich bed of mid Holocene age from a silted up river meander in northern Ireland, they recorded acorns (fruits of *Quercus*) and 19 different mosses, mainly species of woodlands. The two mosses designated as "abundant" were *N. complanata* and *Thamnobryum alopecurum* (Hedw.) Gang. A second example from much more recent times concerns riverine deposits of Medieval age (about 1100 AD) at Brno, the Czech Republic. These sediments contained both *N. complanata* and *N. crispa* as well as some 40 other mosses including many of woodland origin (Rybnicek, Dickson and Rybnicikova, 1998).

Both hypothesis 1 and hypothesis 2 seem very likely to be correct.

The *Neckera* spp are widespread and locally abundant in Vinschgau at present and in Trentino Alto Adige as a whole (Entleutner, 1884; Venturi, 1879; 1899 and see also Appendix). There is no reason why they were not just as widespread and even more abundant in the Iceman's time. Sample 91/124 consists not just of a large mass of *Neckera complanata* stems up to 65 mm long but also the following species of moss:

Amblystegium serpens (Hedw.) B, S. and G.	1st; 6 mm; g.
Antitrichia curtipendula (Hedw.) Brid.	1 lf; b.
Hydrogrimmia mollis (B., S. and G.) Loeske	1st; 2 mm; g.
Pohlia sp	Ist; 1 mm; g.
Polytrichum piliferum Hedw.	3st; 10 mm; vb and 4 lf, b.
Racomitrium lanuginosum (Hedw.) Brid.	3st; 5 mm; vg.

Also from this sample there are large fragments of Poaceae to more than 100 mm long, some small twigs of *Picea* and some hair.

The *Hydrogrimmia*, *Pohlia*, *Polytrichum* and *Racomitrium* were growing within the catchment of the hollow and have no relevance to the matter of the Iceman's ethnobotany or his provenance. With habitats that overlap with that of the *Neckera*, both the *Amblystegium* and the *Antitrichia* were accidental admixtures with the *Neckera*. The very size of the mass of *Neckera* and its association with a large amount of grass, twigs of *Picea* and hair rules out the thought that this moss was merely an accident. The Iceman might have encountered the *Neckera* spp on his travels and a possible explanation is that he took them from shady rocks, not necessarily calcareous, or from tree trunks as he walked through woodland. Did he deliberately select the *Neckera* spp rather than some other mosses and, if so, why? It is hard to see how there can ever be any certainty about that precise point. Perhaps it is more likely that these mosses came from his village where they and other species such as *Antitrichia* and *Leucodon* could well have been in use or in store ready for use.

Whether collected directly from the wild or from his village, why would the Iceman have gathered mosses whatever the species? Insulation, adornment, hygienic wiping, food, wrapping and medicine are all uses worth discussion.

5.1. Padding for the clothes

Mosses might be considered suitable insulatory material but against this there are (a) No ethnographical parallels known to JHD and (b) The clothes would have had to have had double layers to have held the mosses in place. There is no evidence for such a doubleness.

5.2. Decoration for the clothes

The Maoris of New Zealand are known to have had capes decorated with robust mosses (Polytrichaceae) and New Guineans also use mosses as part of their headdresses. Both *Neckera* spp and especially *N. crispa* can be several cm long and much branched and so might have been considered striking enough to have been decoration. There is no evidence of deliberate attachment of the mosses to the cape or other clothes.

5.3. Hygiene

Mosses have often been recovered from Medieval and Viking cesspits in Ireland, Scotland and Norway and have been observed to have eggs of human intestinal parasites attached, notably *Trichuris trichiura* (whip worm). The Roman fort at Bearsden, Scotland had a defensive ditch full of sewage-impregnated silt and mosses such as *Hylocomium splendens* and eggs of *Trichuris*. Mosses make entirely suitable toilet paper. It is plausible that the Iceman carried mosses for such a purpose. However, this cannot be proved; used mosses would have been discarded.

5.4. Food

Unlike the occasional report of the use of such unexpected plants as lichens and ferns, there are no enthnographical reports from any place at present or at any time in the recent past of mosses having been used as food, not even as famine food, nor is there any archaeological evidence to suggest such a use in the distant past. Not just humans but no other large mammals deliberately ingest mosses as food. Though

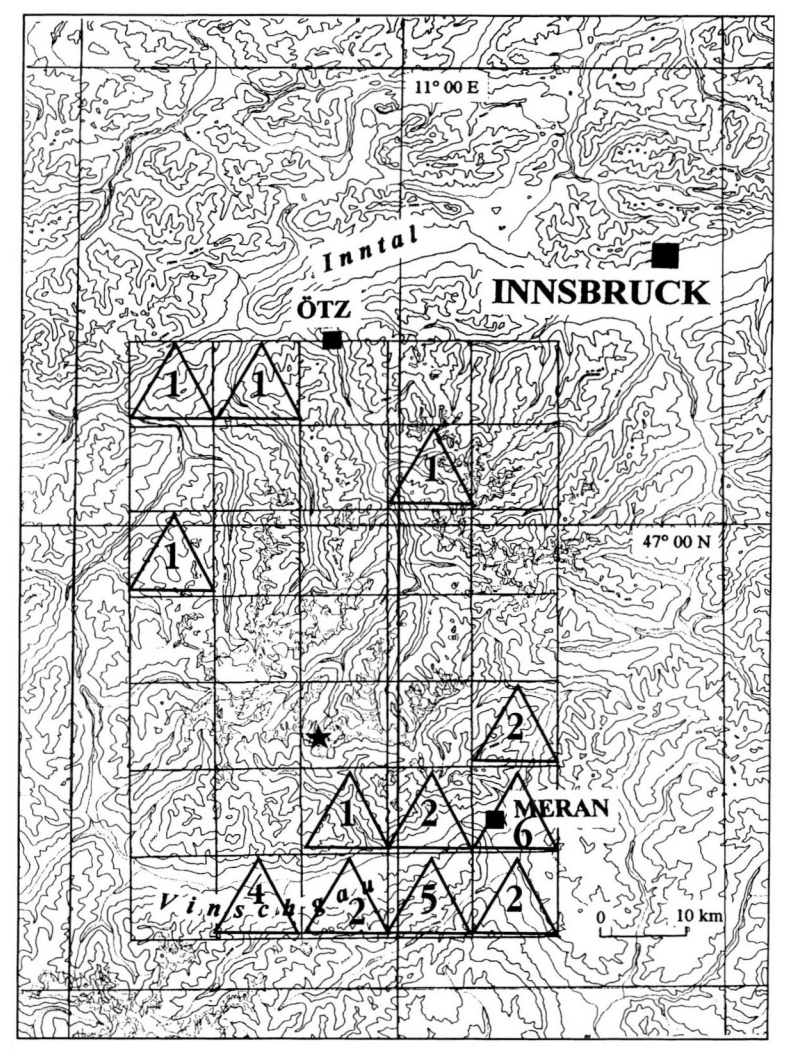

Fig. 5. Map of the present distribution of *Neckera complanata* in the Tyrol. The digits indicate the number of localities in each 10 km square. The star indicates the Iceman site

not poisonous, mosses are neither palatable nor nutritious.

5.5. Wrapping

Large mosses such as the *Neckera* spp and *Hylocomium splendens* which can be gathered easily in large, clean masses could well have been used as wrapping material. Perhaps the Iceman had used the *Neckera* spp to cover food in his rucksack (if there had been any provisions to wrap); this is the explanation put forward by JHD (1997) to account for the leaf of *Neckera complanta* found in the colon sample. The mosses such as *Sphagnum* found in the intestines of the famous Danish and English bogbodies (Dickson, 1997) and in human coprolites from the Orcadian Iron Age (Bell and Dickson, 1989) are likely to have been ingested with drinking water. The habitats of *Neckera complanata* makes the ingestion of a leaf in such a way improbable.

5.6. Medicine

There are exceedingly few reports of mosses being used internally as medicine from any part of the world; none refer to *Neckera* spp. The use of *Sphagnum* externally as wound dressing is well known and comes well into the present century. There is a claim for such use in prehistoric Scotland.

However, the one minute fragment of *Sphagnum* attached to the Iceman's clothes is hardly a good basis for deducing an intended medicinal use.

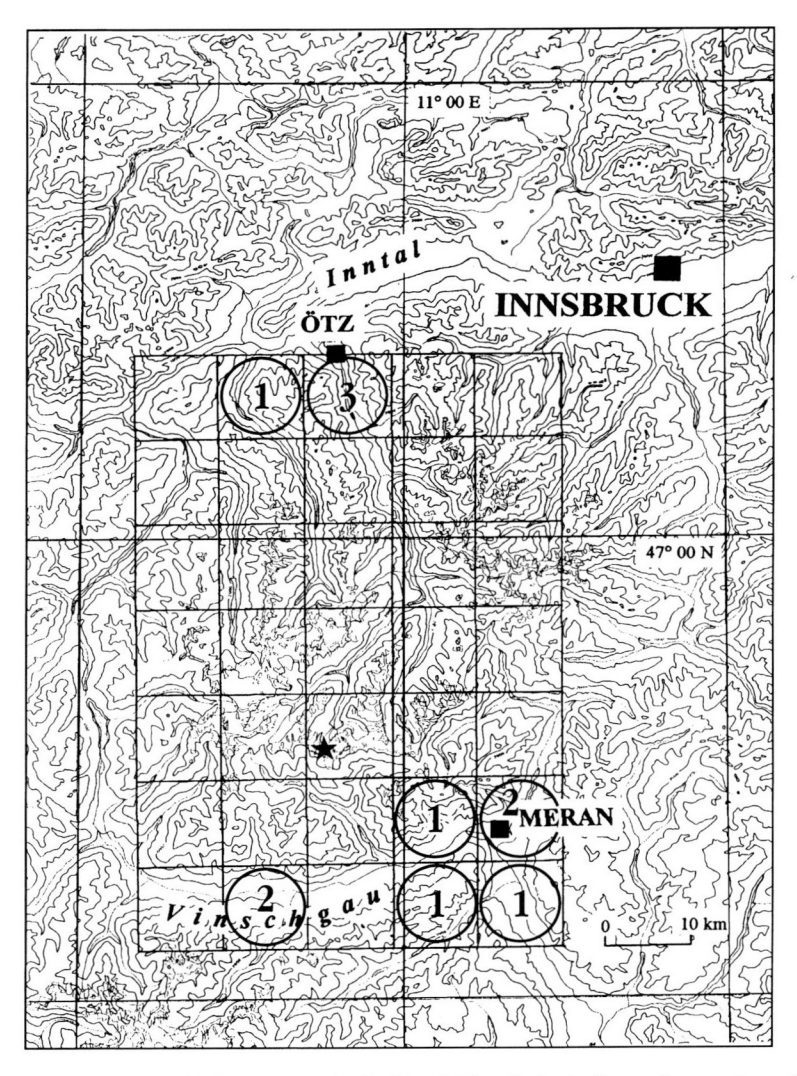

Fig. 6. Map of the present distribution of *Neckera crispa* in the Tyrol. The digits indicate the number of localities in each 10 km square. The star indicates the Iceman site

The most likely explanations for the Iceman having deliberately carried the large mass of *Neckera complanata* are hygiene or wrapping. The other explanations are improbable.

While it seems impossible that the large clump of *Neckera complanata* could have been merely sticking to the Iceman's clothes, perhaps other species had done so. Unlike many flowering plants, mosses do not have any adaptations for adhesion; they are not viscous nor do they bear hooks. Nevertheless, many of the species are represented by tiny fragments or detached leaves which by their very minuteness (a few mm or less) would have readily adhered to clothing especially if wet.

However high the timber-line may have risen in the Holocene, nobody claims that in the Tyrol it reached an altitude anywhere close to 3210 m, the height of the death site. The *Neckera* spp do not grow in the nival zone of the Alps (Pitschmann et al., 1954) and never would have grown at such heights. Even if they had reached the timber-line in the Ötztal Alps they would have occurred in abundance only at much lower altitudes, just as is the case now. Therefore the fragments of the *Neckera* spp represent clues to the Iceman's journeys, perhaps solely his last journey. This is true even if the mosses had not been deliberately gathered and had been merely adhering to his clothing.

In the summers of 1994, 1995 and 1997 six weeks of bryological recording have been carried out with the principal aim of producing distribution maps of the species found attached to the Iceman's clothes, especially the low to moderate altitude species. A rectangle of 3500 square km made up of 35 ten km squares was surveyed; the grid used is that on the Austrian maps of 1:50,000 and on the Italian maps of 1:25,000. The two maps are reproduced here (Figs. 5 and 6) also show localities taken from the publications of Entleutner (1884), Venturi (1899), Dalla, Torre and Sarthein (1904) and Düll (1991) and from specimens in the Innsbruck herbarium. The two *Neckera* spp are seen to occur in Vinschgau (Val Venosta) very close to Hauslabjoch whereas to the north they are much further away, in and close to Inntal.

The *Neckera* spp are by no means the only low to moderate altitude mosses among the 30 bryophytes but, on the basis of their present and probable past chorology, they are the most important in deducing the Iceman's provenance which is much more likely to have been Southern Tyrol than Northern Tyrol. More than 20 of the bryophytes have no relevance for this argument. They are likely to have been growing at and around the death site and are important in allowing environmental inferences to be made; such deductions are not discussed here.

Appendix

List of localities of the Neckera spp within the surveyed rectangle

The 10 km squares are numbered from the northwest corner to the southeast corner.
JHD = James Dickson, AMcM = Andrew McMullen, RP = Ronald Porley.

Neckera besseri

8 and 14. Düll 1991. "Bei Sölden (ca. 1400 m) und bei Lägenfeld (ca. 1200 m) beide leg. Berggren 1867."
30. Near Vellau. JHD 1997. On the underside of small boulder of micaschist, shaded by *Corylus avellana* in woodland on south-facing slope, along track 22, c. 1000 m.
30/35. Meran. Düll 1991. "so um Meran und Bozen noch mehrfach, z.B. auch an der Mauer einer Bachschlucht östlich Schloss Rametz, ca. 500 m, D.91."

Neckera complanata

1. Piller. Düll, 1991; "nur im untersten PT [Pitztal]· ·bei Piller, 1290 m."
2. Jerzens. Düll, 1991; " nur im untersten PT an der Pitzebrücke unterh. Jerzens, 960 m."
9. Gries in Sulztal. Düll, 1991; "Auch OT [Ötztal] s [selten] so bei Gries bis 1500 m leg." F. Koppe 1957.
11. Tösens. JHD, 1995; In considerable quantity on heavily shaded rock face and boulder of mainly east and north aspects, Tösner Bach, close to waterfall near Giggl, c. 1150 m.
25. Moos in Passeier. RP et al., 1995. In small quantity on backside rock of west aspect along Passer, a little north of Moos, c. 1,050 m.
25. St. Leonhard in Passeier. JHD et al., 1995. In small quantity on trunk of roadside *Fraxinus*, minor road east of Breiteben, c. 750 m.
28. Schnalstal. RP et al., 1994. Heavily shaded schistose boulders at foot of gorge of Schnalser Bach downstream of Platthaus at c. 850 m.
29. Partschins. AMcM 1994. Near the waterfall, shaded, southeast-facing schistose cliff with *N. crispa*, at c. 1150 m.
29. Partschins. RP 1994.
30. Gratsch. A.F. Entleuner 1884.
30 Vellauer Thal. A.F. Entleuner 1884.
30. Algund. A.F. Entleuner 1884.
30. Between Leiter Alm and Hochganghaus. RP, 1995. In small quantity on rock, c. 1,700 m
30. Meraner Hochweg, near Leiter Alm. JHD, 1997. On shady, east-facing quartz muscovite schist and biotite schist in woodland, c. 1520 m.

30. Above Toll. JHD, 1994. Near Quadrat, in large quantity with *N. crispa* on heavily shaded vertical rockface of northeast aspect and also on shaded boulders and as epiphyte on *Larix* twigs, beside stone quarry at c. 1100 m.
30. Meran. JHD, 1995. Via Labers, heavily shaded, west-facing roadside wall, at c. 500 m.
32. Laas. JHD, 1994. Large quantity on east-facing, shady rock in Valdaunbach, at c. 1100 m.
32. Laas. JHD et al., 1994. In large quantity with *N. crispa* on vertical, heavily shaded rockface of west aspect, in Valdaunbach, at 1200 m.
32. Laas. JHD et al., On large, shaded marble boulder near marble quarry, in Valdaunbach, at c. 1300 m.
32. Martelltal. JHD, 1994. Bad Salt, in large quantity on vertical, heavily shaded northeast-facing side of very large boulder, c. 1350 m.
33. Near marble quarry above Morter. JHD, 1997. On north-facing boulders of quartz muscovite and biotite schist in woodland, c. 1180 m.
33. Near Parmant, above Marein. JHD, 1997. On shady outcrop of amphibolite in gully, c. 1300 m.
34. Schnalstal. JHD et al., 1994. On large, shaded concrete waterpipe close to Schloss Juval at c. 850 m.
34. Schnalstal. JHD, 1997. On east-facing augen gneiss outcrop along road below Schloss Juval, c. 900 m.
34. Above Rabland. JHD, 1995. At two places between Aschbach and Rabland, on north-facing, shaded rock, at c. 1100 m.
34. Above Naturns. JHD et al., 1995. Below Zeten Alm, in small quantity with abundant *N. crispa* on north-facing, shaded cliff, at c. 1500 m.
34. Above Naturns. JHD, 1997. Near Lind, on shady marble, c. 850 m.
35. Trauttmanstorff. A.F. Entleuner 1884.
35. Völlan. JHD and AMcM, 1995. Obertalmühle Gasthof, in gorge in considerable abundance with *N. crispa* on north-facing rock, c. 750 m. JHD, 1997. On granite boulder, *Fagus* trunk and cut *Picea* stump.
South of square 35. Saxl, south of Prissian. JHD et al., 1995, in heavy shade on volcanic ash with secondary calcite, c. 650 m.
South of square 35. Prissianer Tal JHD et al., 1995. In abundance with *N. crispa* on shaded, northfacing, shaded rock in gorge, also on dead branch on ground and on *Fagus* trunk, c. 600 m.

Neckera crispa

2. Jerzens. Düll 1991; "im untersten PT, s: beim Fernsturm ob. Arzl, Kalk, cat 960 m und unterh. Jerzens, ca. 960 m."
3. Piburg. E. Bauer 1901. Piburgsee bei Ötz auf Granitfelsen.
3. Piburg. JHD, 1995. In considerable quantity on roadside, unshaded vertical rock of north aspect, below Piburg at c. 900 m.
3. Tumpen. JHD, 1995. In small amount on heavily shaded, vertical, east-facing side of large rock, near road just south of Tumpen, at c. 900 m.
29. Partschins. AMcM, 1994. Above waterfall, in large quantity close to ground on very large boulder of northeast aspect at c. 1450 m.
30. Gratsch. A.F. Entleuner 1884; "zwischen Gratsch und Algund."
30. Above Toll. JHD, 1994. Near Quadrat, in large quantity with *N. complanata*, on heavily shaded, vertical rockface of northeast aspect and also on shaded boulders, beside stone quarry at c. 1100 m.
32. Laas. AMcM et al., 1994. In small quantity on unshaded, east-facing rock, Valdaunsbach, at c. 1100 m.
32. Laas. JHD et al., 1994. In large quantity with *N. complanata* on heavily shaded, vertical rockface of west aspect, c. 1200 m.
34. Above Naturns, JHD et al., 1995. Below Zeten Alm, in abundance on shaded, north-facing cliff, with *N. complanata*, c. 1500 m.
35. Trauttmanstorff. A.F. Entleuner 1884.
35. Völlan. JHD and AMcM, 1995. Obertalmühle Gasthof, in gorge in great abundance on north-facing rock with *N. complanata*, 750 m. JHD, 1997. On *Fagus* trunk and cut *Picea* stump.
South of square 35. Prissianer Tal JHD et al., 1995. In great abundance with *N. complanata* on shaded, north-facing, rock in gorge of north aspect, also both on boulder and dead branch on ground, at 600 m.

Neckera pumila

30. Düll, 1991. "Sonst aus S.Tirol nur von Verdins (leg. Milde, um 1860, in DS p. 441)."

Abstract

Remains of the mosses *Neckera complanata* and *N. crispa* were recovered with the Iceman in 1991. These species are woodland mosses sometimes growing on tree trunks and branches but more often on shady, mostly base-rich, rocks and are most abundant at altitudes below 1,500 m in the Tyrol at present. They were carried by the Iceman on his last journey. Together with some 28 other species of bryophytes, they provide important evidence concerning the Iceman's southern provenance, ethnobotany and environment.

Zusammenfassung

Bei dem Eismann wurden 1991 Reste der Moose *Neckera complanata* und *N. crispa* gefunden. Diese Spezies sind Waldmoose, die manchmal auf Baumstämmen und Ästen, meistens aber auf schattigen, vor allem basenreichen Felsen wachsen und heute in Tirol vorwiegend in Höhen bis zu 1500 m vorkommen. Der Eismann hatte sie auf seiner letzten Reise bei sich. Zusammen mit 28 weiteren Arten von Bryophyten liefern sie wichtige Hinweise auf das südlicher gelegene Herkunftsland des Eismanns, seine Ethnobotanik und Ökologie.

Riassunto

Con l'uomo del Similaun sono stati individuati resti dei muschi *Neckera complanata* e *Neckera crispa*. Si tratta di specie forestali presenti talvolta sui tronchi d'albero e rami, ma più sovente su rocce ombreggiate, spesso ad alta alcalinità che sono molto frequenti ad altitudini sotto i 1500 m s.l.m., nel Tirolo odierno. Essi erano presenti sull'uomo del ghiacciaio nel suo ultimo cammino. Insieme con altre 28 specie di briofiti essi forniscono importanti indizzi per la sua provenienza dal Sud delle Alpi Centrali, nonchè per l'etnobotanica e l'ambiente.

Résumé

L'Homme des glaces, découvert en 1991, portait sur lui des résidus de *Neckera complanata* et *N. crispa*. Ce sont des espèces de mousses des forêts, poussant parfois sur des souches et des branches d'arbres, mais le plus souvent sur des roches alcalines et ombragées. On les trouve actuellement en abondance au Tyrol jusqu'à une altitude de 1500 m. Ces mousses avaient été emportées par l'Homme des glaces lors de son dernier voyage. Avec 28 autres espèces de bryophytes, elles sont des indices importants d'une origine méridionale de l'Homme des glaces, de l'ethnobotanique et l'environnement de sa région d'origine.

Acknowledgements

JHD acknowledges with gratitude Prof. S. Bortenschlager, the University of Innsbruck and the Carnegie Trust for the Universities of Scotland for funding his visits to the Tyrol in 1994 and to the Royal Society of London and the Austrian Academy of Sciences for funding his 1995 sojourn in Innsbruck and Prof. S. Bortenschlager and the University of Glasgow for funding his 1997 visit. Dr. Klaus Oeggl and Messrs R. Porley and A. McMullen were keen, skilful helpers and good companions in the field, as was Dr. J. G. MacDonald. Dr. Ursula Maier kindly informed JHD of unpublished data on mosses from Hornstaad-Hörnle.

References

Ando H. and Matsuo A. (1984) Applied Bryology. In: Schultze-Motel W. (ed.) Advances in Bryology 2: 133–224.
Bates J. W. (1993) Epiphytic bryophytes preserved in a French farmhouse. Journal of Bryology 17: 511–512.
Behre K.-E. (1969) Untersuchungen des botanischen Materials der frühmittelalterlichen Siedlung Haithabu (Ausgrabung 1963–1964). Berichte über die Ausgrabungen in Haithabu 2: 7–46.
Bollinger T. (1994) Samenanalytische Untersuchung der frühjungsteinzeitlichen Seeufersiedlung. Egolzwil 3. Diss. Botanicae 221: 1–172.
Buchner A., Hofbauer W. und Gärtner G. (1993) Beitrag zur Moosflora von Seefeld und Umgebung und des Leutascher Beckens (Nordtirol). Ber. nat.-med. Verein Innsbruck 80: 53–67.
Cappers R. T. J. and van Zanten B. O. (1994) Ecology and representativeness of subfossil mosses from the Heveskesklooster terp (the Netherlands). In: Cappers R. T. J. An ecological characterization of plant macro-remains of Heveskesklooster (the Netherlands). A methodological approach. Thesis, Rijksuniversiteit Groningen, pp 191.
Culicova V. (1995) Rekonstruktion der synanthropen Vegetation des mittelalterlichen Most. Pamatly Archeologicke 86: 83–131.
Dalby D. H. (1971) Mosses and Ferns. In: Greenfield E. The Roman villa at Denton Lincolnshire Part TI. Lincolnshire History and Archaeology 1: 53–55.
Dalla Torre K. W. v. and Sarthein L. v. (1904) Die Moose (Bryophyta) von Tirol, Vorarlberg und Lichtenstein. Innsbruck, Universitäts Buchandlung, pp 671.
Dick M. (1989) Wirtschaft und Umwelt cortailod- und horgenzeitlicher Seeufersiedlungen in Zürich (Schweiz). Diss. Botanicae 132: 1–114.
Dickson C. (1989) Human coprolites. In: Bell B. and Dickson C. Excavations at Warebeth (Stromness Cemetery) Broch, Orkney. Proceedings of the Society of Antiquaries of Scotland 119: 101–131.
Dickson J. H. (1973) Bryophytes of the Pleistocene. Cambridge University Press, pp 256.
Dickson J. H. (1986) Bryophyte Analysis. In: Berglund B. E. (ed.) Handbook of Holocene Palaeoecology and Hydrology 627–643pp.
Dickson J. H. (1997) The Moss from the Tyrolean Iceman's colon. Journal of Bryology 19, 449–451.
Dickson J. H., Bortenschlager S., Oeggl K., Porley R. and McMullen A. (1996) Mosses and the Tyrolean Iceman's southern provenance. Proceedings of the Royal Society of London, Series B 263: 567–571.
Dickson J. H. and Brough D. (1989) Biological studies of a Pictish Midden. Diss. Botanicae 133: 155–166.
Duda J. and Opravil E. (1988) Archeobotanicke nalezy mechu (Musci) v Ceskoslovensku. Cas. Slez. Muz. Opava (A) 37: 207–216.
Düll R. (1991) Die Moose Tirols unter besonderer Berücksichtigung des Pitztals/Ötztaler Alpen, Bd 2. IDH-Verlag, Bad Münstereifel-Ohlerath, pp 216.
Düll R. (1994) Deutschlands Moose, Bd 3. IDH-Verlag, Bad Munstereifel-Ohlerath, pp 256.
Entleutner A. F. (1884) Beiträge zur Laubmossflora von Meran. S. Pötzelberger, Meran, pp 32.
Fraser M. J. (1980) A Study of the Botanical Material from Three Medieval Scottish Sites. Unpublished University of Glasgow M. Sc. Thesis, pp 150.
Grosse-Brauckman G. (1979) Pflanzliche Grossreste von Moorprofilen aus dem Bereich einer Steinzeit Seeufersiedlung am Dümmer. Phytocoenologia 6: 106–117.
Hamilton A. C. and Carter R. W. G. (1983) A Mid-Holocene moss bed from eolian dune sands near Articlave, Co Londonderry. Irish Naturalist's Journal 21: 73–75.
He S. (1997) A revision of Homalia (Musci: Neckeraceae). Journal of the Hattori Botanical Laboratory 81: 1–52.
Heer O. (1865) Die Pflanzen der Pfahlbauten. Neujahrsblatt der Naturforschenden Gesellschaft Zürich für das Jahr 1865: 1–65.

Home P. and Ireland R. R. (1991) Moss and a Guanche Mummy: An Unusal Utilisation. The Bryologist 94: 407–408.

Jorgensen G. (1980) Om kostvaneridet middelalderlige Svendborg. Naturens Verden 1980: 203–209.

Karg S. (1990) Pflanzliche Grossreste der jungsteinzeitlichen Ufersiedlungen Allensbach-Strandbad, Kr. Konstanz. Siedlungsarchäologie im Alpenvorland II. Forschungen und Ber. Vor- und Frühgeschichte Baden-Württemberg 37: 114–154.

Kenward H. K., Hall A. R. and Jones A. K. G. (1986) Environmental Evidence from a Roman Well and Anglian Pits in the Legionary Fortress. The Archaeology of York 14: 241.

Knörzer K.-H. (1980) Römerzeitliche Pflanzenfunde aus Aachen-Burtscheid. Archaeo-Physica 7: 35–60.

Knörzer K.-H. (1995) Planzenfunde aus dem bandkeramischen Brunnen vom Kuckhoven bei Erkelenz. Vorbericht, pp 81–86. In: Kroll H. and Pasternak R. (eds.) Res archaeobotanicae. Oetker-Voges Verlag, Kiel.

Körber-Grohne U. (1979) Einige allgemeine Bemerkungen zu einer pflanzensoziologischen Zuordnung subfossiler Floren des Postglazials. In: Wilmans O. and Tuxen R. (eds.) Werden und Vergehen von Pflanzengesellschaften. Berichte der Internationalen Symposien der Internationalen Vereinigung für Vegetationskunde, 16: 43–59.

Kuc M. (1964) Bryogeography of the Southern Uplands of Poland. Monographiae Botanicae 17: 1–212.

Marstaller R. (1992) Die Moosgeschaften des Verbandes Neckerion complanatae. Herzogia 9: 257–318.

Mitchell G. F., Dickson C. A. and Dickson J. H. (1987) Archaeology and Environment in Early Dublin. Royal Irish Academy, Dublin, pp 40.

Neuweiler E. (1905) Die prähistorischen Pflanzenreste Mitteleuropas mit besonderer Berücksichtigung der schweizerischen Funde. Vierteljahrsschrift der Naturforschenden Gesellschaft in Zürich 50: 23–134.

Ochsner F. (1975) Neckera crispa Hedw. an Pfahlbau-Fundstellen in der Schweiz. Phytocoenologia 2: 9–12.

Okland R. H. (1988) C. Moss remains from latrine samples. In: Schia E. (ed.) De arkeologiske utgravninger i Gamlebyen, Oslo. Bind 5: 109–114.

Pitschmann H. and Reisigl H. (1954) Zur Nivalen Moosflora der Ötztaler Alpen (Tirol). Revue Bryologique et Lichenologique 23: 123–141.

Preston C. D. (1994) Neckera crispa Hedw. and Neckera complanata Hedw. In: Hill M. O., Preston C. D. and Smith A. J. E. (eds.) Atlas of the Bryophytes of Britain and Ireland. Harley Books, Colchester, pp 220 and 222.

Robinson D. and Rasmussen P. (1989) Botanical investigations at the Neolithic lake village at Weier, Northeast Switzerland: leaf hay and cereals as animal fodder. In: Milles A., Williams D. and Gardner N. (eds.) The Beginnings of Agriculture. BAR International Series 496: 149–163.

Rösch M. (1988) Subfossile Moosfunde aus prähistorischen Feuchtbodensiedlungen: Aussagemöglichkeiten zu Umwelt und Wirtschaft. In: Küster H. (ed.) Der prähistorische Mensch und seine Umwelt. Theiss Verlag, Stuttgart, pp 177–198.

Rybníček K., Dickson J. H. and Rybníčekova E. (1998) Flora and vegetation at about A.D. 1100 in the vicinity of Brno, Czech Republic. Vegetational History and Archaeobotany 7: 155–165.

Sjögren E. (1964) Epilithische und epigaische Moosvegetation in Laubwäldern der Insel Oland (Schweden). Acta Phytogeographica Suecica 48: 1–184.

Stevenson R. (1986) Bryophytes from an archaeological site in Suffolk. Journal of Bryology 14: 182–184.

Störmer P. (1969) Mosses with a Western and Southern Distribution in Norway. Universitetsforlaget, Oslo, pp 288.

Van Zeist W. (1974) Palaeobotanical studies of settlement sites in the coastal area of the Netherlands. Palaeohistoria 16: 226–371.

Van Zeist W. (1988) Botanical evidence of relations between the sand and clay districts of the north of the Netherlands in medieval times. In: Küster H. (ed.) Der prähistorische Mensch und seine Umwelt. Theiss Verlag, Stuttgart, pp 381–394.

Venturi G. (1879) Bryineae ex regione italica Tirolis, Tridentina dicta. Revue Bryologique 3: 49–62.

Venturi G. (1899) Le Muscinee del Trentino. Giovani Zippel ea., Trento, pp 107.

Wiethold J. W. (1995) Plant remains from the town moats and cesspits of Medieval and post-Medieval Kiel (Schleswig-Holstein, Germany), pp 359–384. In: Kroll K. and Pasternak R. (eds.) Res archaeobotanicae. Oetker-Voges Verlag, Kiel.

Williams D. (1976) A Neolithic moss flora from Silbury Hill, Wiltshire. J. Arch. Science 3: 267–270.

Wright E. V. (1990) The Ferriby Boats Seacraft of the Bronze Age. Routledge, London.

The diet of the Iceman

K. Oeggl

Institut für Botanik, Universität Innsbruck

1. Introduction

The evaluation of the Iceman's bodily constitution is essential to an understanding of the circumstances of his death and to a reconstruction of the last moments of his life. Thus far medical examinations of the body have illuminated the Iceman's state of health during his life time. Several degenerative deformations of his skeleton have been detected (zur Nedden and Wicke, 1992) and the positions of the tattoos found on his body have been connected with some kind of therapeutic treatment (Sjövold et al., 1995). Investigations of a nail plate from a finger indicate three periods of severe stress within the last five months of his life (Capasso, 1995). The third and most serious one occurred two months before his death and lasted at least two weeks. However none of these diseases known so far seriously threatened his life and the circumstances of his death are still unclear. One scenario discusses that the Iceman was in a poor physical condition and perished lost and exhausted in the wilderness (Spindler, 1996). The investigations of the food residue in his alimentary canal can help to solve this question and to reveal his nutritional state. Learning whether and what he had eaten leads to a better understanding of the incidents during the last day of his life.

The residue of food in the intestine depends on what was consumed and reflects the diet of prehistoric people. Certainly the composition of food residues in alimentary canals is complicated and the information gained from gut contents is restricted additionally by the low number of well preserved contemporaneous prehistoric human bodies. The latter prevents a statistical evaluation of the data. But, rare though they may be, micro- and macrofossil analyses of gut contents give direct information about prehistoric diet, complementary to the palaeoethnobotanical investigations on soil samples from archaeological excavations.

Usually the nutrition of prehistoric populations is evaluated by indirect evidence. Remains of edible plants found in samples from excavations tell us something about former diet and crop plants. Heer (1866) had already started these investigations in the last century and made an inventory of cultivated plants of the lake dwelling sites in Switzerland. Since then many more palaeoethnobotanical investigations have been done in the area around the Alps. They have increased our knowledge about Neolithic food supplies and even the agricultural technique of crop breeding used in prehistoric times (Jacomet et al., 1989; Küster, 1995; Biagi and Nisbeth, 1987; Hopf, 1992). However, the occurrence of a cultivated plant does not necessarily imply that it was part of the prehistoric human food supply. Quantitative analyses are essential to separate intentional crop plants from unintentional contamination of fields with cultivated plants. It is well known that in the Neolithic period cultivated plants which were not the main crop often thrived in minor quantities as a weed among the crop plants. Such cultivated plants did not contribute very much to the former diet. Also the different nutritional values of plants have influenced the nature of dietary use. There is evidence that even in prehistoric times specific crop plants were fed mainly to animals and only consumed by man in times of famine (Hopf, 1992). Therefore direct evidence of prehistoric food and food preparation is very important. Occasionally food remains in ancient pottery (Richter, 1987) or finds of prehistoric bread (Behre, 1991; Hansson, 1996; Maurizio, 1916; Währen, 1990; 1995a,b) allow a brief insight into how cultivated plants were prepared for food and consumed. More data are contributed to this topic by studies of excrement.

Palaeofeces, coprolites and gut contents of prehistoric people are preserved only under very specific – waterlogged or arid – conditions. When complementary techniques like micro- and macrofossil analyses are used much can be learned about diet, medicine, the state of health of a person, the cause and even season of death, and the environment in which the person lived (Bryant, 1974a,b; Helbaeck, 1950, 1958; Holden, 1986, 1990, 1995; Holden and Núñez, 1993; Dickson and Dickson, 1988; Dickson, 1989; Bell and Dickson, 1989; Hadorn, 1994; Reinhard and Bryant, 1992). When more than one corpse can be investigated, such studies even allow an insight into the interindividual variability of nutrition (Schoenwetter, 1974, 1996; Gremillon and Sobolik, 1996).

All this background knowledge raises a wide variety of questions for the investigation of the residue in the Iceman's intestine. To gain the maximum information from the colon content both, macrofossil and pollen analyses, were carried out. The main goal of the macrofossil analyses was to detect what kind of plant species are represented, how they reflect the Neolithic cultivated plants known from the inner Alpine area, and what the man consumed for the last meal. The pollen analysis provided additional evidence for useful plants. Furthermore, pollen analysis can be a good indicator of seasonality of meal consumption. By such means in this case the season of death has been determined. Lastly, pollen analysis can indicate the habitat in which the Iceman lived.

2. Methods

Data were obtained from two different sources. One is derived from the Iceman's body and gives direct evidence of his nutrition within the last days. The second came from soil samples from the find spot as well as from archaeological excavations at Neolithic sites in North- and South Tyrol.

The sample from the body consists of a 40 mg food residue from the transverse section of the colon (Plate I, Fig. 1). It was removed endoscopically in order to limit damage to the body itself. After documentation the colon sample was divided evenly into four subsamples A–D (Fig. 1), as independent replicates used for verification of internal consistency of the sampling and analytical procedures. Three subsamples (A–C) were prepared for light microscopy analysis of pollen- and macroremains and one (subsample D) for ultrastructural investigations by a scanning electron microscope (Leitz). Since the chemical treatment on samples for pollen and for plant macrofossil analysis conflict with each other and therefore a specific sample rescue technique had to be developed.

2.1. Sample preparation techniques

Each subsample was rehydrated in a 0.5% solution of trisodium phosphate for 72 hours (Pearsall, 1989; Holden, 1994), and screened through 500 μ, 250 μ, 125 μ and 63 μ steel meshes; at this stage samples (spectra) were numbered 2 to 5 with decreasing mesh size (Fig. 1). The filtered solution (outwash) was also kept for further investigation (spectrum 6). To evaluate the efficiency of the preparation technique and any problems arising from it, sample screening was carried out

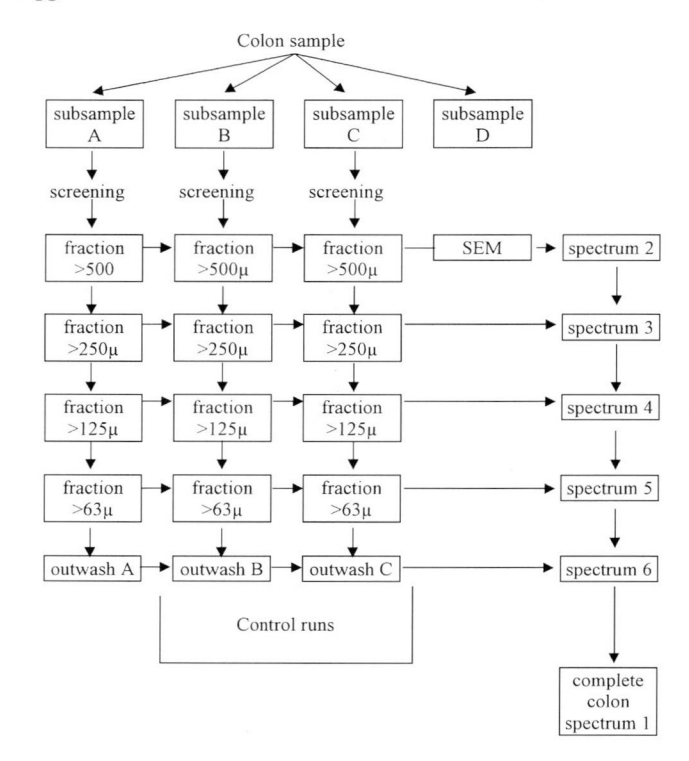

Fig. 1. Flow chart of the sample recovering procedures

initially on subsample A, prior to the preparation of samples B–D. After preparation of subsample A small random sample was then taken from the 500 μ residue and investigated for its microfossil content and its state of preservation. Then each subsample was boiled for 10 min in 10% potassium hydroxide to remove the amorphous gel, consisting mainly of starch, protein and fat enclosing the recovered material, washed in distilled water, and then treated with ethanol (Gassner et al., 1989). For final analysis samples were stained with fuchsin and mounted in glycerine. The light microscope investigations were carried out under magnifications of 750x and 1200x with the aid of phase contrast and interference contrast.

2.2. Soil samples

Charred plant remains from soil samples of archaeological excavations from Neolithic sites within the Tyrol as well as soil samples from the find spot were rescued by flotation technique. The fraction of the sieve set corresponds with that one mentioned above.

2.3. Quantification and presentation of the results

Because of the small sample size of the colon residue the application of volumetric methods was not possi-

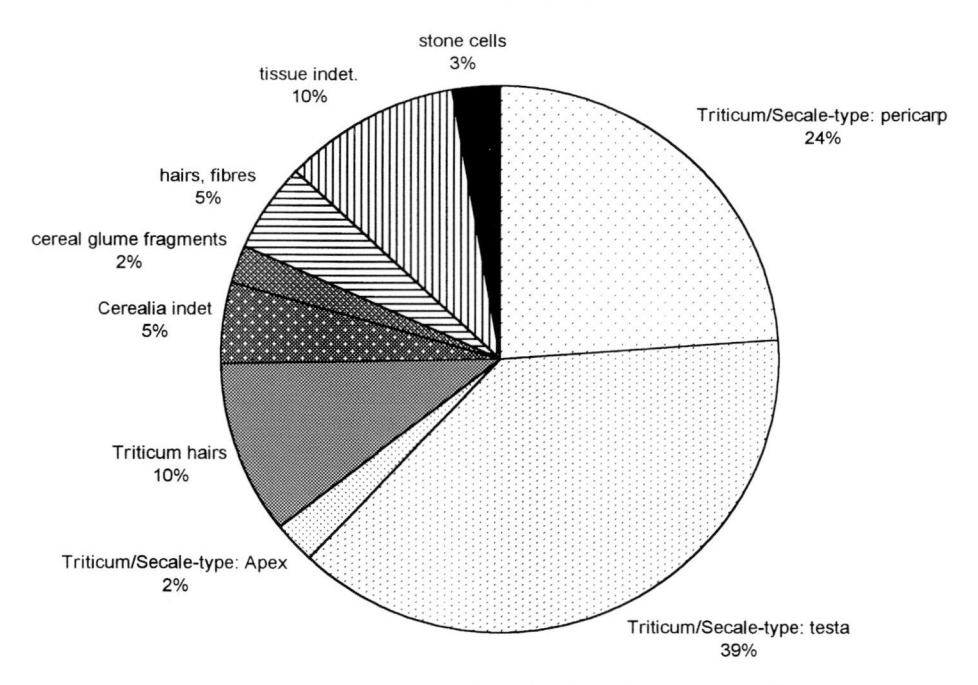

Fig. 2. Relative abundancies of the various remains in the food residue from the Iceman's colon

Pie chart values:
- stone cells 3%
- tissue indet. 10%
- hairs, fibres 5%
- cereal glume fragments 2%
- Cerealia indet 5%
- Triticum hairs 10%
- Triticum/Secale-type: Apex 2%
- Triticum/Secale-type: testa 39%
- Triticum/Secale-type: pericarp 24%

ble, and a classification in "very abundant", "abundant" and "rare" seemed inappropriate. To have a more precise estimation of the quantities of each single component, each type of remain was counted and represented by absolute numbers. Results of macro fossil analyses are presented in tables and pie charts. The results of pollen analyses are given in percentage diagrams in which the size fractions from all 4 subsamples (spectra 2–5) and the outwash (spectrum 6) are averaged and presented as a single spectrum (spectrum 1). The total pollen sum of the colon sample was then calculated from the averaged values.

3. Results

3.1. Colon sample

The plant remains in the food residues of the intestinal tract are in a good state of preservation. Single components like cereal bran, hairs and fibres were distinguishable, when viewed under a stereo microscope. The investigation by light microscopy confirmed the first impression. Different types of material, primarily plant macroremains, alongside carbon particles, mineral particles, pollen, spores, diatoms, muscle fibres and eggs of a human parasite, the human whipworm (*Trichuris trichiura*) were identified (Fig. 2; Plates I–IV).

3.2. Plant macroremains

The plant macrofossil contents are derived mainly from bran. The pericarp and testa tissues are well preserved. The testa remains are characterised by a layer of elongate cells similar to the transverse cells of the pericarp (Dickson, 1987). In a right angle underneath them follow light brown coloured, long cells which are characteristic of wheat (*Triticum*) and rye (*Secale*). Fortunately the pericarp layers are also preserved and the diagnostic features – longitudinal cells, transverse cell layer and tube cells – are easily recognisable to enable a detailed taxonomic classification. Again the cell pattern of the bran fragments corresponds with the *Triticum/Secale*-type (Plate II, Figs. 2, 3). Both, wheat (*Triticum*) and rye (*Secale*) have pitted longitudinal cells, transverse cells in rows and underlying tubular cells in the longitudinal direction (Körber-Grohne and Piening, 1980). Rye (*Secale*) can be separated from wheat (*Triticum*) by the sickle-shaped thickening of the unpitted sidewalls. Because such thickenings are missing, rye (*Secale*) can be discounted. According to the occurrence of the tube cells all over the surface of the pericarp tissues, Einkorn (*T. monococcum*) is identified (Körber-Grohne and Piening, 1980; Dickson, 1987). Given that rye (*Secale*) was not cultivated in central Europe during the Neolithic period and that several spikelets of Einkorn (*Triticum monococcum*) were found on the body, all the bran fragments belong to Einkorn (*Triticum monococcum*).

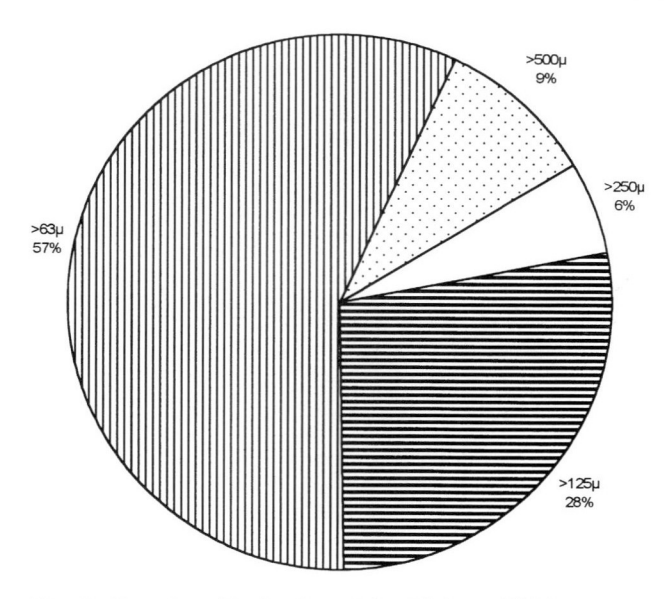

Fig. 3. Size class distribution of the Einkorn (*Triticum monococcum*) pericarp and testa tissues from the food residue in the colon

Several other tissues and cell types, which are difficult to determine to the species level, were also identified. Most of the tissues originate from wheat (*Triticum*). Apices of wheat grains are identified by the attached hairs (Plate III, Fig. 4). The unicellular hairs of wheat (*Triticum*) are characterised by thick cell walls and a small lumen (Gassner et al., 1989). Small amounts of glume fragments were also found. They show the typical Gramineae cell pattern with long and short-cells and most probably belonging to wheat (*Triticum*). In a few cases the bran fragments were too small for detailed determination; so they are listed as Cerealia indeterminata (Fig. 2). Also 13% of the fragments are represented by cell types which are very common in plant tissues, e.g., stone cells or vessels with ring-like thickenings. The latter are thought to derive from the vascular bundles of Einkorn (*Triticum monococcum*). Such thickenings occur in the proto- and metaxylem of all vascular plants and not all of them may belong to Einkorn (*Triticum monococcum*). Two vessel elements were detected with spiral thickenings and simple perforations, features that occur in the secondary xylem of many families with edible plants, e.g., Rosaceae, Leguminosae, Chenopodiaceae, Tiliaceae. Together with the stone cells they document that the Iceman had consumed other vegetable food in addition to Einkorn (*Triticum monococcum*).

As a whole, the cereal remains add up to 75% of the plant macro remains. The size distribution of the pericarp and testa remains (Fig. 3) show that particles of $> 500\,\mu$ and $> 250\,\mu$ size add up to only 15%, whereas the smaller size classes $> 63\,\mu$ dominate. Therefore it is presumed that the corn used for his last meal was finely ground.

It was surprising that a single moss leaf occurred in the food residues. It belongs to the species *Neckera complanata*. This moss species is distributed in woodlands of low to moderate attitudes and grows on rocks and tree trunks. It was unintentionally incorporated in the food residue (Dickson, 1997).

The charcoal remains (Plate III, Fig. 5), which add up to a quarter of the remains in the colon content, were also accidentally ingested. They belong to conifers but are too tiny for species determination. Most probably they stuck to the food and indicate that the food was in direct contact with an open fire.

3.3. Pollen

With 30 different pollen types and two spore types, the pollen flora (Figs. 4, 5; Plate II, IV) in the food residues has turned out to be far more diverse than the macrofossil content. The AP (arboreal pollen) represent 84%, and the NAP (non arboreal pollen) 16% of the total pollen. This is of considerable interest given that an NAP dominance was to have been expected due to all the edible herbaceous species in the diet.

The composition of the colon pollen spectrum (spectrum 1 in Fig. 4) reflects a deciduous forest vegetation dominated by *Ostrya* with an admixture of *Betula*, *Corylus*, *Picea* and *Pinus*. All these arboreal pollen types represent forest-forming tree-taxa characteristic of the valley bottoms of the investigation area. The NAP represent a mixture of taxa from arable fields, as well as taxa commonly distributed in meadows and tall forb communities.

The diversity of pollen types in the colon needs explanation. One way much pollen becomes ingested is by adherence to the swallowed food plants. Such pollen types are classified as "economic pollen" and they yield information about the diet or other ethnobotanical usage. On the other hand, the pollen content of the atmosphere is unintentionally incorporated in the food residue by breathing or drinking water. They are defined as "background pollen". This background pollen reflects the general composition of the vegetation in which the man lived (Bryant, 1974a, Bryant and Holloway, 1983). For a better understanding of the taphonomy of the Iceman's colon pollen spectrum it seems sensible to separate the pollen types identified into "economic" and "background" pollen.

3.4. Background pollen

The category of background pollen is composed of 23 types. Most of them are dispersed by the wind and are abundant in the air. They are part of the pollen rain that normally settles to the ground. Such pollen became incorporated accidentally and to some extent this pollen component of the spectrum must reflect the vegetation the Iceman traversed just before his death.

Ostrya-type: The presence of the Hop Hornbeam (*Ostrya*-type) pollen was most surprising (Plate II, Figs. 7, 8, 10; Plate IV, Figs. 1,2). Only two tree species contribute to the Ostrya-type pollen: *Ostrya carpinifolia* and *Carpinus orientalis*, but the latter is not native in Central Europe. In particular the numerically predominant *Ostrya*-type pollen indicates the thermophilous Hop Hornbeam (*Ostrya carpinifolia*), the only indigenous species of this genus in the area. This small tree establishes only in forests along the valleys south of the main range of the Alps. The Hop Hornbeam (*Ostrya carpinifolia*) produces abundant, medium-sized, sphaeroidal pollen grains and sheds them in the spring. Pollen is only found within the distribution range, and its presence in the pollen spectrum of more than 5% indicates local occurrence in the vegetation (Huntley and Birks, 1983). For this reason the *Ostrya*-type pollen is a good indicator of the Iceman's provenance and this discovery had to be treated with greatest rigour, the more so given the fact that the cellular gametophytes were still visible in the pollen grains (Plate II, Figs. 7, 8, 10). Although sample preparation and analyses were carried out in sterile laboratory conditions outside the flowering season, the presence of pollen with intact gametophytes lead to an initial concern that subsample A had been contaminated in some way, and the small absolute quantities of this pollen type in the fraction $> 500\ \mu$ lent weight to these concerns. However, a control run of the sampling and the extraction procedures with the subsamples B and C gave the same results indicating that no contamination had taken place and that the Hop Hornbeam (*Ostrya carpinifolia*) pollen effectively originated from the colon sample.

One of the most exciting things was that the gametophyte was still visible in what were clearly pollen grains of Birch (*Betula*) as well as in approximately 80% of the Hop Hornbeam (*Ostrya carpinifolia*) pollen. The preservation of cellular microgametophytes occurs rarely in fossil grains, although such survival since the Palaeozoic, ca. 300 million years ago has been reported (Taylor and Millay, 1979). Further investigation was therefore required to highlight those rare conditions which could guarantee gametophyte preservation in such ancient pollen. It is well-known empirical knowledge that the cell content of pollen grains is decomposed after several weeks in oxidising conditions in water or soil. Desiccation or freezing may, however, preserve such structures. Reference herbarium material was therefore investigated to see if some indication was forthcoming concerning the survival of gametophytes in pollen grains which had been air-dried. In all the investigated material dating as far back as 1875, gametophytes could clearly be observed in the Hop Hornbeam (*Ostrya carpinifolia*) pollen grains. It therefore seems possible for vegetative cells within the pollen to be preserved for more than 100 years in air-dried conditions, which are similar for storing food. In contrast with American *Ostrya* species, Hop Hornbeam (*Ostrya carpinifolia*) is not known to have been eaten or used for ethnobotanical purposes in Central Europe; this makes it likely that this pollen was incorporated unintentionally. It seems possible that Hop Hornbeam (*Ostrya*-type) pollen contaminated the food supply.

Corylus avellana: Hazel (*Corylus avellana*) is the second dominant pollen type in the colon spectrum. Its pollen is sphaeroidal, small, abundantly produced in catkins and released by the wind in spring. Values between 2% and 25% indicate that Hazel (*Corylus avellana*) is present in the forest as an understory shrub (Huntley and Birks, 1983). The ovoid nuts of the Hazel (*Corylus avellana*) have been collected and consumed by man since earliest prehistoric times (Knörzer, 1971), and the rapid spread of Hazel (*Corylus avellana*) in Europe may indicate that man contributed to its fast dispersal after the last deglaciation (Firbas, 1949). The very palatable seeds are protected by thick woody shells, which are removed before consumption. This makes it impossible for larger quantities of pollen to be ingested by eating hazel nuts. Therefore the occurrence of hazel pollen is interpreted as accidental ingestion also.

Betula: Birch (*Betula*) pollen is similar to the above-mentioned types. It has nearly the same shape and size. Two Birch (*Betula*) species are native and grow in the investigation area: Downy Birch (*Betula pubescens*) and Silver Birch (*Betula pendula*). Both trees are high pollen producers and their pollen is well distributed by the wind. Usually in pollen diagrams local presence is assumed from values $> 10\%$ of the total pollen sum (Huntley and Birks, 1983). In this case the low percentages of Birch (*Betula*) pollen found are interpreted as local occurrence, which is confirmed by plant macroremains found associated with the body (Oeggl and Schoch, 1995). A certain amount of Birch (*Betula*) pollen retains the cellular microgametophyte; this indicates an ingestion soon after the pollen release from the flowers.

Pinus: Pine (*Pinus*) pollen type is characteristic of four Central European species: Dwarf Mountain Pine (*P.*

mugo), Scots Pine (*P. sylvestris*), Swiss Stone Pine (*P. cembra*) and Black Pine (*P. nigra*). The pollen can be separated by minor morphological differences into Diploxylon and Haploxylon species (Klaus, 1975). Pollen of the Diploxylon subgenus lack distal verrucae on the corpus and are represented by the native Scots Pine (*P. sylvestris*), Dwarf mountain pine (*P. mugo*) and Black Pine (*P. nigra*). The Haploxylon subgenus has verrucae on the distal side of the pollen and indicates the Swiss stone pine (*P. cembra*). In the colon residue only pollen from the Diploxylon type were found (Plate II, Figs. 6, 11; Plate IV, Fig. 4), in quantities below 5%. They refer to the montane distributed Scots Pine (*P. sylvestris*) as well as to the subalpine Dwarf Mountain Pine (*P. mugo*). The interpretation of Diploxylon pine pollen values is difficult, because the big sized pollen has two air bladders attached (sacci) and is well adapted for long distance transport. Even in areas without pine high percentages of this kind of pollen can be registered. Although in forested regions values between 20 and 30% indicate local occurrence (Lang, 1994), it is suggested that the pollen quantities found in the food residue are of regional origin. This suggestion is confirmed by remains of *Pinus* found in connection with the body.

Picea: Pollen of Norway spruce (*Picea abies*), the only native species of the genus in the Alps, adds up to 1,5% of the total pollen sum in the colon spectrum. Compared with modern vegetation spruce (*Picea*) pollen is well represented. Its local presence can be assumed even by small pollen quantities of 0,5% (Lang, 1994). Needles and wood remains of Norway spruce (*Picea abies*) recovered with the body of the Iceman confirm its local occurrence. In any case, Norway spruce (*Picea abies*) is a dominant canopy tree in the investigation area.

Several pollen types of forest-forming tree species in Central Europe are encountered in the food residue in quantities below 1%, e.g., European Silver Fir (*Abies*), Alder (*Alnus*), European Beech (*Fagus*), Willow (*Salix*) and Lime (*Tilia*). The potential information gained from these pollen taxa is restricted. European Silver Fir (*Abies*), Alder (*Alnus*), European Beech (*Fagus*) and Willow (*Salix*) are all wind pollinated trees or shrubs and the values found reflect possible regional occurrence in the investigation area.

Lime (*Tilia*) produces moderate amounts of large and heavy pollen (Plate II, Fig. 9). However, the shape of Lime (*Tilia*) pollen is also well adapted to wind-distribution, but is poorly dispersed because the flowers are insect-pollinated (Huntley and Birks, 1983). Two species of Lime (*Tilia*) grow as natives in the investigation area: Small-leaved Lime (*Tilia cordata*) and Broad-leaved Lime (*Tilia platyphyllos*). The pollen of the two species can be distinguished morphologically by certain holes (fovae) in the surface structures (Christensen and Blackmore, 1988). Only pollen of the Broad-leaved lime (*Tilia platyphyllos*) was recognised, from which it is assumed that they are of local origin, because macroremains of Lime (*Tilia*) were found.

Caltha-type: Pollen of the *Caltha*-type includes Columbine (*Aquilegia vulgaris*), Marsh Marigold (*Caltha palustris*) and Mousetail (*Myosurus minimus*). The three species cannot be distinguished from one another by pollen morphology. All of them are insect-pollinated and under-represented in local pollen rain. In former times Marsh Marigold (*Caltha palustris*) is known to have been used as a caper substitute and young leaves can be eaten as salad though the plant is slightly poisonous. For medical purposes it was given against liver and gall bladder diseases (Pahlow, 1985). Columbine (*Aquilegia vulgaris*) was used as medicine for scurvy and jaundice (Hegi, 1935).

All these species prefer mesic sites; Marsh Marigold (*Caltha palustris*) and Mousetail (*Myosurus minimus*) grow along rivulets and this habitat makes unintentional ingestion by drinking water possible.

Primulaceae: Taxa of this family are widely distributed in the investigation area from valley bottoms up to alpine regions. Ethnobotanical use of Primulaceae species was common in former times. Some species are eaten. The young leaves of Cowslip, Oxlip and Primrose (*Primula veris, elatior, vulgaris*) are consumed as salad or added to soups. The roots of Sowbread (*Cyclamen europaeum*) are eaten dry or roasted. Several species are used for various diseases, e.g., jaundice, rheumatism, diarrhoea, scurvy, urinary problems, etc. (Hegi, 1935).

The pollen of all these species is distributed by insects, and in pollen spectra this pollen type indicates local occurrence. Amounts of 4% may indicate intentional consumption. Nevertheless, it has to be considered that the montane species prefer mesic conditions and often grow near open water, which makes also accidental ingestion by drinking water possible.

The source of the zoophilous taxa Compositae s. l., *Polygonum persicaria*-type, Rosaceae (Plate IV, Fig. 6), Rubiaceae, *Saxifraga oppositifolia*-type, Umbelliferae, *Vaccinium*-type, and Valerianaceae are also difficult to trace. Although entomophilous, some of the taxa are well distributed by the wind and contribute in low numbers to the pollen rain without growing at the site. They are over-represented in the local pollen spectrum, e.g., Rubiaceae, Umbelliferae (Jochimsen, 1986), and yield only minor information.

On the contrary even low abundancies of Compositae s.l., *Saxifraga oppositifolia*-type and *Vaccinium*-type (Plate II, Fig. 12) indicate local occurrence. They all include many edible and medicinal plants, but because mainly vegetative parts are used, only minor quantities of pollen is consumed by that means. Species of these families grow commonly from valley bottoms up to the mountain summits. *Saxifraga oppositifolia*-type includes the three native species *S. aizoides, S. oppositifolia* and *S. paniculata*. They thrive in rock crevices, on scree, near springs and along rivulets in the Alpine regions. South of the main range of the Alps they are often found transported by avalanches or rivulets down the mountains to the valley bottoms (Dalla Torre and Sarnthein, 1900). *Saxifraga oppositifolia* is a species flowering early in spring when the snow melts.

According to Beug (1961) *Vaccinium*-type represents eight taxa thriving in the montane forests as well as in the high alpine zones, e.g., Whortleberry (*Vaccinium*), Spring Heath (*Erica herbacea*),Wild Rosemary (*Andromeda polyfolia*), Bearberry (*Arctostaphylos*), Trailing Azalea (*Loiseleuria procumbens*), Alpenrose (*Rhododendron*), Wintergreen (*Pyrola*) and Indian Pipe (*Monotropa*). They have a wide ecological range and occur on dry soils as well as on moist places.*Vaccinium, Erica, Pyrola* and *Monotropa* are characteristic understory elements of boreal montane conifer-forests of the investigation area, whereas *Arctostaphylos, Loiseleuria procumbens* and *Rhododendron* are distributed at and above the timberline. They all are entomophilous species and produce large pollen, which are mainly locally deposited. Their presence in modern surface spectra means local occurrence at the site.

Polygonum persicaria-type includes *P. hydropiper, P. lapathifolium, P. minus, P. mite, P. orientale, P. persicaria* (Hedberg, 1946). The first two species indicate a moist habitat, whereas the rest of the species are commonly associated with crop plants in arable fields or are part of rural vegetation. Their pollen is large and transported by insects. Even low abundancies of this kind of pollen in modern surface samples indicate their existence at the site.

Spores: Monolete and trilete spores of ferns (Pteridophyta) are traced in abundancies of more than 2%. In general ferns are prolific spore producers, and the monolete wind-borne spores are regularly found in modern pollen surface samples in values below 1%, where they reflect a regional occurrence. Trilete fern spores were recorded in low abundancies and indicate more their local occurrence (Jochimsen, 1986; Huntley and Birks, 1983). Ferns tend to grow in moist habitats all over the investigation area. Extracts of the rhizomes of some species, e.g., *Dryopteris filix-mas*, were used against tapeworms in former times (Endlicher, 1842).

Trilete microspores of *Selaginella selaginoides* were also detected in low frequencies. In the Alps this plant occupies a wide range of communities, soils and altitudes from 900–2600 m. In the pollen rain its occurrence is indicative for local growth (Jochimsen, 1986).

3.5. Economic pollen

This category encompasses pollen types of edible plants, either cultivated or collected, and of species associated with crop plants, indicating cultural activities, e.g., ruderals. Only 2 types occur in percentage values: *Cerealia*-type and Papilionaceae. Associated herb pollen belong to synanthropic species commonly growing in cereal fields or at dwelling places, e.g., Chenopodiaceae, *Plantago major*-type, *Polygonum persicaria*-type, *Rumex acetosella*, Urticaceae. Most probably they were also trapped in the bran during crop processing and were picked up unintentionally with the cereal meal.

Cerealia-type: Cereal pollen is the only pollen type indicating crop plants in the Iceman's colon content. The *Cerealia* pollen identified belong to the *Triticum*-type (Figs. 4, 5; Plate II, Figs. 4, 5; Plate IV, Figs. 7, 8). Even registered near a field, this pollen type is not a large contributor to local pollen rain, because cereals are self-pollinators. Therefore most possibly this pollen derives from the Einkorn (*Triticum monococcum*) bran he had eaten.

Small amounts of grass (Gramineae) pollen were detected, which are smaller in size than the cereal pollen types found and do not show the distinct annulus around the aperture. They are listed separately as Gramineae.

Papilionaceae: Pollen of the legume family reach values above 1%. The legume family is represented by many species in the area. They are all pollinated by insects. In modern surface samples Papilionaceae pollen does not contribute much to local pollen rain and scarcely registers above 1%. Within this family several crop plants, e.g., pea (*Pisum sativum*), lentil (*Lens culinaris*), were already in use during Neolithic times.

Several wind-born pollen species, e.g., Goosefoot family type (Chenopodiaceae), Sheeps Sorrel (*Rumex acetosella*-type), Greater Plantain (*Plantago major*-type), Nettle (*Urtica*), occur in low quantities in the colon spectrum. This low pollen recovery rate corresponds to modern soil samples, because these pollen are commonly dispersed in the air. All these taxa identified indicate disturbed vegetation and are associated with fields or other rural sites. Their pollen may settle on cereal remains eaten by man. On the other hand it is well known that the young leaves of these species are consumed as salad or spinach. However, investigations have proved that unintentionally ingested pollen by consuming vegetative parts of plants results also in very low numbers of pollen (Bryant, 1974b).

4. Various accessory remains

Diatoms: In addition to pollen, 24 different diatom species were also recognised in the food residue. The dominant species identified are *Achnanthes minutissima, Gomphonema olivaceum, G. pumilum, Diatoma ehrenbergii, Fragillaria arcus* and *Navicula radiosa*. They characterise a freshwater diatom flora typical of mountain rivulets within the investigation area (Rott, this volume).

Zoological remains: According to all the plant remains in the Iceman's food residue it might be argued that he ate mainly vegetable food, but muscle fibres (Plate III, Fig. 8) as well as charred bone fragments, were also found and indicate that meat was also part of his diet. Probably this meat was from ibex (*Capra ibex*), just as the finds of charred ibex bones at the find spot indicate (Driesch and Peters, 1995).

Parasites: Eggs of the whipworm (*Trichuris trichiura*, Plate III, Fig. 7) in a substantial amount among the colon content confirm that the Iceman was infested by this human parasite (Aspöck et al., 1995, 1996 this volume). Infection with the whipworm (*Trichuris trichiura*) was very common in prehistoric times in Central Europe (Herrmann, 1993) and sheds light on the sanitary conditions in former times.

Mineral particles: The inorganic component of the food residue is dominated by mineral particles, which were most probably unintentionally consumed as a byproduct of the milling process and alongside the food preparation (Plate III, Fig. 6).

4.1. Provenance of the Colon Pollen

As briefly outlined above, the pollen taphonomy of the food residue in the colon is complex and may have derived from different sources. There are several ways in which pollen can be incorporated into the intestinal residue: diet, drinking water, breathing and swallowing bronchial mucus, etc. Furthermore it has to be considered whether pollen is incorporated intentionally as a dietary component by consuming plants (economic pollen) or unintentionally as a contamination of food, water or air (background pollen). Nevertheless, pollen accumulated by these means in coprolites can be very informative about diet, seasonality of site occupation and palaeoenvironmental conditions (Bryant, 1974a; Sobolik, 1988).

The classification of pollen into background and economic categories is artificial and may differ according to ecological circumstances. In this case it is based on anemophily and pollen occurrence in local or regional modern rain. On the other hand economic pollen includes anemo- and entomophilous pollen of crop plants, of edible and medicinal plants as well as pollen of weeds closely associated with arable fields.

It is unexpected that the economic pollen content in the food residue of the Iceman is low, indicating fairly highly processed cereal. This is confirmed by the size classes of the bran remains recognised (Fig. 3). Sizes smaller than 63 μ are predominating with 57%. Only 6% of the testa remains are bigger than 250 μ, whereas 28% are smaller than 125 μ.

The high amount of background pollen in the food residue is very surprising. Most of the taxa found are wind-born like the predominant Hop Hornbeam (*Ostrya carpinifolia*) and Hazel (*Corylus avellana*). Although they are detected in remarkable proportions it seems unlikely that these pollen were intentionally consumed. Whereas American *Ostrya* species are used as a remedy against rheumatics, diarrhoea and toothache, very little information exists about medicinal use of pollen from European catkin bearing plants. In former times in some areas of the inner Alps, catkins of Hazel (*Corylus avellana*) were fed to animals within the hay to prevent diseases (Hegi, 1935). Any application for humans appears to be unknown. Therefore, this kind of pollen is classified as background pollen.

In normal cases a limit of 2% entomophilous pollen in the food residue may indicate intentional consumption (Bryant, 1974b; Sobolik, 1988). From the colon pollen taxa this threshold is exceeded by *Caltha*-type and Primulaceae; Papilionaceae occur in values near this limit. Intentional consumption of the insect-pollinated taxa seems likely, and there exists a probability that all this entomophilous pollen derives from consumed honey. However, a comparison of the colon NAP spectrum with pollen spectra of honeys from the Italian Alps (Vorwohl, 1972) shows no relationship and makes it unlikely that the NAP pollen was incorporated in that way.

It is quite clear that pollen is recruited into coprolites differently than it is into soil samples or Burkard traps but nevertheless, a comparison of these samples can be attempted in order to obtain a more precise indication of the location of the deciduous vegetation type represented by the background pollen types. Therefore the colon spectrum was compared with (Fig. 5):

- a pollen spectrum derived from material extracted from the ice thought to be contemporaneous with the deposition of the body (site 2). This would test the possibility that the Iceman drank water from the gully in which he was discovered.

- five pollen spectra from Burkard-traps collecting airborne pollen over a five-year period were set up along a transect of the Alps within the investigation area. One single trap was placed at the northern alpine border zone (site 3), at the northern inneralpine montane zone (site 4), at the inneralpine subalpine regions (site 5) and at two sites (sites 6–7) in the southern inneralpine regions (Schlanders and Bozen).

Visual comparison, supported by numerical evaluation of the data, clearly indicates that the colon pollen spectrum differs greatly from those of the transect and of the ice samples. The pollen spectrum from the ice at the find spot contains more Alder (*Alnus*) and Pine (*Pinus*), but no Hop Hornbeam (*Ostrya*-type), and therefore water from the find spot can be ruled out as a pollen source for the colon spectrum. Neither do the spectra presented from the Burkard traps show any harmony with the colon spectrum. One underlying reason is that the pollen spectra from the Burkard traps represent a regional flora whereas the colon spectrum seems to show specifically local features. The only thing that can be deduced from the Burkard pollen spectra is that the Hop Hornbeam (*Ostrya*) pollen type is distributed south of the main Alpine ridge. This finding would, however, appear to add further weight to the evidence that the colon spectrum is representative of a southern alpine flora.

Pollen taxa in the colon spectrum (Figs. 4, 5) belonging to the ecologically more demanding deciduous forests (*Querco-Fagetea*) predominate. In particular the numerically predominant *Ostrya*-type indicates the thermophilous Hop Hornbeam (*Ostrya carpinifolia*), the only indigenous species of this genus in the area, which establishes only in forests along the valleys south of the main ridge of the Alps. There, in the lower Vinschgau and the Schnalstal this species is a dominant representative in Manna ash-Hop Hornbeam forests (*Orno-Ostryetum*), which thrive up to 1100 m altitude (Dalla Torre and Sarnthein, 1900; Peer, 1981), and recent stands of Hop Hornbeam (*Ostrya carpinifolia*) are only ca. 10 km away from the find spot. The typical montane vegetation of the inner alpine area, acid spruce forests (*Vaccinio-Piceion*), is also represented by minor occurrences of the pollen grains belonging to Pine (*Pinus*), Spruce (*Picea*), and Whortleberry (*Vaccinium*-type). In general, therefore, this pollen spectrum would appear to reflect the transition zone between a warmth demanding vegetation with Hop Hornbeam-forests (*Orno-Ostryetum*) and the inneralpine Spruce forests (*Vaccinio-Piceion*) characteristic for the montane vegetation in the valleys of the southern Alpine regions.

Similar conclusions concerning the provenance of the Iceman have also been reached on the basis of dendrological analyses (Oeggl and Schoch, this volume) and chorology of the bryophytes (Dickson et al., 1996). The dendrological analyses carried out showed that the majority of the woody plants which have been found belong to species of the montane region, although some subalpine and alpine species were also represented. In particular the ecological requirements of the identified woody species point to the transition zone between thermophilous mixed-oak forest communities (*Quercetalia pubescenti-petreae*) and the montane spruce forest (*Piceetum montanum*). Norwegian Maple (*Acer platanoides*), European Yew (*Taxus baccata*), Ash (*Fraxinus excelsior*), Lime (*Tilia*) and Elm (*Ulmus*) allow us to infer a humid habitat with a mineral-rich, free-draining soil and a mild winter climate, similar to present-day conditions in woodlands along the slopes and in the gorges of the lower Schnalstal and Vinschgau in South Tyrol.

Alongside this increasingly clear picture of the type of environment in which the Iceman spent the hours preceding his death, which emerges from an overview of the botanical material recovered in the colon, further and more detailed information can be obtained from an evaluation of single components of this complex picture. The most fascinating among these is the preserved gametophyte within the grains of the *Ostrya*-type pollen. It was initially assumed that the *Ostrya*-type pollen became associated with the cereals during the preparation of the farinaceous Einkorn-based (*Triticum monococcum*) food. However food preparation phases, such as milling, cooking or baking – especially the heating – would destroy the proteins, damage the cell content irreversibly and change the colour and proportions of the exine. No damaged pollen grains were observed in the colon sample, therefore making it improbable that this pollen originated through direct association with the cereal remains.

Nevertheless, the high frequency of wind-pollinated pollen types in this sample may also indicate accidental ingestion during respiration or by contaminated water supplies. Especially if food is prepared in an open area, such as in a cave, this can be particularly prevalent during the pollinating season of a specific plant. Furthermore, the diatom flora, as well as the pollen from herbs of mesic habitats found in the colon residue make assimilation by drinking water most probable.

But whatever the mechanisms through which the *Ostrya*-type pollen entered the body it is clear that the grains had only recently been released from the inflorescences. *Ostrya*-type pollen was absorbed within a short time after their deposition, otherwise the gametophytes would have been destroyed by oxidising conditions within a few days or weeks of emission. Furthermore, when the flowering season of the other

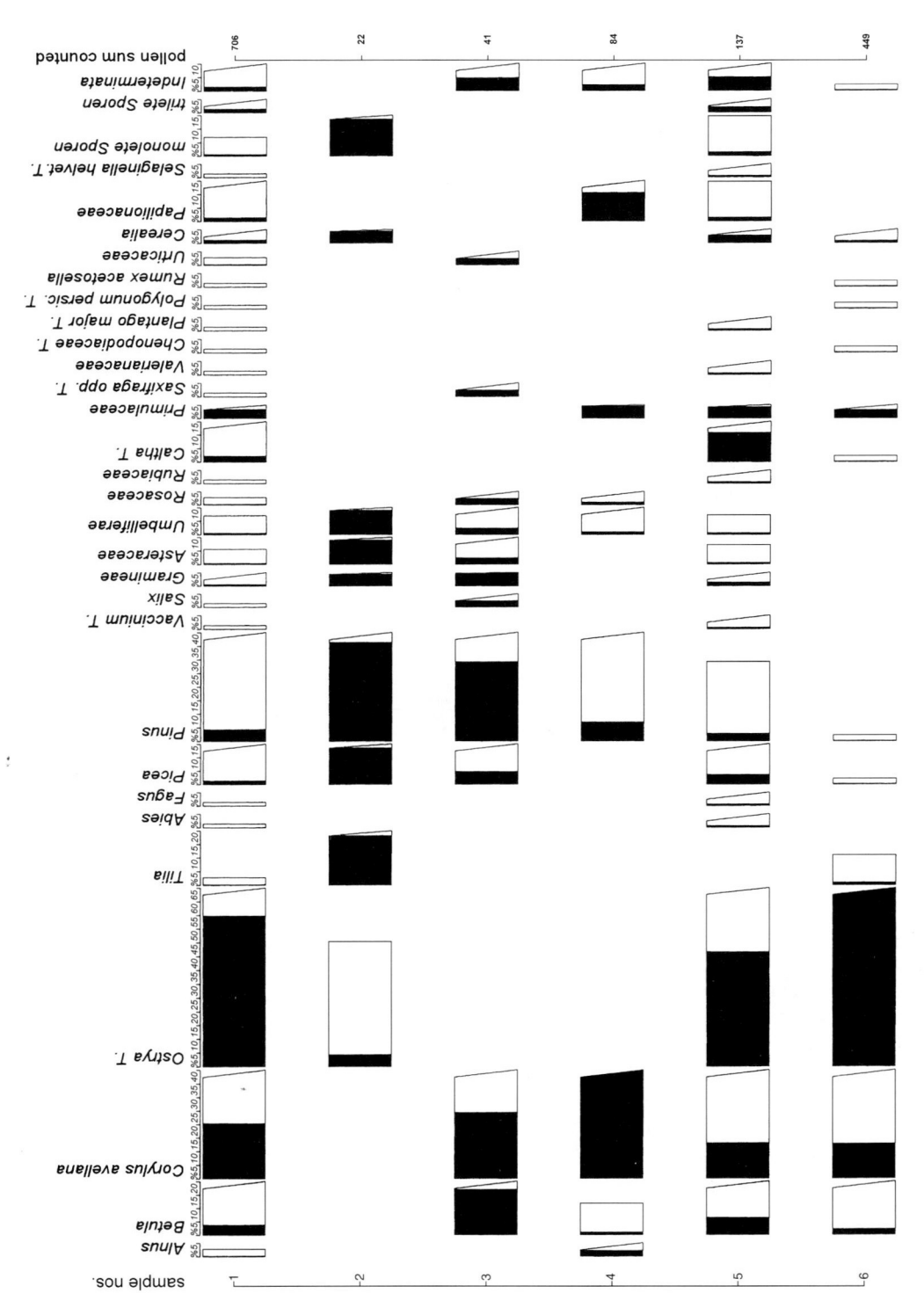

Fig. 4. Pollen spectra of the food residue from the colon: spectrum 1 = colon complete, spectrum 2 = fraction > 500 μ, spectrum 3 = fraction > 250 μ, spectrum 4 = fraction > 125 μ, spectrum 5 = fraction > 63 μ, spectrum 6 = outwash < 63 μ (scale reading: 1 graduation mark means 5%)

The diet of the Iceman 99

Fig. 5. Comparison of the pollen spectrum from the colon with recent pollen spectra along a transect from south to north of the Eastern Alps: 1 = colon pollen spectrum complete, 2 = ice sample from the find spot, 3 = northern alpine border zone: montane regions, 4 = northern inneralpine zone: montane regions, 5 = inneralpine zone: subalpine regions, 6 = southern inneralpine zone (Vinschgau): montane regions, 7 = southern inneralpine zone (Bozen): montane regions. μ (scale reading: 1 graduation mark means 5%)

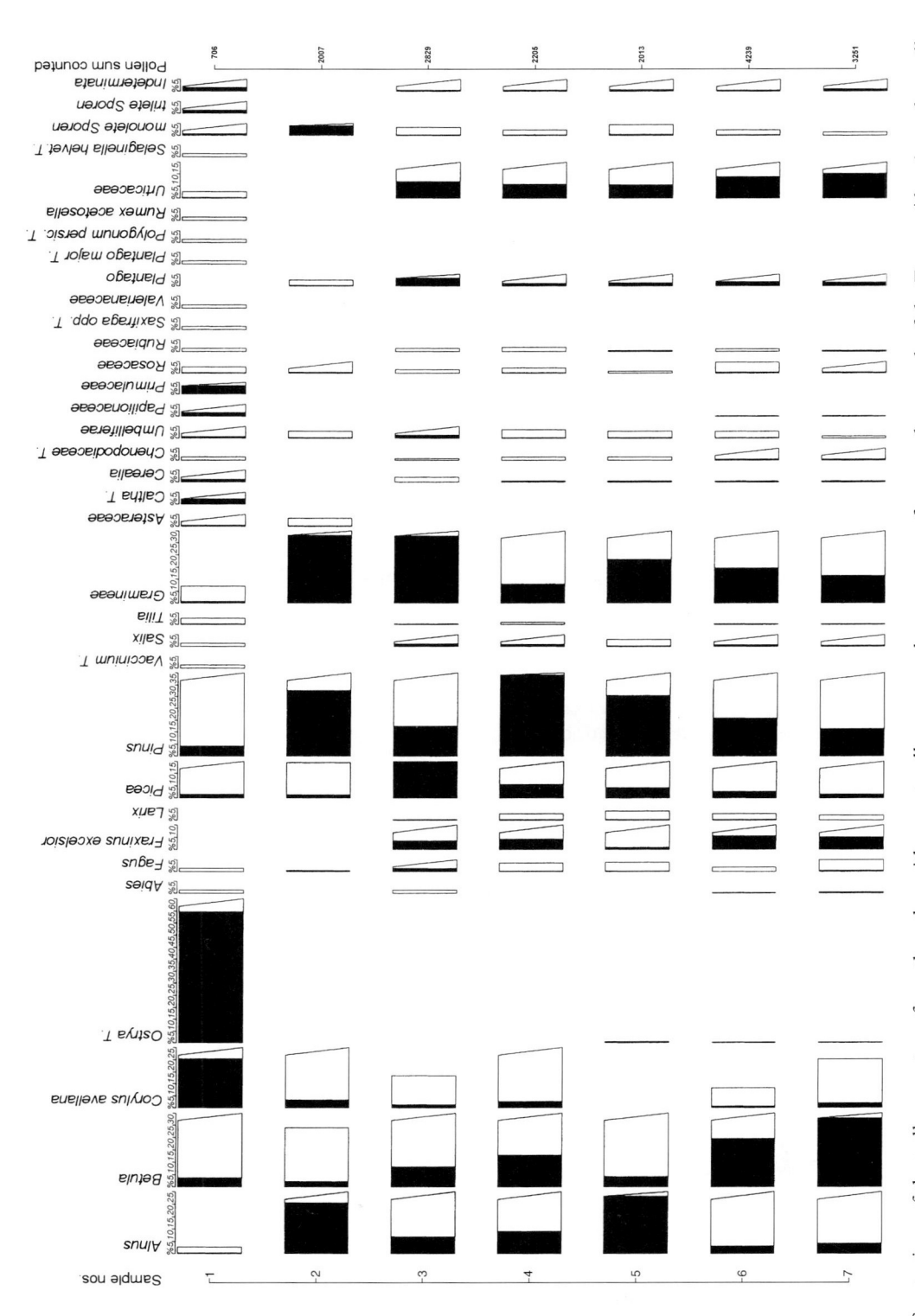

taxa is also taken into consideration, an estimation of the season of his death is possible. It is striking that the predominant arboreal taxa, like Birch (*Betula*), Hazel (*Corylus avellana*) and Hop Hornbeam (*Ostrya carpinifolia*) indicate an early flowering season, and among the NAP spectrum (Figs. 4, 5) mainly early flowering taxa are represented, whereas late flowering taxa such as Mugwort (*Artemisia*) are missing. The preserved cellular gametophyte in Birch (*Betula*) and Hop Hornbeam (*Ostrya carpinifolia*) pollen indicates that the season of his death would appear to be spring. Taking into consideration that pollen, depending on size and shape, can stay within the intestine up to a month (Sobolik, 1988; Gremillion and Sobolik, 1996), the Iceman died in early summer at the latest.

Within this reconstruction one further aspect needs discussion, namely why the pollen content was not affected by enzymes and/or digested in the intestinal tract? Food residue in the colon always contains protein and fat, which are further but only partly digested in this part of the intestinal tract, given that even in human faeces certain amounts of these dietary components can still be found. The exine certainly offers some sort of protection, but this is probably insufficient to enable the preservation of 80% of the pollen in pristine conditions, as is the case of the *Ostrya*-type pollen identified in the colon. Protein and fat content in faeces will increase when the digestive transfer speed is increased. This can be caused for instance by a strong endoparasitic intestinal infection, and the material from the Iceman's colon also includes a large number of eggs of the whipworm (*Trichuris trichiura*), a human parasite (Aspöck et al., 1995; 1996 this volume). Furthermore, anatomical investigations have shown that the intestinal epithelium tissue was affected by a strong parasitic infection, which led to a higher transfer speed in the digestive tract, possibly even manifested with diarrhoea (Platzer, pers. comm.). The link between the forensic and the microfossil data, which together indicate an increase in the digestive transfer speed and a reduction in the catabolic rate and protein assimilation, go a long way to explaining the presence of the remarkably well-preserved *Ostrya*-type pollen in the colon.

5. Macroremain evidence of dietary plants from archaeological excavations in the central part of the Alps

Plant remains discovered in archaeological excavations inform us about prehistoric crop plants. They yield indirect evidence about the food of prehistoric people, because from the existence of a crop plant it cannot be assumed, that it was eaten by man. The use of a cultivar for dietary purposes or for fodder of the life stock depends on the agricultural technology of the prehistoric civilisation. However, edible plants from archaeological excavations yield additional information about the food of prehistoric man. Although only one sample from the food residue in the colon was investigated, it is of interest to compare the results from the colon sample with the cultivar inventory of the investigation area.

Presently crop plants are known from six Neolithic sites in North and South Tyrol (Table 1). First-hand evidence comes from the find spot of the Iceman itself. Several cereal remains were found adhering to the hides of his clothes (Plate I), namely Einkorn (*Triticum monococcum*) and eleven rachis segments of a lax-eared variety of naked six-row Barley (*Hordeum vulgare* var. *nudum*).

The seven spikelets of Einkorn (*Triticum monococcum*) are uncharred, well preserved and almost complete (Plate I, Figs. 3, 4). Only the apices are a little eroded and the awns have broken off. The glumes show the normal curvature. The twin apices were only extant on one specimen. The margins of the glumes run almost parallel to one another. In cross section they are rectangular. The keeled shape conferred by the main nerve is visible and the secondary nerve is also quite prominent. Under low magnification 2–3 further nerves are clearly visible in the space between the main and the secondary nerves. The abscission zone at the internodes is discernible in a distinct oval scar. The measurements of the spikelets are shown in Table 2.

The naked six-row Barley (*Hordeum vulgare* var. *nudum*) is documented by uncharred rachis segments (Plate I, Figs. 5, 6). In most of them the lower parts of glumes are still adhering to the internodes. Three types can be discriminated by the hairs on the lateral rachis margins: specimens with long and straight, with short and woolly or with glabrous rachis margins. Pubescent internodes always show a hairy rachilla too, if preserved. In some rachilla segments the peduncles of the lateral spikelets are preserved. The abscission zone is visible in a distinct oval scar on the ventral side of the internodes. Based on the length of the internodes (limit 2,5 mm) it is a lax-eared variety (Table 3).

These cereal finds are confirmed by charred plant remains recovered from several other archaeological excavations north and south of the main range of the Alps. Because many more sites are known from the south and it is not surprising that the oldest finds of Neolithic crop plants also derive from there. In "La Vela di Trento", a site in the Adige Valley south of Bozen, Einkorn (*Triticum monoccum*) and Emmer (*Triticum dicoccum*) were harvested by Neolithic farmers. The

Table 1. Neolithic finds of cereals and of edible plants from archaeological excavations within the central parts of the Alps. (cal BC = radiocarbon data calibrated to calendar years; bc = radiocarbon data transcribed uncalibrated to sidereal years before Christ as stated by the authors)

Site/County	Radiocarbon data/ archaeological classification	Species	Author
La Vela/Prov. Trento	3420 ± 180 bc 3460 ± 100 bc Vaso Bocca Quadrata II	Einkorn, *Triticum monococcum* Emmer, *Triticum dicoccum* Hazel, *Corylus avellana* Bramble, *Rubus* sp.	Pedrotti, 1990
Mariahilfbergl-Brixlegg/Tirol	3650–3534 cal BC Münchshofener	Emmer, *Triticum dicoccum* Bread-wheat, *Triticum* cf. *aestivo-compactum* Hulled Barley, *Hordeum vulgare* Pea, *Pisum sativum* Strawberry, *Fragaria vesca*	Oeggl, unpubl.
Hauslabjoch/Bozen	3300 cal BC	Einkorn, *Triticum monococcum* Barley, *Hordeum vulgare* Sloe, *Prunus spinosa*	Oeggl, unpubl.
Völseraicha/Bozen	2830 ± 100 bc	Barley, *Hordeum vulgare*	Biagi and Nisbeth, 1987
Villanders-Plunacker/Bozen	Neolithic	Bread-wheat, *Triticum* cf. *aestivo-compactum* Barley, *Hordeum vulgare* Lentil, *Lens culinaris*	Oeggl, unpubl.
Villanders-Plunacker/Bozen	ENeolithic	Emmer, *Triticum* cf. *dicoccum* Blackberry, *Rubus fruticosus*	Oeggl, unpubl.
Tolerait/Bozen	2180 ± 100 bc ENeolithic	Einkorn, *Triticum monococcum* Emmer, *Triticum dicoccum* Barley, *Hordeum vulgare* Hazel, *Corylus avellana*	Biagi and Nibeth, 1987

Table 2. Cereal findings with the Iceman: Einkorn (*Triticum monoccum*), dimensions of the spikelets in mm; measurements according to Jacomet (1987)

Sample number	Width of the glume basis L	Width of the internode basis O	Max. Width of the internode P	O/P-Index	Length of the spikelet without the internode M	Max. Width of the spikelet R
91/90	0,65	0,65	1,3	0,5	8,2	2,7
91/90	0,55	1,1	1,5	0,73	–	–
91/105	0,8	1,0	1,4	0,71	7,9	2,9
91/105	0,8	0,6	1,35	0,44	8,3	2,7
91/105	0,65	–	1,3	–	9,3	2,6
91/124	0,65	0,9	1,6	0,56	8,8	3,2
91/124	0,6	–	1,3	–	9,1	3,0
average	0,67	0,85	1,39		8,6	2,85

settlement is dated to 3420 ± 180 bc (Pedrotti, 1990), slightly older than the Hauslabjoch finds. More or less of the same age as the cereals from the Iceman are the Barley (*Hordeum vulgare*) finds from Völseraicha near Bozen (Biagi and Nisbeth, 1987) and the finds from Villanders-Plunacker. The latter one contains also Bread-wheat (*Triticum* cf. *aestivo-compactum*).

The only Neolithic cultivars from North of the main range of the Alps were recently found near Brixlegg in the Inntal. This Neolithic dwelling site was located on the top of a hill. The pottery found around a fireplace shows close relations to the lower Bavarian Münchshofener culture group (Krauß and Huijsmans, 1996). Besides ceramics the layer contained animal bones and charred plant remains. Five soil samples, with a volume of 10 l each, taken from the cultural layer yielded a spectrum of crop plants. The cereals are dominated by barley (*Hordeum vulgare*), but also Bread-wheat (*Triticum* cf. *aestivo-compactum*) and Emmer (*Triticum* cf. *dicoccum*) are recorded in low quantities. Legumes are documented

Table 3. Cereal findings with the Iceman: naked, lax-eared six-row Barley (*Hordeum vulgare* var. *nudum*), dimensions of the rachis segments in mm; measurements according to Jacomet (1987). *) dimensions uncertain because fragmented

Sample number	Length S	Width U	Width of the internode basis T	Peduncles of the lateral spikelets	Hairs	Species
91/90	3,0	–	0,8	–	long haired	Barley (*Hordeum vulgare*); lax-eared variety
91/90	2,1	1,6	1,1	+	long haired	six-row Barley (*Hordeum vulgare* var. *nudum*); naked variety
91/90	2,6	1,7	0,9	+	long haired	six-row Barley (*Hordeum vulgare* var. *nudum*); naked, lax-eared variety
91/90	3,7	1,7	1,0	–	short haired	Barley (*Hordeum vulgare*); lax-eared variety
91/95	3,0	2,2	1,1	+	long haired	six-row Barley (*Hordeum vulgare* var. *nudum*); naked, lax-eared variety
91/95	3,0	2,2	1,1	+	long haired	six-row Barley (*Hordeum vulgare* var. *nudum*); naked, lax-eared variety
91/105	3,4	1,9	1,2	+	long haired	six-row Barley (*Hordeum vulgare* var. *nudum*); naked, lax-eared variety
91/105	2,8	2,0	1,1	+	glabrous	six-row Barley (*Hordeum vulgare* var. *nudum*); naked, lax-eared variety
91/105	3,7	2,2	1,1	+	long haired	six-row Barley (*Hordeum vulgare* var. *nudum*); naked, lax-eared variety
91/105	2,8	1,5	0,8*)	+	short haired	six-row Barley (*Hordeum vulgare* var. *nudum*); naked, lax-eared variety
91/105	1,9*)	1,5	0,6*)	+	short haired	Barley (six-row Barley (*Hordeum vulgare* var. *nudum*); naked variety)

by pea (*Pisum sativum*). Radiocarbon analysis on charcoal from the cultural layer yielded a date of 4820 ± 40 BP (cal BC 3650–3534; GrN-20981) and places the settlement more or less contemporaneous to the finds of Völseraicha and Villanders-Plunacker.

In addition to crop plants, fruits and seeds of wild plants were also collected and contributed to the food of Neolithic man. With the finds of Hazel (*Corylus avellana*), Strawberry (*Fragaria vesca*), Sloe (*Prunus spinosa*) and Brambles (*Rubus* sp.) only a small part of edible plants in the area has been documented. Nevertheless, they indicate that gathered wild plants constituted a considerable part of the Neolithic diet.

6. Discussion

The food remains from the Iceman's colon provide rare direct evidence of prehistoric diet. Although only a single specimen is investigated it is a substantial contribution to our knowledge of Neolithic diet. The Iceman's last meal, consisting of cereals, vegetables and meat, was well-balanced. The cereal remains found in and with the Iceman correspond well with the information gained from archaeological excavations. Einkorn (*Triticum monococcum*), Emmer (*Triticum dicoccum*) and Barley (*Hordeum vulgare*) were grown in the inneralpine regions during the Neolithic, whereas it is known from the excavations that Barley (*Hordeum vulgare*) was the most important cultivar. It is interesting that the cereal inventory of the inneralpine regions is dominated by Barley (*Hordeum vulgare*). This feature shows close relations to the Neolithic cereal inventory of the Swiss Alpine Foreland and probably indicates a relationship to the Mediterranean regions, because Küster (1995) assumes that Barley (*Hordeum vulgare*) was introduced as a Mediterranean element in South Germany.

The plant part of his last meal consisted mainly of Einkorn (*Triticum monococcum*). The flour of Einkorn (*Triticum monococcum*) is nutritious, has a high protein content, but gives bread poor rising qualities. Therefore Einkorn (*Triticum monococcum*) was consumed primarily as porridge or gruel (Zohary and Hopf, 1993). Nevertheless, both Einkorn (*Triticum monococcum*) and Barley (*Hordeum vulgare*) were used in bread in the Neolithic. Barley bread was baked by pressing a thick porridge or grain-paste over a heated stone (Maurizio, 1916; Hansson, 1994). The grain-paste or porridge therefore was made by whole, crushed and partly ground cereals. The predominantly small size of the bran and mineral particles in the Iceman's colon as well as the low amount of economic pollen indicate a fairly highly processed Einkorn (*Triticum monococcum*) flour that was more likely consumed as a kind of bread than as gruel or soup. Investigations on Neolithic bread and recent experiments by Währen (1995a) show that Neolithic men were already able to produce fine flour with their primitive mills for baking bread of good quality.

Stone cells and vessel elements show that his last meal consisted of additional vegetable ingredients. Unfortunately for taxonomic classification both these cell types are commonly distributed and found in tissues of roots and of many fruits recorded from Neolithic settlements, e.g., Strawberry (*Fragaria vesca*), Brambles (*Rubus* sp.), Apple (*Malus sylvestris*), and Elderberry (*Sambucus nigra*). Future sample may aid in species determination and depends on the occurrence of additional characters.

The accessory remains give additional information about his last meal and his environment. A noticeable amount (25%) in the food residue comes from charcoal fragments derived from cooking on an open fire. Diatoms give evidence of the freshwater he drank or which he used to prepare his meal. The ecological demands of the different species point to several water sources from valley bottoms as well as from high altitudes (Rott, this volume). The pollen describes a habitat in the transition zone of the thermophilous deciduous forest rich in Hop Hornbeam (*Ostrya carpinifolia*) to the boreal Spruce (*Picea abies*) forest of the valleys in the lower Vinschgau.

The high amount of the *Ostrya* pollen in the colon raises the question of intentional ingestion of this pollen. Although the consumption of pollen has never been part of the European tradition of eating habits, this fact needs to be considered. Pollen is a concentrated energy- and vitamin-rich foodstuff that in contemporary times is consumed as a medicinal treatment and food supplement. The nutritive value of pollen is comparable to that of dried legumes and it is rich in vitamins (Stanley and Linskens, 1984). It is well known that American Indian cultures in prehistoric as well as in recent times made extensive use of pollen as food and in ritual (Linskens and Jorde, 1997). They gathered pollen and added it to the flour of bread or used it unalloyed as a powerful remedy to prevent exhaustion. It is quite reasonable that the use of pollen as a food supply in European cultures is forgotten or not recorded, but there are good reasons to assume that the Iceman consumed the pollen unintentionally and therefore it is classified as background pollen. This classification is based on the fact that (1) the Hop Hornbeam (*Ostrya carpinifolia*) is wind pollinated, (2) it is never a dominant component in honeys of the region and (3) the pollen concentration is low. Calculated to one gram the pollen concentration of the total sample is below 20 000 grains. One half of it (9700 grains/g) is allotted to the Hop Hornbeam (*Ostrya carpinifolia*). According to Sobolik (1988) pollen in such a concentration indicate the accidental ingestion of pollen (background pollen) during respiration, or through contaminated food and water supplies.

Nevertheless, however the pollen was swallowed, it was incorporated immediately after the opening of the flowers, because of the preserved cellular microgametophyte. This would also be the case if Hop Hornbeam (*Ostrya carpinifolia*) pollen was ingested intentionally for medicinal or ritual purposes. Further investigations will focus on this topic of the origin of the intestinal Hop Hornbeam (*Ostrya carpinifolia*). Was it incorporated by breathing air or by drinking water or was it consumed as a medicine? If high amounts of pollen were consumed they would be absorbed from the epithelial cells of the intestinal wall and transported by the blood stream. Then they could be detected in other tissues of the body (Linskens and Jordi, 1997).

The pollen spectrum from the colon also indicates the season of the Iceman's death. Based on the flowering seasons of the pollen taxa, the Iceman died in the late spring or in the early summer. This conflicts with the opinion that the Iceman died in the autumn, more or less at the same period in which he was discovered (Spindler, 1996). This assumption is based on the good preservation of the body, the find of a sloe (*Prunus spinosa*) as well as spikelets of Einkorn (*Triticum monoccum*) and the find of female ectoparasitic insect remains in the fur and hides of the Iceman (Gothe and Schöl,

1992). Spindler (1996) argues that immediately after the Iceman died, the body was covered with snow, otherwise predators would have damaged or destroyed the body. We know now from the lipid composition of the Iceman's skin that the body was submerged in water for several months and desiccated afterwards (Bereuter et al., 1996), which points to an earlier season of his death. His good preservation is primarily due to the temperature regime in such high altitudes. Even during summer, temperatures at the find spot vary around the freezing-point and are comparable to a refrigerator that minimises bacterial growth and thus the decay of organisms. Still, there is other evidence which indicates an autumn death. The harvest-time of the sloe (*Prunus spinosa*) and of the Einkorn (*Triticum monococcum*) is considered the end of summer. This assumption is based on the occurrence of fresh fruits and neglects the fact that both fruits can easily be stored for longer periods. Nowadays in Central Europe sloes (*Prunus spinosa*) ripen from July till September and remain on the shrub over winter. Frost reduces their high tannin content and makes them tastier. South of the main Alpine ridge they mature earlier (July) than in the north (end of September or October). The author himself collected and tasted ripe sloes at the end of July on the hills below Juval Castle at the entrance of the Schnals Valley. Jacomet et al. (1989) cite November-December as the harvest-time of sloes in the Swiss Neolithic lake dwellings. On the other hand as mentioned above dried sloes can easily be stored and are found year round. Neuweiler (1935) states that in Austria dried sloes were still stored for medicinal purposes into recent times. Morphological investigations of the sloe (*Prunus spinosa*) found with the body did not reveal whether it was fresh or dried when deposited.

Einkorn (*Triticum monococcum*) is harvested in July and August depending on how this primitive wheat was grown. It has to be considered that winter cereals mature earlier than summer cereals. From detailed palaeoethnobotanical investigation on the lake dwelling sites in Switzerland Jacomet et al. (1989) conclude that both cultivation techniques were already common in the Neolithic. At the beginning of plant domestication in the Near East Einkorn (*Triticum monococcum*) was cultivated in a manner closely followed the life cycle of its wild ancestor. It was sown in autumn and harvested in early summer. This practice was continued during the spread of agriculture to Europe, where the crop plants adapted to different climate conditions (Willerding, 1996). Einkorn (*Triticum monococcum*) is very cold resistant, well adapted to mountain areas and was grown as a winter and summer crop in the Alps up to recent times (Mayr, 1936). Therefore it has to be considered that Einkorn (*Triticum monococcum*) was already available from stored reserves or from the winter crop in July.

The find of the haematophagous fly *Lipoptena cervi* supports however, an autumnal demise (Gothe and Schöl, 1992). This ectoparasitic insect lives in the fur of red dear and sucks blood. The imago creeps out of the pupa in autumn and attacks its host. After three weeks, the wings are shed and the insect remains on the host for the rest of its life. The discovery of wing fragments make it plausible that the Iceman died in autumn, but already Gothe and Schöl (1992) assume, that the fly did not live as a parasite on the Iceman's body. More probably, the fly derives from the hide of which the Iceman's clothes were made, and this ectoparasite lived in the fur of the animals. It was already dead when the hides were sewn into clothes and the Iceman ventured up to the Tisenjoch.

Nevertheless, all the evidences for an autumn death are indirect, derive from the find spot and may not have been gathered and deposited contemporaneously with the demise of the Iceman. All these plant remains especially could have been collected earlier. By contrast, the pollen from his intestines comes directly from the body and represents more reliable data about the time of death. The preservation of the microgametophyte in the Hop Hornbeam (*Ostrya carpinifolia*) and Birch (*Betula*) pollen is strongly indicative that the pollen was ingested immediately after its release from the flowers. Normally pollen protein, polymer carbohydrates and lipids are digested after entering the intestine tract, but digestion is time dependent. Substances located at the surface and in the pollen wall (exine) cavities are easily leached out and digested by the gastric enzymes, whereas the cytoplasm (gametophyte) in the interior of the pollen is protected by the pollen wall (exine). The disintegration of the cytoplasm starts from the pollen aperture region and goes on to the inner parts, but it takes more time (Franchi, 1987).

Furthermore the good gametophyte preservation speaks against the retention of the Hop Hornbeam (*Ostrya carpinifolia*) pollen in the folds of the intestines for a longer time. Modern faecal studies determined the time it takes for pollen to pass through the digestive system. They show that spheroidal pollen of medium size with smooth exine like Hop Hornbeam (*Ostrya carpinifolia*) pass more quickly through the intestines (Sobolik, 1988). Therefore the extraordinarily good gametophyte preservation is explained by a reduction in the catabolic rate and seems to be caused by a higher digestive transfer speed. This increase in the digestive transfer rate results from the strong infestation with the human whipworm (*Trichiuris trichuria*).

The Norway maple (*Acer platanoides*) leaves all lacking stalks also hint at the season the Iceman climbed to the high mountain regions. When found they still contained chlorophyll; an indication that they did not derive from the usual leaf fall of deciduous trees in autumn (Oeggl and Schoch, 1995). Experiments by the author have shown that green maple leaves are easily stripped off by hand but the petioles remain on the twigs in spring or early summer. Later the leaf ribs strengthen and are more strongly attached to the petioles. So the entire leaves come cleanly from the twigs e.g., lamina and petioles. This supports the results of the pollen analysis. The gametophyte preservation as well as the flowering season of all the other pollen types found indicate that the Iceman died in late spring or at the latest early summer.

Abstract

Plant macrofossil and pollen analyses were investigated in a 40 mg specimen of food residue from the Iceman's colon. The results show the composition of his last meal, and contribute to knowledge of Neolithic diet, the reconstruction of his environment and the season of his death. His last meal consisted of cereals, vegetables and meat. The farinaceous dish was made mainly of Einkorn (*Triticum monococcum*), but stone cells and vessel elements prove that he ate also other vegetables. This is in congruence with the palaeoethnobotanical investigations of soil samples from archaeological excavations, which have shown that Einkorn (*Triticum monococcum*) and Barley (*Hordeum vulgare*) were grown in the inneralpine areas during the Neolithic. Besides these crop plants wild fruits and seeds were gathered and eaten. The microfossil content of the colon residue is rich in pollen (30 types, 2 types of spores), diatoms, mineral particles, charcoal fragments and ova of intestinal parasites. Taxa from the more demanding deciduous forests (*Querco-Fagetea*) predominate among the pollen. The most registered *Ostrya*-type indicates the warmth demanding Hop Hornbeam (*Ostrya carpinifolia*) and gives unequivocal evidence the he came from the valley-bottoms south of the main range of the Eastern Alps. The taxa-rich pollen flora from the colon residue was ingested most probably by drinking water, as the occurrence of diatoms shows. From the flowering times of the pollen taxa and from the unique preservation of cellular gametophytes in the pollen of both Hop Hornbeam (*Ostrya carpinifolia*) and Birch (*Betula*) the deduction is made that the Iceman's last journey took place in spring or early summer at the latest.

Zusammenfassung

Speisereste aus dem Dickdarm (Colon) des Eismannes wurden auf ihren Gehalt an Pflanzengroßresten und Blütenstaub untersucht. Die Ergebnisse der Analysen liefern die Zusammensetzung seiner letzten Mahlzeit und tragen zur Kenntnis der neolithischen Diät, sowie der Rekonstruktion seines Lebensraumes und der Todeszeit bei.

Die mehlartige Speise setzt sich hauptsächlich aus Einkorn (*Triticum monococcum*) zusammen. Steinzellen und Gefäßelemente beweisen, daß zusätzlich auch andere Pflanzen gegessen wurden. Dies stimmt gut mit den Ergebnissen paläoethnobotanischer Untersuchungen von Bodenproben aus archäologischen Grabungen überein. Dadurch ist bekannt, daß während des Neolithikums Einkorn (*Triticum monococcum*) und Gerste (*Hordeum vulgare*) in den Inneralpen angebaut wurde. Außer diesen Kulturpflanzen wurden Wildfrüchte gesammelt und gegessen.

Der Mikrofossilgehalt des Darminhaltes ist reich an Blütenstaub (30 Pollentypen, 2 Sporentypen), Kieselalgen, Mineralteilchen, Holzkohlenfragmenten und Eier eines Darmparasiten. Taxa der thermophiler Laubwälder (*Querco-Fagetea*) überwiegen unter den Pollen. Am häufigsten wurde der *Ostrya*-Typ registriert, der die wärmeliebende Hopfenbuche (*Ostrya carpinifolia*) anzeigt. Da die Hopfenbuche nördlich des Alpenhauptkammes nicht natürlich verbreitet ist, liefert dieser Pollentyp den eindeutigen Beweis, daß der Eismann aus den Tallagen südlich des Alpenhauptkammes stammt. Die artenreiche Pollenflora des Darminhaltes wurde aller Wahrscheinlichkeit nach durch Trinkwasser aufgenommen, was sich an den gefundenen Kieselalgen ableiten läßt. Eine Auswertung der Blühphasen der Pollentaxa aus den Speiseresten zusammen mit der einzigartigen Erhaltung des zellulären Mikrogametophyten im Blütenstaub der Hopfenbuche (*Ostrya carpinifolia*) und der Birke (*Betula*) führt zum Ergebnis, daß der Eismann im Frühjahr, spätestens im Frühsommer, zu seiner letzten Reise aufgebrochen ist.

Résumé

Les restes alimentaires prélevés dans le gros intestin (côlon) de l'Homme des glaces ont été examinés en vue de détecter d'éventuels résidus de plantes et de pollens. Les résultats des analyses nous renseignent sur la composition de son dernier repas tout en contribuant à élucider le régime alimentaire néolithique et à établir l'environnement et le moment de la mort de cet homme.

Le plat farineux retrouvé est constitué principalement d'engrain (*Triticum monococcum*). Les cellules de sclérenchyme et les éléments vasculaires contenus indiquent la consommation d'autres plantes. Ce résultat concorde avec les analyses paléoethnobotaniques de carottes prélevées à l'occasion de fouilles archéologiques. Celles-ci avaient révélé la culture de l'engrain (*Triticum monococcum*) et de l'orge (*Hordeum vulgare*) pratiquée dans les Alpes centrales à l'époque néolithique. En plus des plantes cultivées, les hommes du néolithique mangeaient des fruits sauvages dont ils faisaient la cueillette.

Le contenu microfossile de l'intestin examiné est riche en pollens (30 types de pollens, 2 types de spores), diatomées, particules minérales, fragments de charbon de bois et oeufs d'un parasite intestinal. Parmi les pollens, les taxa des forêts caducifoliées thermophiles (*Querco-Fagetea*) sont prédominants, le plus fréquent étant le type *Ostrya* qui signale la présence de l'ostryer à feuilles de charme (*Ostrya carpinifolia*), une espèce thermophile. Cette essence n'étant pas naturellement présente sur le versant nord des Alpes centrales, son pollen apporte la preuve irréfutable que l'Homme des glaces a été originaire d'une des vallées du sud de la chaîne principale des Alpes. Les pollens, très variés, retrouvés dans son bol intestinal, avaient très probablement été absorbés avec l'eau de boisson, hypothèse appuyée par la présence de diatomées. L'exploitation des périodes de floraison des pollens contenus dans les résidus alimentaires, ainsi que la parfaite conservation des microgamétophytes cellulaires dans les pollens de l'ostryer (*Ostrya carpinifolia*) et du bouleau (*Betula*), amènent l'auteur à la conclusion que l'Homme des glaces a entrepris son dernier voyage soit au printemps, soit au début de l'été au plus tard.

Riassunto

Si sono ricercati resti grossolani di piante nonchè pollini nel cibo contenuto nell'intestino crasso dell'uomo del ghiaccio. In base ai risultati delle analisi si è individuata la composizione dell'ultimo pasto dell'uomo del ghiaccio contribuendo alla conoscenza della dieta neolitica nonchè alla ricostruzione del suo ambiente vitale e dell'ora della sua morte.

Il cibo consistente in un impasto farinoso è composto prevalentemente di spelta minore (*Triticum monococcum*). Le schlerencime ed elementi di vasi dimostrano che sono presenti nella dieta anche altre piante. Tale fatto viene confermato anche dai risultati di analisi paleo-etno-botaniche dal suolo di scavi archeologici. Così si sa che durante il neolitico venivano coltivati spelta minore (*Triticum monococcum*) ed orzo (*Hordeum vulgare*) nella area alpina interna. Oltre a queste piante coltivate venivano però anche raccolte e consumate piante selvatiche.

Il contenuto di microfossile intestinale è composto di pollini (30 tipi diversi, 2 tipi di spore) alghe silicee, particelle minerali, frammenti lignei carbonizzati, uova di un parasita intestinale. Prevalgono tra i pollini le taxa delle foreste termofili (*Querco-Fagetea*). Il polline piú frequente è quello dell' *Ostrya* che è un indicatore dell' *Ostrya carpinifolia* termofila. Non essendo questa diffusa naturalmente a Nord delle Alpi centrali la presenza di questo tipo di polline è la dimostrazione inequivocabile dell'origine dell'uomo del ghiaccio nelle valli a Sud delle Alpi centrali. I pollini presenti in una molteplice flora intestinale furono assunti, molto probabilmente, attraverso l'acqua potabile e ciò in base alla presenza di alghe silicee. Una valutazione delle fasi di fioritura delle taxa dei pollini presenti nei resti di cibo nell'intestino, insieme con la singolare conservazione dei microgametofiti cellulari nei pollini dell' *Ostrya carpinifolia* e della betulla portano alla conclusione che l'uomo del Similaun ha intrapreso il suo ultimo cammino in primavera o al più tardi nella prima estate.

Acknowledgements

This research was funded by the Austrian Science Fund (Project No. 10151-SOZ). The author would like to thank: Univ. Prof. Dr. Werner Platzer, Univ. Prof. Dr. Othmar Gaber, and Univ. Prof. Dr. Karl-Heinz Künzel, Anatomical Institute of the University of Innsbruck, for fruitful discussions and for inclusion in the sample recovery procedures. Prof. Vaughn M. Bryant, Texas A & M University, Prof. Owen K. Davis, University of Arizona, and Prof. James Schoenwetter, Arizona State University, for presenting literature and for helpful suggestions in the preparation of this manuscript. Prof. Dr. James H. Dickson, Institute of Biomedical and Lifescience of the University of Glasgow, made a critical review of the manuscript and Mrs. Brenda K. Fowler improved the English language.

References

Aspöck H., Auer H. and Pichler O. (1995) The mummy from the Hauslabjoch: A medical parasitology perspective. Alpe Adria Microbiology Journal 2: 105–114.

Aspöck H., Auer H. and Pichler O. (1996) *Trichuris trichuria* eggs in the neolithic Glacier mummy from the Alps. Parasitology Today 12: 255–256.

Behre K.-E. (1991) Zum Brotfund aus dem Ipweger Moor, Ldkr. Wesermarsch. Berichte zur Denkmalpflege in Niedersachsen: 9.

Bell B. and Dickson C. (1989) Excavations at Warebeth (Stormness Cemetery) Broch, Orkney. Proceedings of the Society of Antiquaries of Scotland 119: 101–131.

Bereuter T. L., Reiter C., Seidler H. and Platzer W. (1996) Postmortem alterations of human lipids - part II: lipid composition of a skin sample from the Iceman. In: Spindler K., Wilfling H., Rastbichler-Zissernig E., zur Nedden D., Nothdurfter H. (eds.): Human Mummies. The Man in the Ice, vol. 3: 275–278.

Beug H.-J. (1961) Leitfaden der Pollenbestimmung für Mitteleuropa und angrenzende Gebiete. Stuttgart: pp 63.

Biagi P. and Nisbeth R. (1987) Ursprung der Landwirtschaft in Norditalien. ZfA. Z. Archäol. 21: 11–24.

Bryant V. M. (1974a) The role of coprolite analysis in Archaeology. Bulletin of the Texas Archaeological Society 45: 1–28.

Bryant V. M. (1974b) Prehistoric diet in Southwest Texas: The coprolite evidence. American Antiquity 39: 407–420.

Bryant V. M. and Holloway R. G. (1983) The role of palynology in Archaeology. Advances in Archaeological Method and Theory 6: 191–224.

Capasso L. (1995) Unguel morphology and pathology of the Iceman. In: Spindler K., Rastbichler-Zissernig E., Wilfling H., zur Nedden D., Nothdurfter H. (eds.): Der Mann im Eis. Neue Funde und Ergebnisse. The Man in the Ice, vol. 2. Wien: 231–239.

Christensen P. B. and Blackmore S. (1988) The Northwesteuropean Pollen Flora, 40. Tiliaceae. Review of Palaeobotany and Palynology 57: 33–43.

Dalla Torre und Sarnthein (1900) Flora der gefürsteten Grafschaft von Tirol, des Landes Vorarlberg und des Fürstenthumes Lichtenstein. 6. Bände. Innsbruck.

Dickson C. (1987) The identification of cereals from ancient bran fragments. Circea 4: 95–102.

Dickson C. and Dickson J. H. (1988) The diet of the Roman army in deforested central Scotland. Plants Today: 121–126.

Dickson C. (1989) The Roman army diet in Britain and Germany. In: Küster H. (ed): Archäobotanik: 135–154.

Dickson J. H. (1997) The Moss from the Tyrolean Iceman's Colon. Journal of Bryology 19: 449–451.

Dickson J. H., Bortenschlager S., Oeggl K., Porley R. and McMullen A. (1996) Mosses and the Iceman's Provenance. Proceedings of the Royal Society London, Ser. B, 263: 567–571.

Driesch A. van den and Peters J. (1995) Zur Ausrüstung des Mannes im Eis. Gegenstände und Knochenreste tierischer Herkunft. In: Spindler K., Rastbichler-Zissernig E., Wilfling H., zur Nedden D., Nothdurfter H. (eds.): Der Mann im Eis. Neue Funde und Ergebnisse. The Man in the Ice, vol. 2. Wien: 59–67.

Endlicher S. (1842) Die Medicinal-Pflanzen der österreichischen Pharmakopöe. Ein Handbuch für Aerzte und Apotheker. Wien, pp 608.

Firbas F. (1949) Waldgeschichte Mitteleuropas. Bd I. Stuttgart, pp 480.

Franchi, G. G. (1997) Researches on pollen digestability. Atti Società Toscana di Scienze Naturale Memorie 94: 43–52.

Gassner G., Hohmann B. und Deutschmann F. (1989) Mikroskopische Untersuchung pflanzlicher Lebensmittel. Stuttgart, pp 414.

Gothe R. and Schöl H. (1992) Hirschlausfliegen (Diptera, Hippoboscidae: *Lipoptena cervi*) in den Beifunden der Leiche vom Hauslabjoch. In: Höpfel F., Platzer W. und Spindler

K. (eds.): Der Mann im Eis, Bd 1. Veröffentlichungen der Universität Innsbruck 187: 299–306.

Gremillon K. J. and Sobolik K. D. (1996) Dietary variability among prehistoric Forager-Farmers of Eastern North America. Current Anthropology 37: 529–539.

Hadorn P. (1994) St. Blaise/Bains des Dames 1. Palynologie d'un site néolithique et histoire de la vegetation des derniers 16 000 ans. Archéologie neuchâteloise 18: pp 121.

Hansson Ann-Marie (1994) Grain paste, porridge and bread. Ancient cereal-based food. Laborativ Arkeologi 7: 5–20.

Hansson Ann-Marie (1996) Bread in Birka and on Björkö. Laborativ Arkeologi 9: 61–78.

Hedberg O. (1946) Pollen Morphology in the Genus Polygonum L. s. lat. and ist Taxonomical Significance. Svensk Botanisk Tidskrift 40: 371–404.

Heer O. (1866) Die Pflanzen der Pfahlbauten. Naturforschende Gesellschaft 68: 1–54.

Hegi G. (1935) Illustrierte Flora von Mitteleuropa. Wien, 6 Bände.

Helbaek H. (1950) Tollund mandens sidste maaltid et botanisk bidrag til belysning af oldtidens kost. Årbog f. Nordisk Oldkyndighed og Historie: 311–341.

Helbaek H. (1958) The Last Meal of Grauballe Man. Kuml: 111–116.

Herrmann B. (1993) Parasitologische Untersuchungen mittelalterlicher Kloaken. In: Hermann B. (ed.): Mensch und Umwelt im Mittelalter. Frankfurt a.M.: 159–169.

Holden T. (1994) Dietary evidence from the intestinal contents of ancient humans with particular reference to desiccated remains from Northern Chile. In: Hather J. G. (ed.): Tropical Archaeobotany. Applications and new developments. London: pp 65–85.

Holden T. (1995) The last meals of the Lindow Bog Men. In: Turner R. C. and Scaife R. C. (eds.): Bog bodies- new discoveries and new perspectives.

Holden T. G. (1986) Preliminary report on the detailed analyses of the macroscopic remains from the Gut of Lindow Man. In: Stead I. M., Bourke J. B. and D. R. Brothwell (eds.): Lindow Man - the Body in the Bog. British Museum Publications: 116–126.

Holden T. G. and Núñez L. (1993) An Analysis of the Gut Contents of Five Well-Preserved Human Bodies from Tarapacā, Northern Chile. Journal of Archaeological Science 23: 595–611.

Hopf M. (1992) South and Southwest Europe. In: Van Zeist, Wasylikowa and Behre (eds.): Progress in Old World Palaeoethnobotany. Rotterdam: 241–247.

Huntley B. and Birks H. J. B. (1983) An atlas of past and present pollen maps for Europe: 0–13000 years ago. Cambridge University Press, pp 667.

Jacomet S. (1987) Prähistorische Getreidefunde. Eine Anleitung zur Bestimmung prähistorischer Gersten- und Weizenfunde. Botanisches Institut der Universität Basel. pp 70.

Jacomet S., Brombacher Ch. und Dick M. (1989) Archäobotanik am Zürichsee. Ackerbau, Sammelwirtschaft und Umwelt von neolithischen und bronzezeitlichen Seeufersiedlungen im Raum Zürich. Züricher Denkmalpflege, Monographien: pp 344.

Jochimsen M. (1986) Zum Problem des Pollenfluges in den Hochalpen. Diss. Botanicae 90: pp 249.

Klaus W. (1975) Über bemerkenswerte morphologische Bestimmungsmerkmale an den Pollenkörnern der Gattung Pinus L. Linzer Biologische Beiträge 3: 329–369.

Knörzer K.-H. (1971) Genutzte Wildpflanzen in vorgeschichtlicher Zeit. Bonner Jahrb. 171: 1–8.

Körber Grohne U. and U. Piening (1980) Microstructure of the Surfaces of Carbonized and Non-Carbonized Grains of Cereals as Observed in Scanning Electron and Light Microscopes as an Additional Aid in Determining Prehistoric Findings. Flora 170: 189–228.

Krauß R. und Huijsmans M. (1996) Die erste Fundstelle der Münchshöfener Kultur in Nordtirol. Ein Vorbericht. Archäologisches Korrespondenzblatt 26: 43–51.

Küster H. (1995) Postglaziale Vegetationsgeschichte Südbayerns. Geobotanische Studien zur prähistorischen Landschaftskunde. Berlin: pp 372.

Lang G. (1994) Quartäre Vegetationsgeschichte Europas. Stuttgart, pp 462.

Linskens H. F. and Jorde W. (1997) Pollen as Food and Medicine – A Review. Economic Botany 5: 78–87.

Maurizio A. (1916) Die Getreide-Nahrung im Wandel der Zeiten. Zürich: pp 237.

Mayr E. (1936) Die Ausbreitung des Getreidebaues, die Anbau- und Erntezeiten und die Fruchtfolge in Nordtirol und Vorarlberg. Veröffentlichung Museum Ferdinandeum Innsbruck 15: 3–27.

Nedden zur D. and Wicke K. (1992) Der Eismann aus der Sicht der radiologischen und computertomographischen Daten. In: Höpfel F., Platzer W. and Spindler K. (eds.): Der Mann im Eis, Bd. 1. Veröffentlichungen der Universität Innsbruck 187: 131–147.

Neuweiler E. (1905) Die prähistorischen Pflanzenreste Mitteleuropas mit besonderer Berücksichtigung der schweizerischen Funde. Vierteljahrschr. Naturf. Ges. Zürich 50: 23–134.

Oeggl K. and Schoch W. (1995) Neolithic plant remains discovered together with a mummified corpse ("Homo tyrolensis") in the Tyrolean Alps. In: Kroll H. and Patsernak R. (eds.): Res archaeobotanicae. Kiel: 229–238.

Pahlow M. (1985) Das große Buch der Heilpflanzen. Gesund durch die Heilkräfte der Natur. München: pp 499.

Pals J. P. and Voorips A. (1979) Seeds, Fruits and Charcoal from two Prehistoric sites in Northern Italy. In: Körber-Grohne U. (ed.): Festschrift Maria Hopf, Köln: 217–235.

Pearsall D. M. (1989) Palaeoethnobotany. A Handbook of Procedures. San Diego: pp 470.

Pedrotti A. (1990) L'abitato neolitico di "La Vela" di Trento. In: Die ersten Bauern, Bd. 2: Pfahlbaufunde Europas. Schweizerisches Landesmuseum Zürich: 210–224.

Peer T. (1981) Die aktuellen Vegetationsverhältnisse Südtirols am Beispiel der Vegetationskarte 1:200.000. Angewandte Pflanzensoziologie 26: 151–168.

Reinhard K. J. and Bryant V. M. (1992) Coprolite Analysis. A Biological Perspective on Archaeology. In: Schiffer M. B. (ed.): Archaeological Method and Theory, Vol. 4. Tucson & London: 245–288.

Richter B. (1987) Mikroskopische Untersuchungen an Speiseresten. In: Suter R. (ed.): Zürich, Kleiner Hafner. Tauchgrabungen 1981–1984. Berichte der Züricher Denkmalpflege, Monographien 3: 180–183.

Schoenwetter J. (1974) Pollen analyses of Human Palaeofeces from Upper Salts Cave. In: Watson P. J. (ed.): Archaeology of the Mammoth Cave. New York: 49–58.

Schoenwetter J. (1996) Palynological evidence for an early woodland ritual meal (in press).

Sjövold T., Berhard W., Gaber O., Künzel K.-H., Platzer W. und Unterdorfer H. (1995) Verteilung und Größe der Tätowierungen am Eismann vom Hauslabjoch. In: Spindler K.,

Rastbichler-Zissernig E., Wilfling H., zur Nedden D., Nothdurfter H. (eds.): Der Mann im Eis-Neue Funde und Ergebnisse. The Man in the Ice, vol. 2. Springer, Wien: 279–286.

Sobolik K. D. (1988) The importance of pollen concentration values from coprolites: an analysis of southwest Texas samples. Palynology 12: 201–214.

Spindler K. (1996) Iceman's last weeks. In: Spindler K., Wilfling H., Rastbichler-Zissernig E., zur Nedden D., Nothdurfter H. (eds.): Human Mummies. The Man in the Ice, vol. 3. Springer, Wien: 252–263.

Stanley, R. G. und Linskens H. F. (1984) Pollen, Biologie, Biochemie, Gewinnung und Verwendung. Greifenberg/Ammersee.

Taylor T. N. and Millay M. A. (1979) Pollination Biology and Reproduction in Early Seed Plants. Review of Palaeobotany and Palynology 27: 329–355.

Vorwohl G. (1972) Das Pollenspektrum von Honigen aus den italienischen Alpen. Apidologie 3: 309–340.

Währen M. (1990) Teig und Feingebäck in der Jungsteinzeit. Neuidentifizierungen. Helv. Archeol. 21: 130–152.

Währen M. (1995) Die Urgeschichte des Brotes und Gebäcks in der Schweiz. Helv. Archaeol. 25: 75–89.

Währen M. (1995) Haushaltskonserven der Jungsteinzeit. Helv. Archaeol. 25: 90–116.

Willerding U. (1996) Fruchtfolge. Reallexikon der Germanischen Altertumskunde 10: 133–138.

Zohary D. and Hopf M. (1993) Domestication of Plants in the Old World. 2nd edn. Oxford.

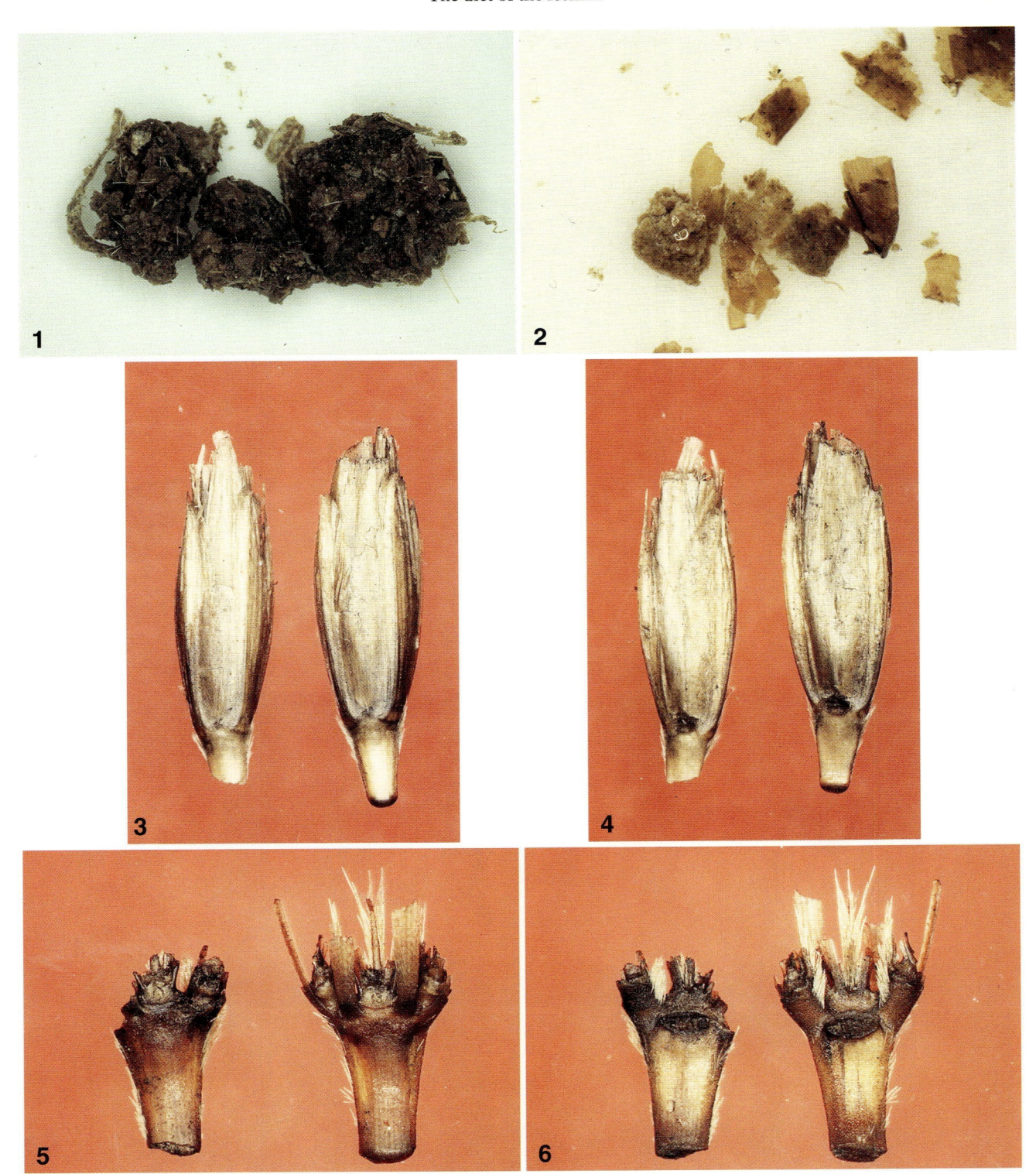

Plate I. Cereal remains found with the Iceman: **1** the food residue (6x), **2** rehydrated food residue with bran fragments (10x), **3** spikelets of Einkorn (*Triticum monoccum*) ventral, and **4** dorsal view (10x), **5,6** naked six-row Barley (*Hordeum vulgare* var. *nudum*) ventral and dorsal view (10x)

110   K. Oeggl

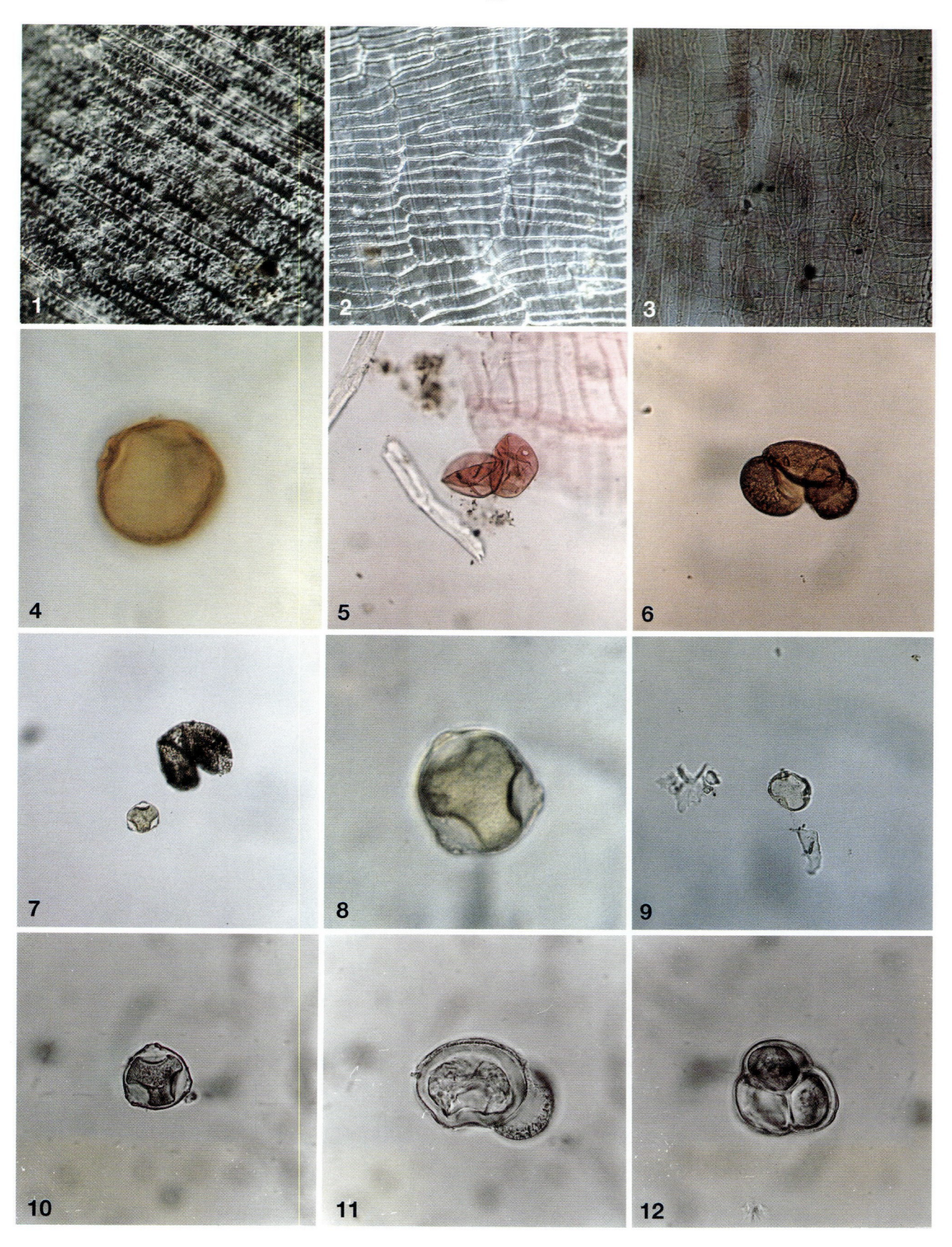

Plate II. Remains found in the food residue from the Iceman's colon (light microscope): **1** glume epidermis of Einkorn (*Triticum monococcum*) 1000x; **2** pericarp of Einkorn (*Triticum monococcum*) transverse cells 1000x, **3** pericarp of Einkorn (*Triticum monococcum*) tube cells 1000x; **4** pollen of the *Triticum*-type 1000x, **5** pollen of the *Triticum*-type 750x, **6** Pine (*Pinus sp.*) pollen 600x, **7** Hop Hornbeam (*Ostrya carpinifolia*) pollen grain with cellular microgametophyte (left) with spruce (*Picea abies*) pollen (right) 400x; **8** Hop Hornbeam (*Ostrya carpinifolia*) pollen grain with cellular microgametophyte 1200x, **9** Lime (*Tilia* sp.) pollen 400x, **10** Hop Hornbeam (*Ostrya carpinifolia*) pollen grain with cellular microgametophyte 750x, **11** Pine (*Pinus diploxylon*-type) pollen with cellular microgametophyte 750x, **12** Whortleberry (*Vaccinium*-type) pollen with cellular microgametophyte 750x

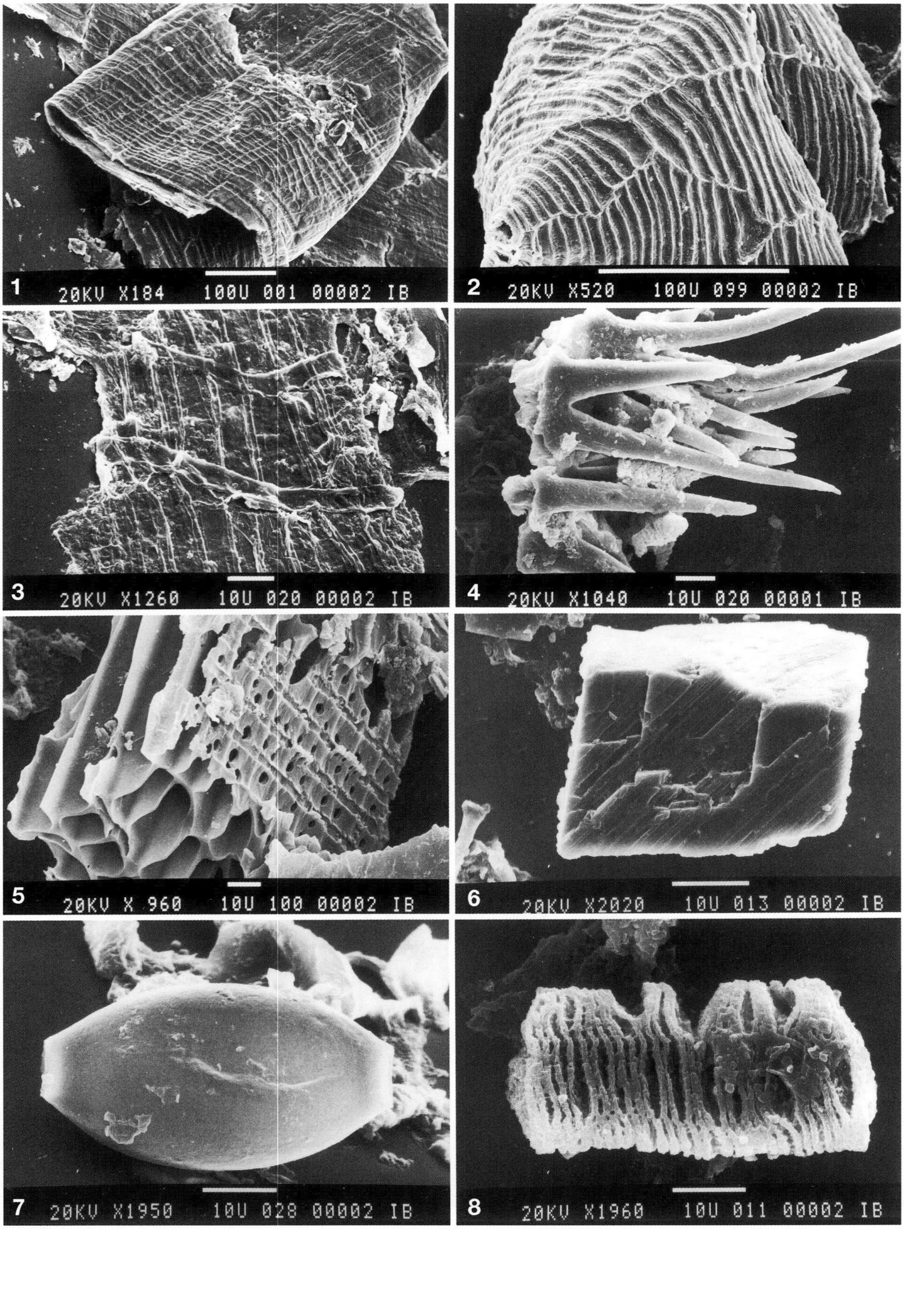
1
20KV X184 100U 001 00002 IB
2
20KV X520 100U 099 00002 IB
3
20KV X1260 10U 020 00002 IB
4
20KV X1040 10U 020 00001 IB
5
20KV X 960 10U 100 00002 IB
6
20KV X2020 10U 013 00002 IB
7
20KV X1950 10U 028 00002 IB
8
20KV X1960 10U 011 00002 IB

Plate III. Remains found in the food residue from the Iceman's colon: **1** pericarp of Einkorn (*Triticum monococcum*) SEM (180x); **2** pericarp of Einkorn (*Triticum monococcum*) transverse cells, SEM (520x); **3** pericarp of Einkorn (*Triticum monococcum*) tube cells, SEM (1260x); **4** pericarp of Einkorn (*Triticum monococcum*) hairs from the grain apex, SEM (1040x); **5** carbon particle of conifer wood, SEM (960x); **6** mineral particle, SEM (2020x); **7** egg of a Human Whipworm (*Trichuris trichiura*) SEM (1950x); **8** muscle fibre, SEM (2000x)

114 K. Oeggl

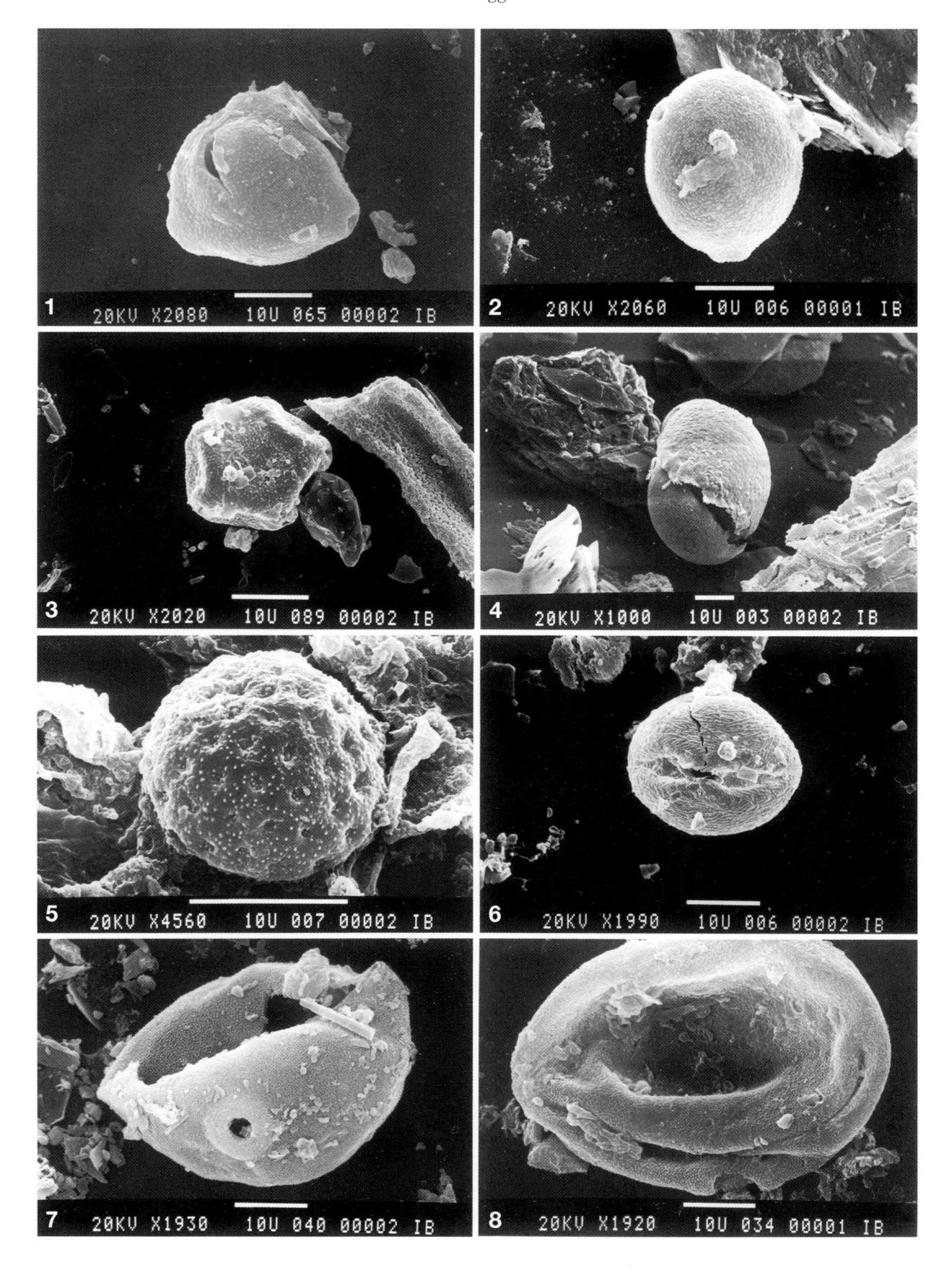

Plate IV. Pollen found in the food residue from the Iceman's colon: **1** Hop Hornbeam (*Ostrya carpinifolia*) SEM (2100x); **2** Hop Hornbeam (*Ostrya carpinifolia*) SEM (2060x); **3** Alder (*Alnus* sp.) SEM (2000x), **4** Pine (*Pinus* sp.) Diploxylon type SEM (1000x); **5** Goosefoot family (Chenopodiaceae) SEM (4560x); **6** Rose family (Rosaceae) SEM (2000x), **7** cereal pollen of wheat (*Triticum*) type, SEM (1930x); **8** cereal pollen of wheat (*Triticum*) type SEM (1920x)

Diatoms from the colon of the Iceman

E. Rott

Institut für Botanik, Universität Innsbruck

1. Introduction

Diatoms are autotrophic organisms of microscopic size. Their persistent silica structures (frustules, valves) are easily dispersed and found almost everywhere, even in the air (Geißler and Gerloff, 1965). The majority of the diatom frustules are persistent over very long time periods under non-acid conditions. Therefore it was not astonishing that with the findings of the Iceman diatoms could be sampled (Oeggl, this volume). Although diatoms have been used for a long time in forensic medicine to identify causes of drowning (e.g., Peabody, 1980; Ludes and Coste, 1996), this is the first time that diatoms are used to identify the diet and living conditions of a prehistoric human being.

From earlier investigations we know that diatoms are incorporated into the human body during life by both inhalation and ingestion (Ludes and Coste, 1996). Assimilated diatoms do not accumulate in the body over time, since the diatom content is independent of age (Geißler and Gerloff, 1965). Furthermore, a very rapid turnover of diatom frustules in the human body is likely, since experiments with dogs supplied with diatom suspensions lead to the first excretion in the urine after one hour (Spitz and Schmidt, 1966). Passive postmortem incorporation was only observed in dead bodies immersed in several metres water depth (Ludes and Coste, 1996). So it seems promising to use the diatom spectrum in the colon of the Iceman to reveal his living conditions during the few hours before his death.

2. Methods

A content sample (40 mg) was taken from the transversal section of the colon by an endoscope (for details see Oeggl, this volume). The subsample for scanning electron microscopy (Leitz AMR 1000) with a total weight of 10 mg was hydrated, acid cleaned and dried at 80 °C on cleaned cover glasses. The sample was sputtered with gold. Due to the small sample size no light microscopical mounts could be made from the colon material. At first the SEM sample was screened for diatom species (Table 1). In a later counting procedure of the same sample the relative portions of the species were worked up to identify the community composition (Table 2). In addition recent diatom samples were taken from eight streams between 800 and 2600 m altitude in the Tisental and the Schnalstal close (maximum 10 km distance) to the find spot. The samples were scraped from stones with a knife, acid cleaned and mounted in Naphrax. Two replicates were made and used for screening of the species spectrum at 400- and 1000-times magnification under an Olympus BX 50 microscope. For taxonomic identification Krammer and Lange-Bertalot (1986–1991), Lange-Bertalot (1993), and Reichardt and Lange-Bertalot (1991) were used.

3. Results

From the colon content sample of the Iceman altogether 24 taxa of diatoms were identified (Table 1), 13 of which were redetected within the later counting process (Table 2). Together with complete diatom frustules many fragments were found. The majority of the specimens observed are small (between 10 and 15 µm length). The longest dimensions of the largest taxa *Fragilaria arcus* (Fig. 16), *Diatoma ehrenbergii* (Fig. 12), *Navicula radiosa* (Fig. 28) and *Gomphonema olivaceum* (Figs. 22, 23) were less than 50 µm.

Almost all common taxa are known to occur mainly in freshwater habitats. This is especially the case for the species of the four most common genera *Achnanthes*, *Diatoma*, *Fragilaria* and *Gomphonema*. Just one species, *Hantzschia subrupestris*, with a taxonomic position close to *H. amphioxys*, is most likely an aerophyte (species preferring subterrestrial habitats, such as moist soil and wet places).

The general preference of freshwater habitats can also be seen from the moisture classification according to van Dam et al. (1994) in Table 1. The classification of the species according to pH places the majority of the species into the circumneutral and alkaliphilous range.

Table 1. List of all diatom species found in the SEM preparation of the colon content sample of the Iceman and some ecological preferences of the taxa according to van Dam et al. (1994) (R = pH: 3 = circumneutral, 4 = alkaliphilous, 5 = alkalibiontic; T = Trophic state: 1 = oligotraphentic, 2 = oligo-mesotraphentic, 3 = mesotraphentic, 4 = meso-eutraphentic, 5 = eutraphentic, 7 = oligo- to eutraphentic (hypereutraphentic); M = Moisture: 1 = never, or only very rarely, occurring outside water bodies, 2 = mainly occurring in water bodies, sometimes on wet places, 3 = mainly occurring in water bodies, also rather regularly on wet and moist places)

Diatom species	R	T	M
Achnanthes biasolettiana Grunow var. *subatomus* Lange-Bertalot (Figs. 4, 5, 7)	4	3	–
Achnanthes minutissima Grunow	3	7	3
Type 1 "lineare" (Figs. 1, 2)			
Type 2 "with bent raphe endings" (Figs. 3, 6)			
Type 3 "microcephalum" (Fig. 8)			
Achnanthes sp. (Psammothidium)			
Caloneis sp. ad *C. bacillum* (Grunow) Cleve	4	4	2
Cocconeis placentula Ehrenberg var. *euglypta* (Ehrenberg) Grunow (Fig. 9)	4	5	2
Cyclotella sp.			
Cymbella affinis Kützing (Fig. 10)	4	5	2
Cymbella sp. (Fig. 11)			
Diatoma ehrenbergii Kützing (Fig. 12)	5	4	1
Diatoma mesodon (Ehrenberg) Kützing (Figs. 13, 14)	3	3	2
Fragilaria arcus (Ehrenberg) Cleve (Figs. 16, 17)	4	2	3
Fragilaria capucina Desmazières sensu stricto	3	3	–
Fragilaria virescens Ralfs (Figs. 15, 18)	3	2	3
Gomphonema angustum Agardh (Figs. 19, 20)	4	1	–
Gomphonema olivaceum Hornemann (Brébisson) sensu stricto (Figs. 22, 23)	5	5	1
Gomphonema olivaceum Hornemann (Brébisson) var. *minutissimum* Hustedt (Figs. 21, 24)	3	2	3
Gomphonema procerum Reichardt & Lange-Bertalot (Fig. 25)	–	–	–
Gomphonema pumilum (Grunow) Reichardt & Lange-Bertalot (Figs. 26, 27)	–	7	–
Hantzschia subrupestris Lange-Bertalot (Fig. 29)	–	–	–
Navicula radiosa Kützing (Fig. 28)	3	4	3
Nitzschia sp. (*paleacea-group*) (Fig. 30)	–	–	–

Table 2. Diatom counts for the colon content sample of the Iceman based on the SEM preparation made to identify community composition

Achnanthes minutissima sensu lato	62
Gomphonema olivaceum sensu stricto	7
Gomphonema pumilum	5
Achnanthes biasolettiana	4
Diatoma ehrenbergii	4
Gomphonema olivaceum var. *minutissimum*	4
Fragilaria arcus	2
Fragilaria virescens	2
Cyclotella sp.	1
Cymbella sp.	1
Diatoma mesodon	1
Hantzschia subrupestris	1
Nitzschia sp.	1
Total	95

For the classification of the taxa in relation to trophic status the situation is not so clear: when all taxa are considered, the range is scattered between oligotrophic and eutrophic conditions. However the classification system by van Dam et al. (1994) is made for data from the Netherlands and needs some modification for the application in alpine areas.

For many of the taxa found autecological information from the Alps is available. *Achnanthes biasolettiana* (Figs. 4, 5), for example, is known to prefer pristine mountain streams (e.g., Pipp and Rott, 1993), whereas *Achnanthes minutissima*, a broad and taxonomically heterogeneous taxon with many ecomorphs, is found in many environments at all altitudes. However, even the latter is known to dominate in many glacial high mountain streams (Kawecka, 1974, 1981; Pfister, pers. comm., own unpubl. observ.) and subalpine springs in the Alps (Cantonati, 1997). The genus *Diatoma* comprises two species of divergent ecology: *Diatoma ehrenbergii* (Fig. 12), a common representative of nutrient rich mountain streams and rivers from lower altitudes (normally below 1500 m), and *Diatoma mesodon* (Figs. 13, 14), a widespread form of springs and streams, frequently found in the high mountain areas. The genus *Fragilaria* comprises two species of interest: *Fragilaria arcus* (Figs. 16, 17), a common representative of fast flowing mountain streams, and *Fragilaria virescens* (Figs. 15, 18), a species preferring spring habitats. The majority of the species within the genus *Gomphonema* (*G. angustum*, Figs. 19, 20, *G. olivaceum* var. *minutissimum*, Figs. 21, 24, *G. procerum*, Fig. 25 and *G. pumilum*, Figs. 26, 27) are common in pristine streams and springs, whereas *Gomphonema olivaceum s.s. (Figs. 22, 23) is generally missing at higher altitudes (above 1500 m, rarely 1800 m) and prefers enriched habitats.

The comparison of the SEM results with the light microscopy of the recent diatom samples from August 1996 showed a general coincidence, since almost all recent samples contained several common taxa found in the SEM, mainly *Achnanthes minutissima*, *A. biasolettiana*, *Diatoma mesodon* and *Fragilaria arcus*. However, the common species of the genus *Cymbella* (around *C. minuta*, *C. silesiaca*) found in the recent samples were missing from the SEM (colon content) preparations, whereas no specimen of the two taxa *Diatoma ehrenbergii* and *Gomphonema olivaceum* well represented in the colon sample was found in the recent light microscopical mounts.

4. Discussion

The diatom spectrum found in the colon of the Iceman indicates that the Iceman had actively ingested the algae from mountain streams or springs. Post-mortem incorporation of diatoms can be excluded, since all species except one are incapable of living on glacial ice or soil near the finding place of the Iceman. It is likely that the Iceman had incorporated diatoms from both low altitude and high altitude places, since in his colon species originating from pristine high mountain habitats and from nutrient richer low altitude habitats were found. This observation parallels the findings for the pollen flora made by Oeggl (this volume). In addition to the probable incorporation of diatoms by drinking water from streams and/or springs, diatoms could also have been incorporated from a meal (bread ?) prepared with stream/spring water. It is likely that diatom frustules are not destroyed by heating over a fire in a bread paste. The low number of diatom specimens found in the SEM material do not allow conclusions concerning the season of the Iceman's death.

Abstract

From an endoscopic sample of the colon of the Iceman 24 diatom taxa could be identified under the SEM. The majority of the species belong to the genera *Achnanthes*, *Diatoma*, *Fragilaria* and *Gomphonema* and are known to be common in pristine mountain streams and springs. Two species found in the quantitative evaluation (*Diatoma ehrenbergii*, *Gomphonema olivaceum*) are common in lower altitude and nutrient richer streams and only one species (*Hantzschia subrupestris*) is occurring mainly in terrestrial habitats. The results indicate that the Iceman, during the last day of his life, had incorporated diatoms by drinking water originating from different places situated at lower (below 1500 m) and higher altitudes and/or by using water for the preparation of his meal.

Zusammenfassung

Aus einer endoskopischen Probe des Darms des Eismannes konnten 24 Kieselalgentaxa im SEM identifiziert werden. Die Mehrzahl der gefundenen Arten (Gattungen *Achnanthes*, *Diatoma*, *Fragilaria* und *Gomphonema*) ist in reinen Bergbächen und Quellen oft sehr hoher Lagen weit verbreitet. Zwei Arten, die auch quantitativ aufschienen (*Diatoma ehrenbergii*, *Gomphonema olivaceum*), haben ihren Verbreitungsschwerpunkt in nährstoffreicheren Bergbächen tieferer Höhenlagen (unterhalb 1500 m), und lediglich eine Art (*Hantzschia subrupestris*) ist für subterrestrische Lebensräume typisch. Dieser Befund der Kieselalgen deutet darauf hin, daß der Eismann am letzten Lebenstag Kieselalgen aus Wasser unterschiedlicher Höhenlagen (sowohl oberhalb als unterhalb von 1500 m) inkorporiert hatte, entweder durch das Trinken des Wassers und/oder durch die Verwendung von Bachwasser zur Herstellung seiner Mahlzeit.

Résumé

24 éspèces de diatomées pouvaient être identifiées à base d'un petit échantillon obtenu par endoscope du colon du "Iceman". La plupart des éspèces trouvées sont bien connues des ruisseaux et sources propres de la région de montagne et de haute montagne. Seulement trois éspèces sont mieux connues d'autres environnements. Une est une espèce subaerienne (*Hantzschia subrupestris*), les deux autres (importants dans le comptage) sont trouvées plus fréquemment dans les rivières de basse altitude (*Diatoma ehrenbergii*, *Gomphonema olivaceum*). Les résultats indiquent que la provenance des diatomées vient de différents endroits, voire que le "Iceman" avait incorporé les diatomées soit en buvant de l'eau, soit en utilisant de l'eau pour préparer son repas sur différents sites, situés en haute montagne et au dessous de 1500 mètres d'altitude.

Riassunto

Da un campione endoscopico dall'intestino dell'uomo del ghiacciaio si sono identificate nel SEM 24 taxa di alghe silicee. La maggioranza delle alghe riscontrate (*Achnantes*, *Diatoma*, *Fragilaria*, *Gomphonema*) è molto diffusa in limpidi torrenti e sorgenti di alta quota. Due tipi altrettanto presenti (*Diatoma Ehrenbergii*, *Gomphonea olivaceum*) sono diffuse prevalentemente in torrenti dalle acque più ricche di sostanze nutritive a quote meno elevate (fino a 1.500 m.s.l.m.) e soltanto una specie (*Hantschia subrupestris*) è tipica di un habitat subterrestre. Il presente ritrovamento di alghe silicee fa supporre che l'uomo del Similaun abbia assunto, nell'ultimo giorno della sua vita, acqua attinta a diversi livelli (al di sotto e al di sopra di quota 1.500 m) sia avendola bevuta sia avendola utilizzata per la preparazione di un pasto.

Acknowledgements

First I thank Doz. K. Oeggl for the idea and the initiative to realize this unusual work with algae. The help of Mr. S. Tatzreiter is kindly acknowledged in making the preparations and taking the photographs of the samples and assisting with the counting.

References

Cantonati M. (ed.) (1997) Le sorgenti del Parco Adamello-Brenta. II: Popolamenti vegetali e animali. Parco Documenti, Strembo (Trento), Italy, manuscript, pp 148.

van Dam H., Mertens A. and Sinkeldam J. (1994) A coded checklist and ecological indicator values of freshwater diatoms from the Netherlands. Neth. J. Aquat. Ecol. 28: 117–133.

Geißler U. and Gerloff J. (1965) Das Vorkommen von Diatomeen in menschlichen Organen und in der Luft. Nova Hedwigia 10: 565–577.

Kawecka B. (1974) Effects of organic pollution on the development of diatom communities in the alpine streams Finstertaler Bach and Gurgler Ache (Northern Tyrol, Austria). Ber. nat. -med. Ver. Innsbruck 61: 71–82.

Kawecka B. (1981) The ecological characteristics of diatom communities in the mountain streams of Europe. In: Ross R. (ed.): Proceedings of the 6th Diatom-Symposium 1980, Koeltz, Königstein: 425–434.

Krammer K. and Lange-Bertalot H. (1986–1991) Bacillariophyceae. 2/1: Naviculaceae, 876pp.; 2/2: Bacillariaceae, Epithemiaceae, Surirellaceae, 596pp.; 2/3 Centrales, Fragilariaceae, Eunotiaceae, 576pp.; 2/4: Achnanthaceae, 437pp. In: Ettl H., Gerloff J., Heynig H. und Mollenhauer D. (eds.): Süßwasserflora von Mitteleuropa, Fischer, Jena.

Lange-Bertalot H. (1993) 85 neue Taxa. Bibliotheca Diatomologica 27: pp 454.

Ludes B. and Coste M. (1996) Diatomées et médecine légale. Lavoisier Technique & Documentation, Paris, pp 258.

Oeggl K. The diet of the Iceman (this volume).

Peabody A. J. (1980) Diatoms and drowning - A review. Med. Sci. Law 20: 254–261.

Pipp E. and Rott E. (1993) Bestimmung der ökologischen Wertigkeit von Fließgewässern in Österreich nach dem Algenaufwuchs. Blaue Reihe des Bundesministeriums für Umwelt, Jugend und Familie 2: pp 147.

Reichardt E. and Lange-Bertalot H. (1991) Taxonomische Revision des Artenkomplexes um *Gomphonema angustum – G. dichotomum – G. intricatum – G. vibrio* und ähnliche Taxa (Bacillariophyceae). Nova Hedwigia 53: 519–544.

Spitz W. U. and Schmidt H. (1966) Weitere Untersuchungen zur Diagnostik des Ertrinkungstodes durch Diatomeen-Nachweis. Dtsch. Z. Ges. Gericht. Med. 58: 195–204.

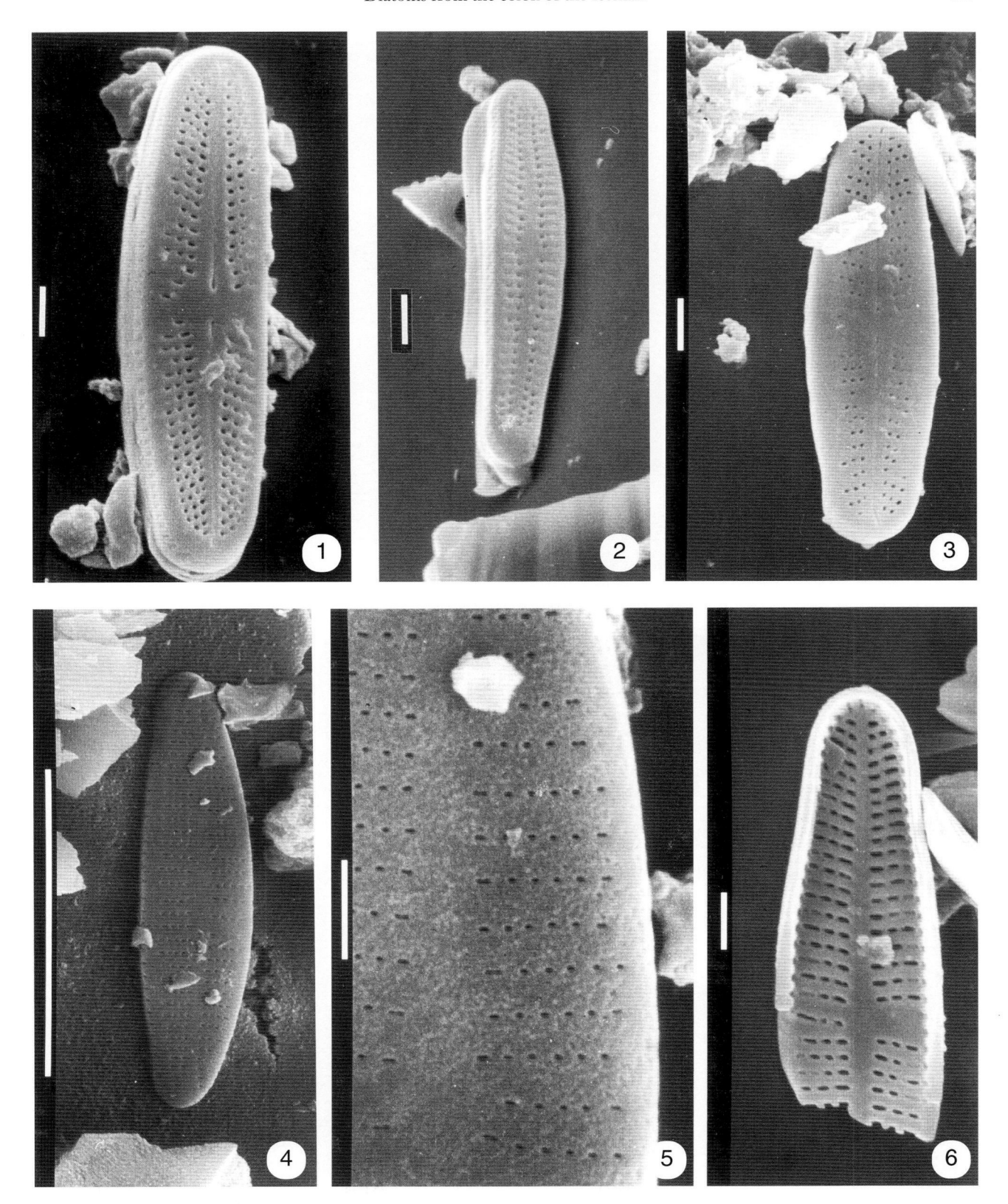

Figs. 1–6. SEM-micrographs of diatoms from the colon of the Iceman: **1** *Achnanthes minutissima*, morphotype "lineare" R – valve, **2** as in 1 RL valve, **3** *Achnanthes minutissima*, unknown morphotype with bent terminal raphe ends, **4, 5** *Achnanthes biasolettiana* RL – valve, **6** *Achnanthes minutissima* morphotype as in 3, RL valve (white bar equals 10 μm in 4, in all other cases 1 μm)

Figs. 7–12. SEM-micrographs of diatoms from the colon of the Iceman: **7** *Achnanthes* sp. (*biasolettiana* ?), **8** *Achnanthes minutissima* morphotype "microcephalum" R – valve, inside, **9** *Cocconeis placentula* var. *euglypta*, RL – valve, **10** *Cymbella affinis* inside view, **11** *Cymbella* sp. (Encyonema ?), **12** *Diatoma ehrenbergii* inside view (all white bars equal 10 m)

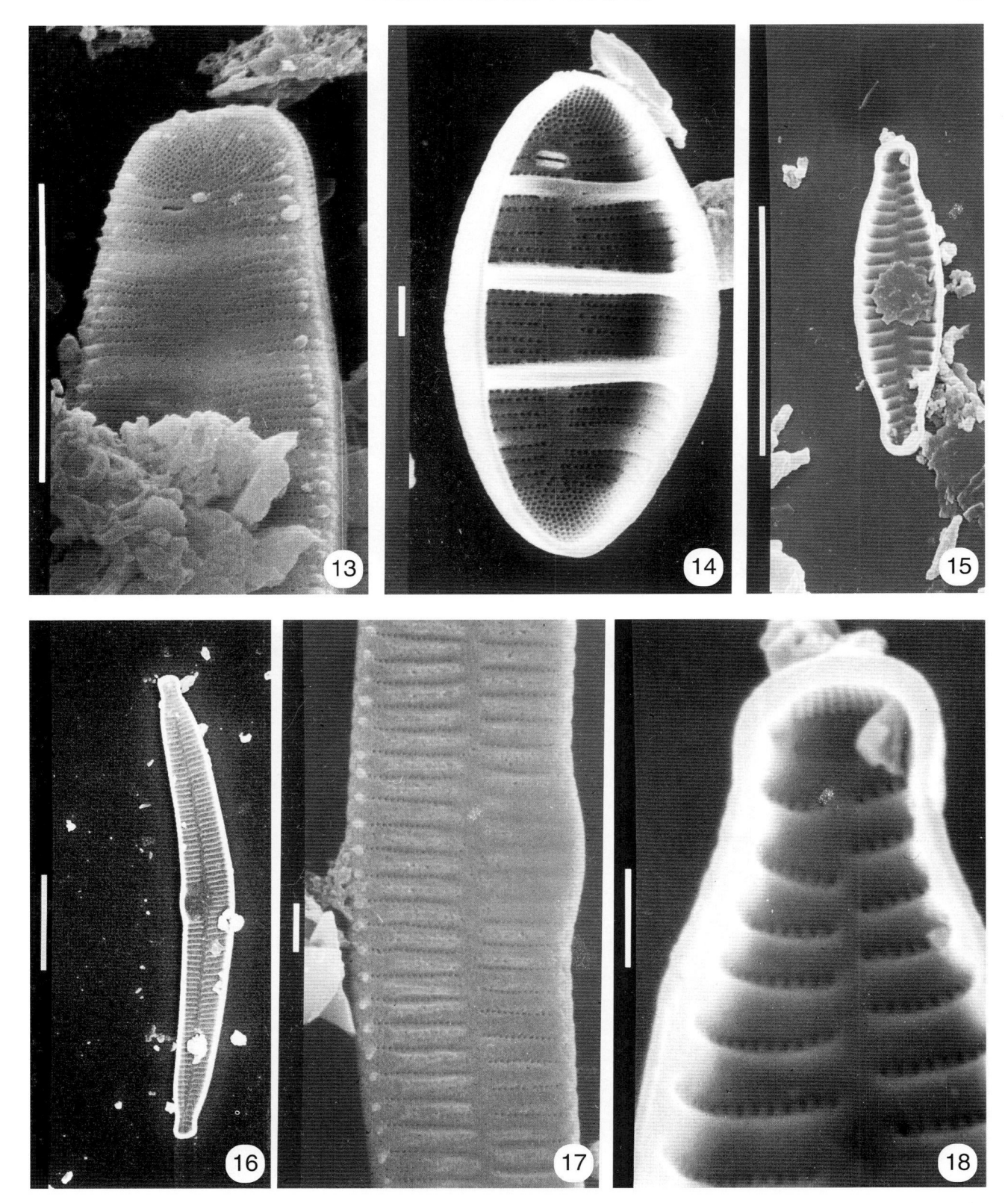

Figs. 13–18. SEM-micrographs of diatoms from the colon of the Iceman: **13, 14** *Diatoma mesodon*, **15, 18** *Fragilaria virescens* inside view, **16, 17** *Fragilaria arcus* (white bars in 14, 17, 18 equal 1 µm, in all other cases 10 µm)

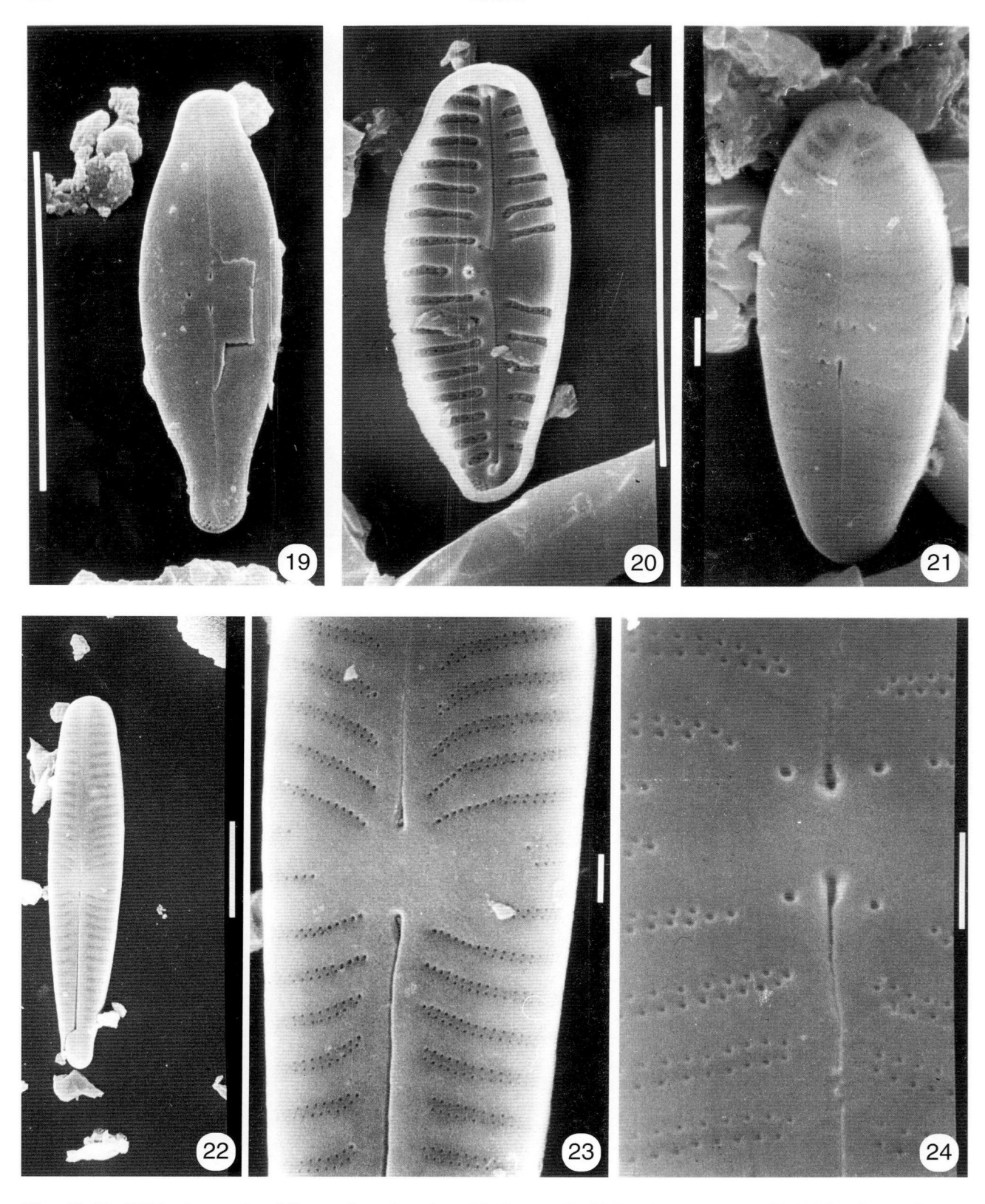

Figs. 19–24. SEM-micrographs of diatoms from the colon of the Iceman: **19, 20** *Gomphonema angustum*, **21, 24** *Gomphonema olivaceum* var. *minutissimum*, **22, 23** *Gomphonema olivaceum* sensu stricto (white bars equal 1 μm in 21, 23, 24 and 10 μm in all other cases)

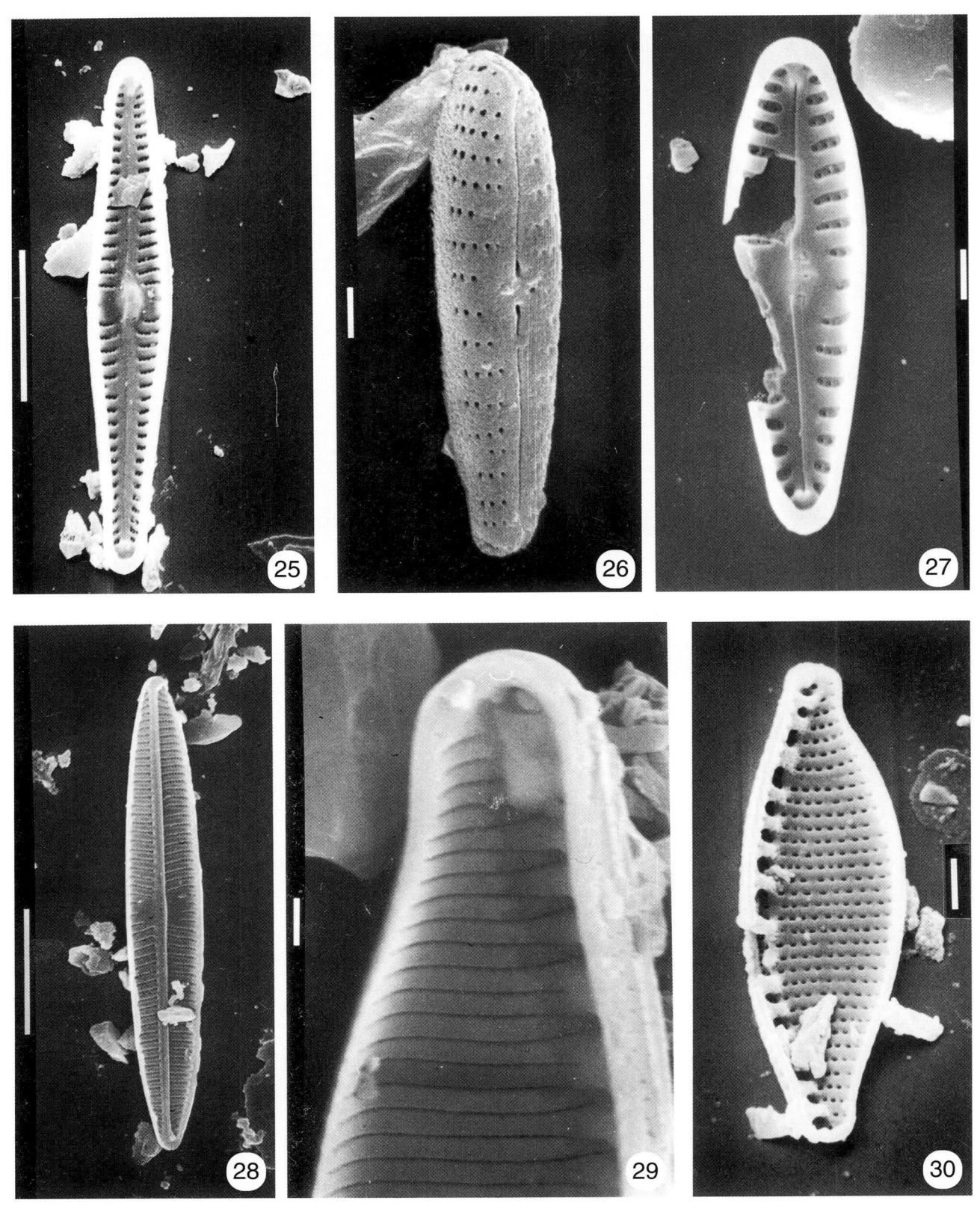

Figs. 25–30. SEM-micrographs of diatoms from the colon of the Iceman: **25** *Gomphonema procerum*, **26, 27** *Gomphonema pumilum*, **28** *Navicula radiosa*, **29** *Hantzschia subrupestris*, **30** *Nitzschia* sp. (paleacea group) (white bars equal 10 μm in 25 and 28, in all other cases 1 μm)

Parasitological examination of the Iceman

H. Aspöck[1], H. Auer[1], O. Picher[1], and W. Platzer[2]

[1] Abteilung Medizinische Parasitologie, Klinisches Institut für Hygiene, Universität Wien
[2] Institut für Anatomie, Universität Innsbruck

1. Introduction

A few weeks after the glacier mummy had been recovered a detailed program for a parasitological examination of the Iceman was established (Aspöck and Auer, 1992; Aspöck et al., 1995). Since October 1991 the body has been investigated with respect to parasites on several occasions, and materials for parasitological examinations have been obtained by necropsies. Furthermore, we have had the opportunity to examine the cap of the Iceman. The most important results have been briefly reported in previous publications (Aspöck and Auer, 1992; Aspöck et al., 1995, 1996). The present paper gives a summarizing overview of the parasitological examinations of the Iceman so far carried out and tries to outline which further investigations should and could be done.

2. Material and Methods

2.1. External inspection for ectoparasites

On March 10th, 1992 two of us (H.A. and H.A.) examined the whole surface of the body for ectoparasites and endoparasites of the skin using magnifying glasses. Moreover, in July 1993 the Iceman's cap – which had been found a year after the body had been recovered and which had been kept frozen in Bolzano (Italy) – was brought to Innsbruck and carefully thawed out overnight. On July 15th, 1993 we spent several hours inspecting the cap (in particular for lice) using a stereomicroscope.

2.2. Examination for endoparasites

The following samples were obtained for parasitological tests:

2.2.1. During the recovery of the mummy the sacral-gluteal region including probably parts of the rectum had been considerably damaged and afterwards heavily covered with dirt. A sample of this material – probably a mixture from soil, dirt and parts of the body, presumably also contents from the intestine – was taken and inspected for endoparasitic helminths and protozoa (see 2.3.1.) on April 16th, 1993 and during the following weeks.

2.2.2. In 1994, after development of special titanium endoscopes, samples from the colon were taken for parasitological examinations; these were processed as described under 2.3.1. and 2.3.2.

2.2.3. In 1997, another sample from the bowels (probably from the small intestine) was taken using a titanium endoscope and processed as sample 2.2.2.

2.2.4. A small sample of lung tissue was obtained in 1994. It was transferred from Innsbruck to Vienna in dry ice and then frozen in liquid nitrogen. So far it has not been examined for anything. The decision about which are the most appropriate PCRs and serological tests will be taken as soon as all samples of other tissues have been obtained. Only then can a careful selection be made. With respect to the tiny samples available this choice is of the utmost importance.

2.3. Procedures of samples

2.3.1. Tests for microscopically visible stages of parasites (eggs of helminths, protozoa)

The sample described under 2.2.1. was immediately put into SAF (= sodium acetate-acetic acid-formalin fixative) solution (Junod, 1972; Marti and Escher, 1990) and processed in the following way:

The vial with the sample in SAF was carefully and slightly shaken and then allowed to stand before microscopy for about 30 minutes. Slides were searched at 100x magnification, each for about 15 minutes and (for any clarification of suspicious structures) at 400x, exceptionally also at 1000x oil immersion. In this way 50 slides (which are about two thirds of the whole sample) were examined. Particular searches were made for eggs of helminths on one hand and for cysts of protozoa on the other. Photographs were taken at both magnifications.

Moreover, an attempt was made to stain the material according to a modified Heidenhain's iron hematoxylin staining. For this purpose a small sample of the size of a pin head was carefully rubbed into some drops of Ringer's solution and smeared on a fat-free slide, afterwards fixed in sublimate alcohol and then processed as usual (Mehlhorn et al., 1995).

The sample described under 2.2.2. was kept in Innsbruck at $-6\,°C$ for 7 weeks, then transferred to Vienna in dry ice and thereafter divided into three parts: One part was immediately put into SAF and processed as described above; about one third was rehydrated with 0,3% trisodium phosphate (Van Cleave and Ross, 1947) and the third part was (and is still) frozen at $-20\,°C$. The material rehydrated with trisodium phosphate was allowed to sediment for about 30 hours without centrifugation. Small amounts of the sediments were then transferred to slides and examined microscopically at 100x and 400x. Further slides were prepared after about 50 hours, 75 hours and later. Thereafter the sediment was transferred to SAF. About 20 slides (about 10 of each kind of sediment) were examined.

The sample described under 2.2.3. was kept at $-6\,°C$ for almost 5 months and then divided into three parts: About one third was put into SAF solution and processed as described above; one part was rehydrated in 0.3% trisodium phosphate; the rest was frozen at $-20\,°C$ for polymerase chain reactions possibly to be carried out later.

2.3.2. Tests for copro-antigens

Samples 2.2.2. and 2.2.3 were tested for antigens of *Giardia* as well as of *Cryptosporidium* using commercial ELISA testkits (Alexon inc., USA). A part of the SAF solution containing the sample was intensively mixed. 200 µl of this suspension were transferred into the wells coated with monoclonal antibodies against *Giardia lamblia* or *Cryptosporidium parvum*, respectively. After incubation and washing, the conjugate – monoclonal antibodies against *Giardia* and *Cryptosporidium*, respectively with peroxidase – was added, and after another incubation and further washing the substrate was added. Change and intensity of colour was measured at 450 nm.

3. Results

3.1. Ectoparasites

No ectoparasites – in particular no ticks, mites, lice or fly maggots – were found on the mummy neither on the body itself nor on the cap. Moreover, no traces of scabies were found on the skin of the man. See, however, Discussion.

3.2. Endoparasites

When we examined the superficial matter taken from the damaged sacral-gluteal region (2.2.1.), we readily found eggs of *Trichuris* in quite high numbers. Due to the contamination of the body with soil, water, polluted ice and simply dirt possibly containing faeces of various mammals and thus perhaps also eggs of species of the genus *Trichuris* which cannot be differentiated with certainty from each other and, in particular, from *Trichuris trichiura*, we did not dare to identify these eggs as those of the human whipworm (Aspöck et al., 1995). The examination of the sample 2.2.2. confirmed, however, our suspicion entirely. We found many splendidly preserved eggs of *Trichuris trichiura* (Figs. 1–6).

Although we spent altogether more than 200 hours inspecting slides with trisodium phosphate rehydrated material and SAF sediments, we did not find any other parasite. A few objects were found which were reminiscent of helminthic eggs at the first glance (Figs. 7–8). We were, however, unable to obtain any convincing evidence of any helminth species, besides *Trichuris trichiura*.

It is particularly worth saying that no cysts of any protozoa were found in the slides with suspended sediments. A number of suspicious objects was found, but none was confirmed to represent a protozoan. Moreover, no protozoa could be detected in Heidenhain's stained smears. We must, however, admit that the method did not work well as the specimens were too small and also too crumbly to yield "stool smears" of a reliable quality.

It is worth mentioning that we did not find any parasites in the sample 2.2.3.

Another observation merits mention: The structure of the rehydrated samples differed not very much from those in SAF, and in particular we could not find any convincing differences in the appearance of the *Trichuris* eggs. For the search for protozoa the trisodium phosphate method should, of course, be given preference.

All tests for copro-antigens of *Giardia lamblia* and of *Cryptosporidium parvum*, respectively with both samples (2.2.2. and 2.2.3.) were negative.

4. Discussion

The discovery of a well preserved mummy of an age of at least 5200 years in an Alpine glacier was and is, of course, a challenge for the parasitologists.

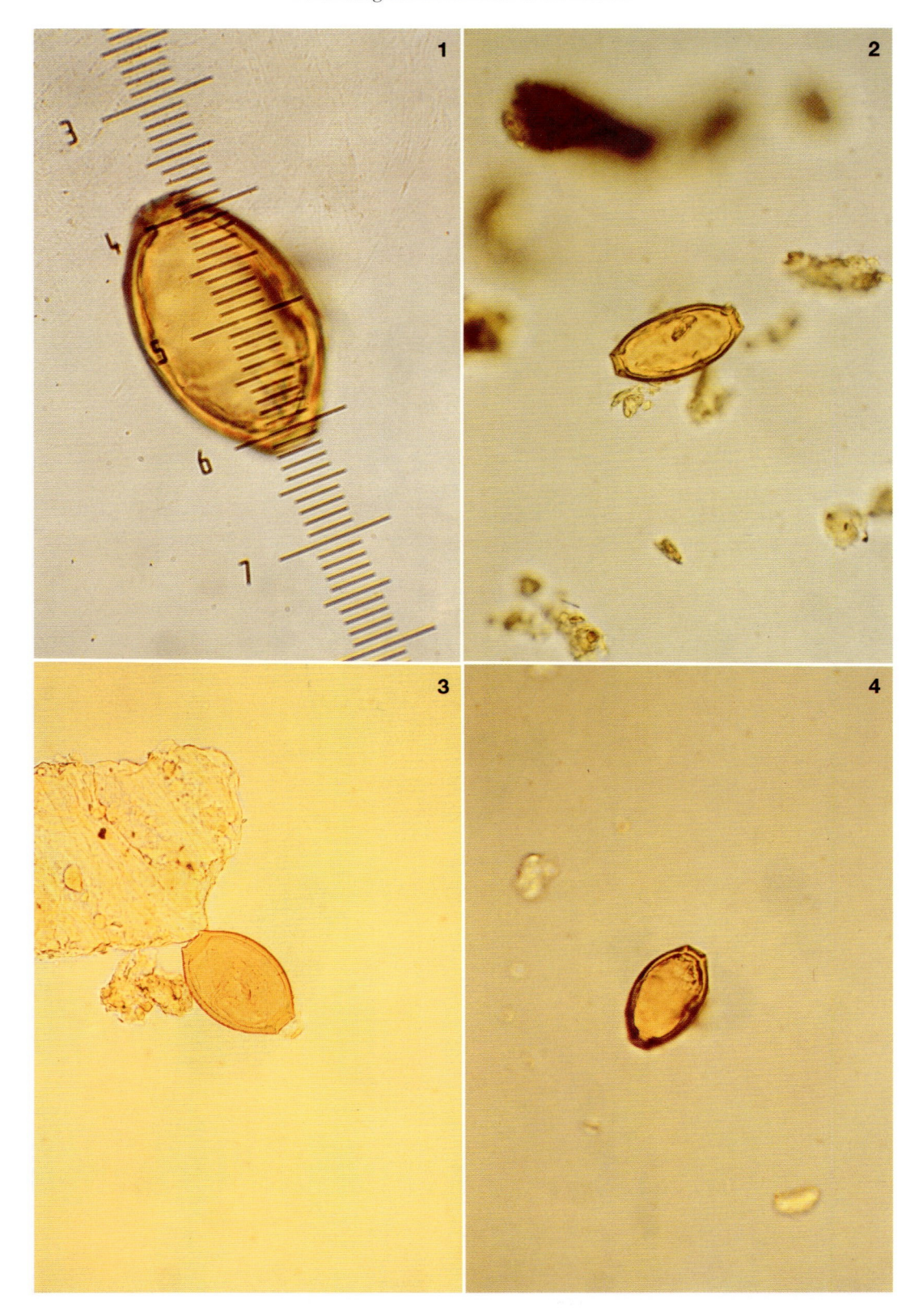

Figs. 1–4. Eggs of *Trichuris trichiura* in various stages of preservation from the colon of the Iceman (SAF-solution) (1 graduation mark=2,5 µm)

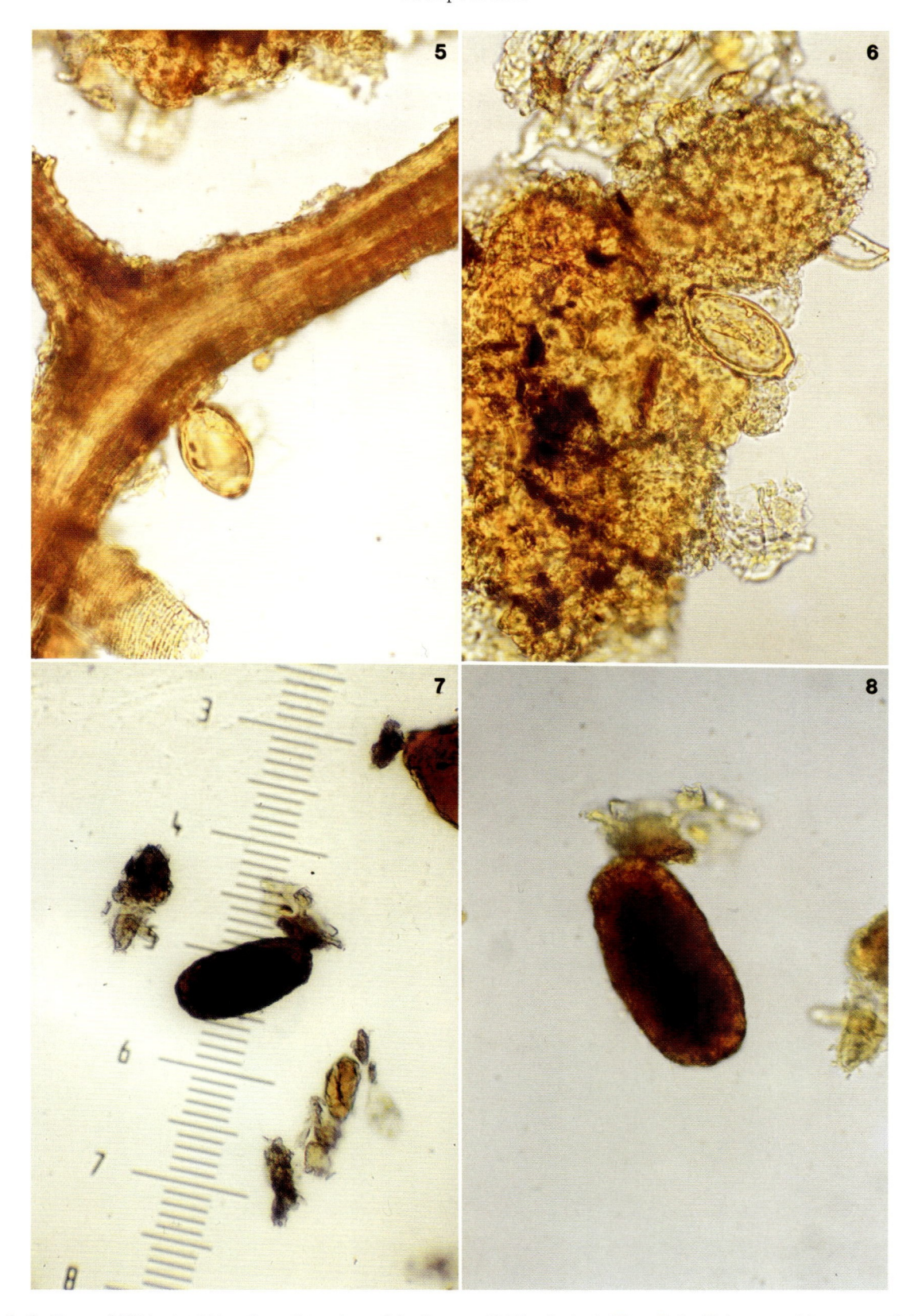

Figs. 5–6. Eggs of *Trichuris trichiura* from the colon of the Iceman (SAF-solution). **Figs. 7–8.** Unknown objects superficially resembling helminthic eggs from the colon of the Iceman (SAF-solution) (1 graduation mark=9,8 µm)

It is no question that the humans of the Neolithic Period in Europe were afflicted by a considerable number of parasites and one can hardly doubt that the spectrum of the species involved was very similar to that which we know today. Nevertheless, the provenance of many human parasites which are distributed throughout large parts of the world or which even occur in any human population in the world is still (and will possibly remain) unknown. Thus, the detection of an agent in a Neolithic human body gives at least evidence of the presence of a certain parasite in a certain geographic area at a certain time.

We did not find any ectoparasites. According to the original hypothesis (Spindler, 1996) the man died late in the year ("shortly before the start of the winter"), "... his body was completely covered during the night of his death ..." and "... it must have remained covered by a layer of light snow, which being in the area of permanent frost, made the dry mummification of the body possible ...". Thus, due to the sudden low temperature, any stationary ectoparasites (in particular ticks or lice) would probably have become inactive and would have remained on the body.

Basing upon histological and chemical investigations of the Iceman's skin, Bereuter et al. (1996) came, however, to the conclusion "... that the body was lying in water for at least several months before the desiccation process started" which led to a loss of the epidermis. If this assumption can be corroborated, one must also conclude that any ectoparasites would have been lost together with the outer layers of the skin. This has to be considered, when discussing the question of ectoparasites of the Iceman.

At any rate, it is of particular interest that the man was apparently not infected with lice. Despite the new findings on the loss of the epidermis and although the mummy is entirely hairless – which makes the detection of lice (*Pediculus* or *Phthirus*) on the body itself unlikely, if not impossible – this statement can be sufficiently corroborated. Many hairs that belonged to the man have been found around the mummy. The Similaun man must have had dense, long, dark hair that fell out shortly after his death. These hairs have been carefully investigated (Wittig and Wortmann, 1992), and no trace of adult lice or nits could be found. Of particular importance for the clarification of the question whether the man had been infested by head lice (*Pediculus humanus capitis*) was the examination of the excellently preserved cap (description and figures see: Goedecker-Ciolek, 1993; Groenman-van Waateringe, 1993). We spent a lot of time investigating this cap made from bear hide. If the man had had head lice, we would certainly have found any remains of them.

It is also worth mentioning that no fly maggots have been found on the corpse. We have not considered myiasis *intra vitam*, of course, that would have been rather unlikely. It is, however, well-known that a dead body attracts a number of Diptera (also at high altitudes in the Alps) within a few hours and that these flies deposit their eggs (or, in certain species, neonate larvae) on the body thus initiating a process which immediately leads to spectacular damages. The fact that the mummy is absolutely free of maggots and does not show any traces of a (pre- or postmortal) myiasis is another indication that the man must either have died late in the year so that coldness and most probably a quickly fallen cover of snow prevented the body from being infested by maggots – or that any other event must have prevented flies from being attracted by the Iceman's body. Even in case of submergence in water and loss of epidermis of the Iceman traces of any myiasis would have remained detectable.

The highlight of our examinations was the detection of eggs of *Trichuris trichiura* in the intestine. To the best of our knowledge this is by far the oldest evidence of whipworm in a human corpse. There are, however, records of *Trichuris trichiura* eggs in human coprolites of almost the same age in France (Bouchet et al., 1995).

The oldest findings of *Trichuris trichiura* in the New World are from coprolites and from a mummy from Brazil about 1600 (1540 ± 120) BC (Ferreira et al., 1980, 1983). Eggs of *Trichuris trichiura* have also been found in human excrements from salt miners of the Hallstatt period (Younger Iron Age), about 800–350 BC. These faeces were found embedded in collapsed prehistoric saltmines in two localities, Hallstatt and Hallein, in the Austrian Alps (Aspöck et al., 1973, 1974). Of about the same age are findings of whipworm eggs in a bog mummy in former Eastern Prussia, about 600 BC (Szidat, 1944). Wei et al. (1981) found eggs of *Trichuris trichiura* in a corpse from the Western Han Dynasty (167 BC) in Jiangling County in China. All other records of *Trichuris* eggs in humans are much younger.

Due to the unusual consistency and the small size of the samples an estimation of the intensity of the whipworm infestation of the Iceman is hardly possible. From the relatively high number of eggs we may conclude that he was rather intensively infested. It is, however, unlikely that he really suffered from this helminthosis – although he may sometimes have had various intestinal problems (abdominal pains, diarrhoea).

Trichuris trichiura is a common parasite of man and has also been recorded in various species of non-human primates (Coombs and Crompton, 1991). The parasite has a world-wide distribution, but is particularly frequent in tropical and subtropical regions with a low standard of hygiene. In some areas more than 90% of

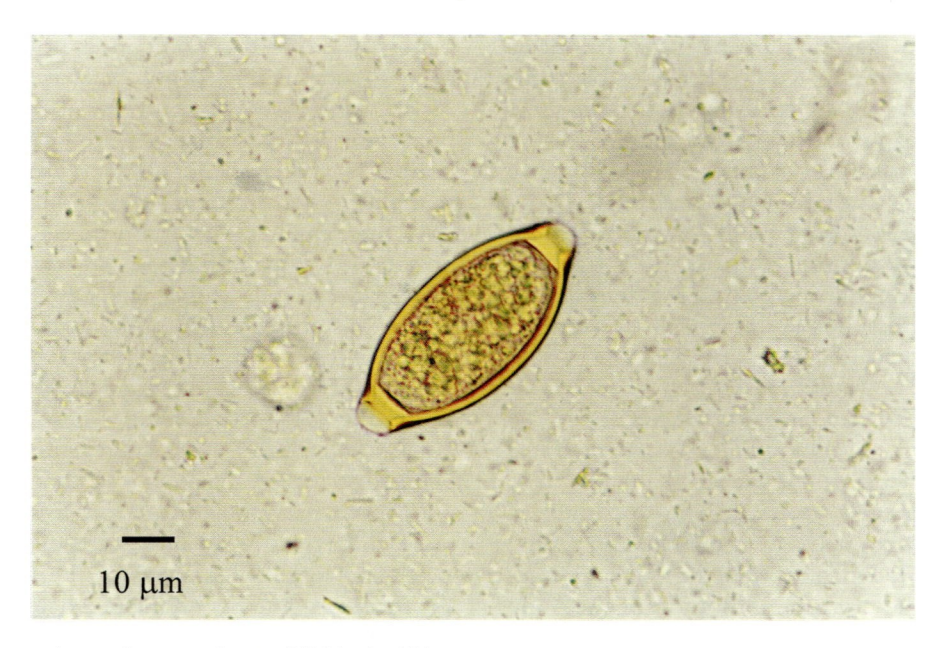

Fig. 9. Freshly passed unembryonated egg of *Trichuris trichiura*

Fig. 10. Life cycle of *Trichuris trichiura* (whipworm). The parasites live in the large intestine, eggs are passed in an unembryonated stage; they develop within a few weeks to months and become infective when ingested, e.g., with contaminated vegetables (reproduction of the Iceman with kind permission of Prof. M. Egg, Römisch-Germanisches Zentralmuseum Mainz, from Egg, 1993)

the human population may be infested, it is estimated that at least 350, possibly 900 millions of the world population suffer from (or at least harbour) whipworms. The adult parasite has a total length of 30 to 50 mm with a very thin hair-like anterior part (which was believed to be the posterior part when the genus was described: cf. the name *Trichuris*). Whipworms live in the large intestine preferably in the caecum, the colon transversum and descending colon. The narrow portion of the body is embedded within the intestinal mucosa (where the worms may also take up blood), the posterior part protrudes into the lumen. Females lay 2000–14000 eggs of typical shape and structure (Fig. 9), 50–60 µm long. The eggs are unembryonated when passed. Depending on temperature and humidity, the infective larva may develop in the egg within a few days up to several months (at 15 °C about 4–6 months, at 26° 3–4 weeks, at 35° about 11 days). If embryonated, infective eggs are ingested, e.g., with soil-contaminated vegetables, the larva will hatch in the small intestine from where it migrates to the colon and reaches maturity there within 2–3 months. Adults may live up to two years (Fig. 10).

Weak whipworm infestations may remain (almost) without clinical symptoms. Eosinophilia is, however, often found. In heavy infestations with hundreds or even thousands of worms diarrhoea, loss of weight, weakness, abdominal pains and anemia occur. Cellular infiltration, inflammation and necrosis of the mucosa and subsequently eczema of the anus, pruritus and urticaria may develop. In cases with heavy colitis and proctitis, particularly in malnourished patients, a rectal prolapse may result (Piekarski, 1987; Eckert, 1998; Katz et al., 1982).

It was surprising that we could not detect any other helminths. In particular, we would have had expected *Ascaris*. Eggs of this worm had been detected in larger numbers in the above mentioned faeces of the salt miners of the Hallstatt period (Aspöck et al. l.c.). It has been discussed that possibly *Ascaris* (and also *Trichuris*) have originally not been genuine parasites of man but may have inhabited him as a new host after the domestication of pigs (Cockburn et al., 1975; Horne, 1985; Hinz, 1990; Kliks, 1990; Anderson et al., 1993; Macko and Dubinský, 1997; Crompton, pers. comm.).

The domestication of wild pigs – independently in many parts of the Old World – dates back to about at least 7000 BC; there is first evidence from Anatolia and in the following millenia in many other parts of the world (Epstein and Bichard, 1984). It is likely that the human populations in the Alps had domesticated pigs, at least at the time when our Iceman lived, as several findings confirm that pigs were kept in Europe in the Neolithic period (Epstein and Bichard l.c.). Thus one may assume that the human population of the 4th millenium BC had already been infested by *Ascaris* (and *Trichuris*), if it is at all true that these helminths are primary parasites of pigs and only secondarily of man. On the other hand, nobody knows when and where this amplification of the host range might have occurred and, after all, we have no solid evidence that the population to which the Iceman belonged had really domesticated pigs.

We could not find any protozoa in the samples from the intestine, but this does not necessarily mean that the Iceman was really free of these microorganisms, some of which are absolutely harmless commensals (e.g., *Entamoeba coli, Entamoeba hartmanni, Jodamoeba buetschlii, Endolimax nana etc.*) showing a high prevalence, at least in the present day population. We do not know, on one hand, how 5200 years in a "lyophilized" and then frozen body may have influenced the structure of protozoa, not only trophozoites but also cysts, so that it may have become difficult or even impossible to detect them in sediments, even after rehydration by trisodium phosphate or after fixation in SAF. On the other hand we did not succeed in producing "stool smears", although the material had been carefully comminuted and stirred into Ringer's solution. Probably this problem mainly results from the low amount of material available.

Also the tests for detection of copro-antigens of *Giardia* and of *Cryptosporidium* were negative. Again we do not know how the antigens may have been altered during these 5200 years. Moreover, it would have been more than surprising to detect *Cryptosporidium*, a parasite which is a typical opportunistic microorganism having gained importance in medicine only since we know AIDS. *Giardia* is, however, a frequent and usually not dangerous parasite, today occurring in a few (1–5) %, depending on the level of hygiene, in the normal population of Central Europe.

We have retained material from all samples from the intestine and frozen at −20 °C. Moreover the small sample of lung tissue has been frozen in liquid nitrogen so that later examinations will be possible using PCR and possibly other molecular biology techniques. It would be easy to do a PCR of this lung sample for *Pneumocystis carinii* and *Toxoplasma gondii* (these tests are carried out routinously in our laboratory) and in co-operation with other laboratories which have experience with PCRs with stool samples – also the rest of the intestinal material could have been tested for *Giardia, Entamoeba histolytica, Cryptosporidium* or Microsporidia. We were, however, reluctant to use up the material for tests which do not seem promising. (It is not likely that *Toxoplasma* or *Pneumocystis* may be found in the lung or *Entamoeba histolytica* or Microsporidia in the stool.) All these tests can be done later and in the meantime improved and enlarged methods may be developed and so yield greater result. The lung tissue has also been retained in case that it will remain the only tissue available for serology.

The most important parasitological examinations still to be carried out with the Iceman are serological tests for (the hopefully possible) detection of specific antibodies. For this purpose we need material with a high amount of dried blood (larger vessels, heart, spleen . . .). Unfortunately so far it has not been possible to gain such materials from the body. We are, however, confident that it will be possible to take suitable samples in near future, rehydrate them and test them for specific antibodies in a number of selected highly sensitive and highly specific tests (indirect immunofluorescence test, enzyme-linked-immunosorbent assays, Westernblot and others) against the most prevalent and/or most important parasites in Central Europe – at least of our days. At any rate, the Iceman should be serologically tested for toxoplasmosis, trichinellosis, toxocarosis and – alveolar echinococcosis, the most famous parasitosis of the Tyrol.

Acknowledgements

We would like to express our grateful thanks to Univ.-Prof. Dr. Othmar Gaber and Dr. Karl-Heinz Künzel (Institute of Anatomy, University of Innsbruck) for various help when examining the mummy for parasites. Moreover, many thanks to Prof. Dr. J. H. Dickson, University of Glasgow, for critical reading and improving the manuscript.

Abstract

So far the following parasitological examinations have been carried out with the Iceman, the 5200–5300 years old Neolithic glacier mummy found in the Ötztaler Alps in September 1991:
1) Inspection for ectoparasites of the whole surface of the body, of the hairs found near the mummy and belonging to the man and, finally, of the cap made from bear hide; 2) Examination of samples taken from the colon and the jejunum, respectively for helminths and protozoa using rehydration by trisodium phosphate, SAF- sedimentation and Heidenhain's staining; 3) Examination of these samples for copro-antigens of *Giardia* and *Cryptosporidium* using enzyme-linked immunosorbent assays.

No ectoparasites were found. In particular it is worth mentioning that the man was apparently not infested by lice. Moreover, no trace of any pre- or postmortal myiasis was found which is of significance for the reconstruction of the circumstances of the Iceman's death.

In the colon many eggs of *Trichuris trichiura* (whipworm) were found. Neither any other helminths nor any infection with protozoa could be detected.

Further parasitological examinations should – besides various PCRs – particularly include serological tests with rehydrated tissues of organs with high amounts of remains of blood.

Zusammenfassung

An der im September 1991 in den Ötztaler Alpen gefundenen 5200–5300 Jahre alten neolithischen Gletschermumie wurden bisher folgende parasitologische Untersuchungen durchgeführt:
1) Inspektion der gesamten Körperoberfläche, der neben der Mumie gefundenen Kopfhaare sowie der Mütze auf Ektoparasiten; 2) Untersuchung von endoskopisch gewonnenen Proben aus dem Colon und dem Jejunum auf Helminthen und Protozoen nach Rehydrierung mit Trinatriumphosphat, mittels SAF-Methode und Heidenhain-Färbung; 3) Untersuchung dieser Proben mittels Enzymimmuntests auf Koproantigene von *Giardia* und *Cryptosporidium*.

Es wurden keine Ektoparasiten gefunden. Besonders sei darauf hingewiesen, daß der Mann offenbar keinen Läusebefall hatte. Ebenso wurden keinerlei Spuren einer prä- oder postmortalen Myiasis festgestellt, was für die Rekonstruktion der Umstände des Todes von gewisser Bedeutung ist.

Im Colon konnten viele recht gut erhaltene Eier von *Trichuris trichiura* (Peitschenwurm) nachgewiesen werden. Andere Helminthen oder Protozoen wurden nicht gefunden.

Weitere parasitologische Untersuchungen sollen – neben verschiedenen PCRs – vor allem serologische Untersuchungen rehydrierter blutreicher Gewebe umfassen.

Résumé

La momie glaciaire néolithique, âgée de 5 200 à 5 300 ans, découverte en septembre 1991 dans les Alpes de l'Ötztal, a fait jusqu'à présent l'objet des examens parasitologiques suivants:
1. recherche d'ectoparasites par l'examen de l'intégralité de sa surface corporelle, des cheveux retrouvés à côté du corps et de son bonnet; 2. recherche d'helminthes et de protozoaires sur des échantillons de côlon et de jéjunum prélevés par endoscopie et réhydratés à l'aide de phosphate trisodique, au moyen de la méthode SAF et de la coloration de Heidenhain; 3. recherche de coprantigènes de *Giardia* et de *Cryptoporidium* par l'application de la méthode immuno-enzymatique à ces échantillons.

Des ectoparasites n'ont pas été détectés. Soulignons que selon toute évidence, le sujet n'était pas atteint de poux. De même, aucune trace de myiasis pré- ou postmortale n'a été découverte, fait qui revêt une certaine importance pour la reconstitution des circonstances de la mort de l'Homme des glaces.

L'examen du côlon a mis en évidence la présence d'un important nombre d'oeufs du trichocéphale (*Trichuris trichiura*). D'autres helminthes ou protozoaires n'ont pas été détectés.

Les examens parasitologiques ultérieurs comprendront - en plus de plusieurs PCR – notamment l'examen sérologique, après réhydratation, des tissus richement irrigués.

Riassunto

Sulla mummia ritrovata nel settembre del 1991 nelle Alpi dell'Ötztal sono stati effettuati i seguenti esami parassitologici:

1) Ricerca di ectoparassiti su tutta la superficie cutanea, dei capelli del capo ritrovati in immediata prossimità della mummia, nonché del beretto della stessa;
2) Ricerca di elminti e protozoi in campioni endoscopici dal digiuno e colon in seguito a reidratazione con trifosfato di sodio, tramite metodo SAF e colorazione Heidenhain;
3) Ricerca di coproantigeni di *Giardia* e *Cryptosporidium* su campioni, tramite test imuno-enzimatico.

Non sono stati identificati ectoparassiti. È importante rilevare che l'uomo non era infestato da pidocchi. Non sono state individuate tracce di una myiasis pre- o postmortale, fatto di una certa rilevanza per la ricostruzione delle circostanze di morte dell'uomo del Similaun.

Nel colon si è riscontrata una certa quantità di uova di *Trichuris trichiura*.

Ulteriori esami parassitologici dovranno interessare – oltre a diversi PCR – soprattutto esami serologici di tessuti di alto tenore ematico, reidratati.

References

Anderson T. J. C., Romero-Abal M. E. and Jaenike J. (1993) Genetic structure and epidemiology of *Ascaris* populations: patterns of host affiliation in Guatemala. Parasitology 107: 319–334.

Aspöck H. und Auer H. (1992) Zur parasitologischen Untersuchung des Mannes vom Hauslabjoch. In: Höpfel F., Platzer W. und Spindler K.(Hrsg.): Der Mann im Eis, Bd. 1. Veröffentlichungen der Universität Innsbruck 187: 214–217. Eigenverl. Univ. Innsbruck, 1992.

Aspöck H., Auer H. and Picher O. (1995) The Mummy from the Hauslabjoch: A Medical Parasitology Perspective. Alpe Adria Microbiol. J. 2: 105–114.

Aspöck H., Auer H. and Picher O. (1996) *Trichuris trichiura* Eggs in the Neolithic Glacier Mummy from the Alps. Parasitology Today 12/7: 255–256.

Aspöck H., Barth F. E., Flamm H. und Picher O. (1974) Parasitäre Erkrankungen des Verdauungstraktes bei prähistorischen Bergleuten von Hallstatt und Hallein (Österreich). Mitt. Anthropol. Ges. Wien 103: 41–47.

Aspöck H., Flamm H. und Picher O. (1973) Darmparasiten in menschlichen Exkrementen aus prähistorischen Salzbergwerken der Hallstatt-Kultur (800–350 v. Chr.). Zbl. Bakt. Hyg. I Abt. Orig. A 223: 549–558.

Bereuter T. L., Reiter C., Seidler H. and Platzer W. (1996) Postmortem alterations of human lipids – part II: lipid composition of a skin sample from the Iceman. In: Spindler K., Wilfing H., Rastbichler-Zissernig E., zur Nedden D. and Nothdurfter H. (eds.): Human mummies. A global survey of their status and the techniques of conservation. In: Moser H., Platzer W., Seidler H. and Spindler K. (eds.): The Man in the Ice, vol. 3. Veröffentlichungen des Forschungsinstituts für Alpine Vorzeit der Universität Innsbruck 3: 275–278. Springer-Verlag, Wien New York.

Bouchet F., Pétrequin P., Paicheler J. C. et Dommelier S. (1995) Première approche paléoparasitologique du site néolithique de Chalain (Jura, France). Bull. Soc. Path. Ex. 88: 265–268.

Cockburn A. and Barraco R. A. (1975) Autopsy of an Egyptian Mummy. Science 187: 1155–1160.

Coombs I. and Crompton D. W. T. (1991) A guide to human helminths. Taylor & Francis, London New York Philadelphia: pp 196

Eckert J. (1998) Parasitologie. In: Kayser H., Bienz K. A., Eckert J. und Zinkernagel R. M. (Hrsg.): Medizinische Mikrobiologie. G. Thieme Verlag, Stuttgart: 483–640.

Egg M. (1993) Die Ausrüstung des Toten. In: Egg M. und Spindler K. (Hrsg.): Die Gletschermumie vom Ende der Steinzeit aus den Ötztaler Alpen. JbRGM 39: 35–100.

Epstein H. and Bichard M. (1984) 17. Pig. In: Mason I. L. (ed.): Evolution of domesticated animals. Longman, London and New York: 145–162.

Ferreira L. F., de Araujo A. J. G. and Confalonieri U. E. C. (1980) The finding of eggs and larvae of parasitic helminths in archaeological material from Unai, Minas Ferais, Brazil. Trans. Roy. Soc. Trop. Med. Hyg. 74: 798–800.

Ferreira L. F., de Araujo A. J. G. and Confalonieri U. E. C. (1983) The finding of helminth eggs in a Brazilian mummy. Trans. Roy. Soc. Trop. Med. Hyg. 77: 65–67.

Goedecker-Ciolek R. (1993) Zur Herstellung von Kleidung und Ausrüstungsgegenständen. In: Egg M. und Spindler K. (Hrsg.): Die Gletschermumie vom Ende der Steinzeit aus den Ötztaler Alpen. JbRGZ 39: 100–113.

Groenman-van Waateringe W. (1993) Analyses of hides and skins from the Hauslabjoch. In: Egg M. und Spindler K. (Hrsg.): Die Gletschermumie. Vom Ende der Steinzeit aus den Ötztaler Alpen. JbRGZ 39: 114–128.

Hinz E. (1990) Zur Herkunft der Helminthen des Menschen in Amerika. In: Hinz E. (Hrsg.): Geomedizinische und biogeographische Aspekte der Krankheitsverbreitung und Gesundheitsversorgung in Industrie- und Entwicklungsländern: 359–406. Peter Lang, Frankfurt M. Bern New York Paris.

Horne P. D. (1985) A review of the evidence of human endoparasitism in the pre-Columbian New World through the study of coprolites. J. Archaeol. Sci. 12: 299–310.

Junod C. (1972) Technique coprologique nouvelle exentiellement destinée a al concentration des trophozoites d'amibes. Bull. Soc. Pathol. Exot. 65: 390–398.

Katz M., Despommier D. D. and Gwadz R. (1982) Parasitic diseases. Springer-Verlag, New York Heidelberg Berlin: pp 264.

Kliks M. M. (1990) Helminths as heirlooms and souvenirs: a review of New World paleoparasitology. Parasitology Today 6: 93–100.

Macko J. K. and Dubinský P. (1997) Taxonomic deliberations on human and pig ascarids. Helminthologia 34: 167–171.

Marti H. P. and Escher E. (1990) SAF – eine alternative Fixierlösung für parasitologische Stuhluntersuchungen. Schweiz. Med. Wschr. 120: 1473–1476.

Mehlhorn H., Eichenlaub D., Löscher T. and Peters W. (1995) Diagnostik und Therapie der Parasitosen des Menschen. 2. Aufl. G. Fischer Verlag. Stuttgart, Jena, New York. pp 452.

Piekarski G. (1987) Medizinische Parasitologie in Tafeln. Dritte, vollst. überarb. Aufl. Springer-Verlag, Berlin. pp 364.

Spindler K. (1996) Iceman's last weeks. In: Spindler K., Wilfing H., Rastbichler-Zissernig E., zur Nedden D. and Nothdurfter H. (eds.): Human mummies. A global survey of their status and the techniques of conservation. In: Moser H., Platzer W., Seidler H. and Spindler K. (eds.): The Man in the Ice, vol. 3. Veröffentlichungen des Forschungsinstituts für Alpine Vorzeit der Universität Innsbruck 3: 249–263. Springer-Verlag, Wien New York.

Szidat L. (1944) Über die Erhaltungsfähigkeit von Helmintheneiern in vor- und frühgeschichtlichen Moorleichen. Z. Parasitenk 13: 265–274.

Van Cleave H. J. and Ross J. A. (1947) A method for reclaiming dried zoological specimens. Science 105: 318.

Wei D. X., Yang W. Y., Huang S. Q., Lu Y. F., Su T. C., Ma J. H., Hu W. X. and Xie N. F. (1981) Parasitological investigation on the ancient corpse of the Western Han Dynasty unearthed from tomb No 168 on Phoenix Hill in Jiangling county. Acta Acad. Med. Wuhan 1: 16–23.

Wittig M. und Wortmann G. (1992) Untersuchungen an Haaren aus den Begleitfunden des Eismannes vom Hauslabjoch. Vorläufige Ergebnisse. In: Höpfel F., Platzer W. und Spindler K. (Hrsg.): Der Mann im Eis, Bd. 1. Veröffentlichungen der Universität Innsbruck 187: 273–298. Eigenverl. Univ. Innsbruck, 1992.

Vivianite from the Iceman of the Tisenjoch (Tyrol, Austria): Mineralogical-chemical data

R. Tessadri

Institut für Mineralogie und Petrographie, Universität Innsbruck

1. Introduction

The discovery in September 1991 of a Late Neolithic man in a glacial field between Austria and Italy offered uniquely preserved archaelogical samples (Seidler et al., 1992). Commonly known as the Iceman (or Ötzi, having been found in the Tyrolean Ötztaler Alps), the body is the oldest to be retrieved from an alpine glacier and is one of the best preserved mummified humans ever discovered (Edwards et al., 1996).

An interesting observation from the mineralogical point of view is the growth of the mineral vivianite on the skin of the Iceman in contact with the surrounding weathered rocks (Tessadri et al., 1996).

The aim of this paper is to present and discuss mineralogical-chemical data of this unique occurrence of vivianite, the accompanying skin and surrounding rocks.

2. Analytical methods

Elemental analyses of the vivianite and the skin has been carried out using TXRFA (Total Reflection X-Ray Analysis), since this is the only available method to quantitatively investigate trace elements in bulk materials with very small sample amounts and sufficient detection limits. An amount of 0.18 mg (!) from the Iceman-Vivianite was dissolved in 1 ml 5M HNO_3 (acid purification by subboiling distillation using a quartz apparatus). The skin was combusted in an oxygen stream (amount: 1.65 mg), the residue was taken again in 1 ml 5M HNO_3.

100 µl portions of the sample solutions spiked with 0.2 µg of gallium as internal standard were pipetted onto a silicon sample support and dried.

Characteristic X-rays of the analyte elements have been generated using a Mo-target and a W-target. Both tubes were operated at 50 kV and variable current (5 to 38 mA), depending on the count rate of the sample. Each sample was measured for 3000 sec. Gallium was used as the internal standard. The known concentration of the Ga-spike together with the instruments' response function allows converting signal intensities directly to concentration values.

The TXRF-spectrometer used was from Atomika Instruments (Extra IIA) at the Institute of Physics/GKSS Forschungszentrum Geesthacht/Germany. The analyzing systems consisted of 80 mm^2 Si(Li)-detectors having an energy resolution of 155 eV at 5.9 keV, the QX 2000 XR-System (Link Analytical) and alternatively a XR-System from Noran Instruments (for a more detailed description see Krivan et al. and operating notes from Atomika Instruments).

Chemical analysis of rocks has been performed using a combination of ICP-OES (solutions obtained using $LiBO_2$-flux technique, standardization with SRMs) and EDXRFA (powder pellet technique, standardization with SRMs) at the Institute of Mineralogy & Petrography/University of Innsbruck.

For crystalline phase analysis of vivianite a SIEMENS D-500 XR-Powder Diffractometer at the Institute of Mineralogy & Petrography/University of Innsbruck was used.

Secondary Electron Microscopy has been performed with a Zeiss DSM940A at the Institute of Geology & Paleontology/University of Innsbruck.

2.1. General Data of Vivianite

In terms of the mineralogical system the mineral vivianite (synonymous names: iron-mica, blue ironstone, native Prussian blue, blue iron earth, glaucosiderite, mullicite or iron-phyllite) belongs to Group 7 (Phosphates, Arsenates, Vanadates).

The ideal stoichiometric chemical formula is $Fe_3(PO_4)_2 \cdot 8H_2O$, whereas iron is in the oxidation state 2^+; crystallographically the mineral is monoclinic.

Different shades of blue colour, ranging from light blue to dark- and even greenish blue are known from vivianite. The changes in colour depend on oxidation of Fe^{2+} to Fe^{3+}. During this the co-ordinative bound water (crystal water) is partly converted to $(OH)^-$ to keep charge balance. Increasing oxidation of vivianite changes colour to a deeper blue. This process does not affect the crystal structure of vivianite.

Another interesting feature of vivianite is the formation condition, which is restricted to a pH > 7 and strongly reducing enviroment.

Chemistry and formation conditions of this mineral determine the occurrences: vivianite is known from leaching zones of poly-metallic ore deposits in lake waters with reducing conditions (St. Agnes/Cornwall, Llallagua/Bolivia, Waldsassen/Germany), from P-rich pegmatites and their weathering products, from pelitic sediments (especially clays) rich in organic material, from swamp-deposits, bog-phosphates and from fossil bones and teeth, where vivianite was formed by reaction of iron-bearing solutions with hydroxyle-apatite (see literature in MinSource Vers. 2.1., 1996).

2.2. The vivianite from the Iceman

Vivianite from the Iceman occurs in small spots (1 mm to about 5 mm) on the skin, which had been in close contact to the surrounding weathered rocks (mostly arms and legs).

Figure 1 shows a photograph of the vivianite, whereas Fig. 2a, b shows secondary electron microscopic sides of the mineral.

The first macroscopic determination of the mineral was done by T. Rowbothan/Institute of Public Health (Leeds) and F. Tiefenbrunner/Institute of Hygiene (Innsbruck), who recognized the blue phase on the skin as vivianite (in analogy to known occurrences of vivianite from Siberian mammoths).

Although identification of the phase from macroscopic data was rather clear a small quantity of the sample was examined by powder X-ray diffraction: the X-ray-pattern obtained clearly shows a reflection at 6.73 Å, which corresponds to the highest reflection (the (020)-reflection) of vivianite (JCPDS-Pattern 30-662). Comparison with other natural vivianites (samples from Waldsassen/Germany, Hagendorf/Germany and Luax Mine/Colorado from the mineral collection of the Institute of Mineralogy & Petrography/University

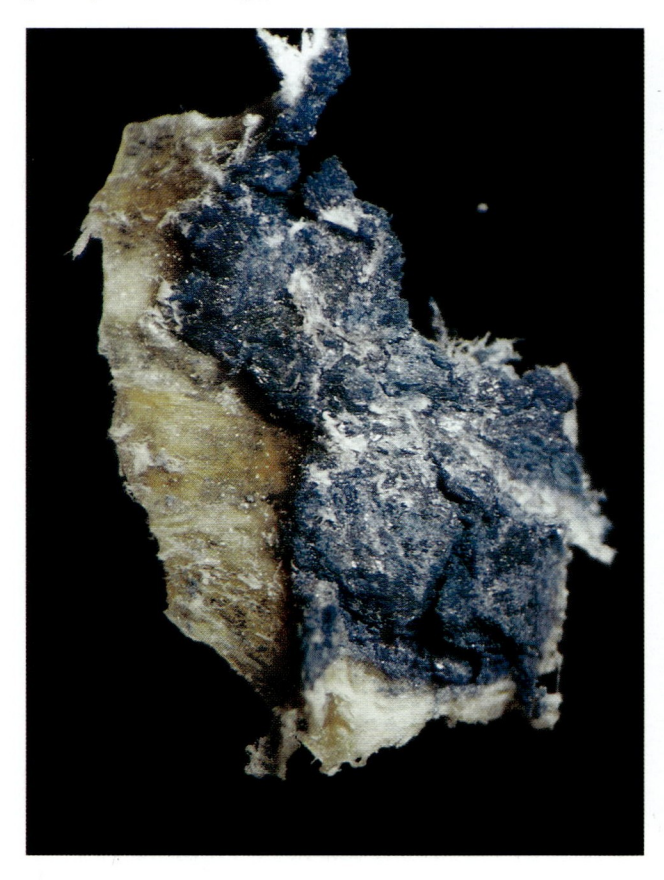

Fig. 1. Vivianite on skin of the Iceman; length about 5 mm (picture from B. Baumgarten, Naturmuseum Bozen)

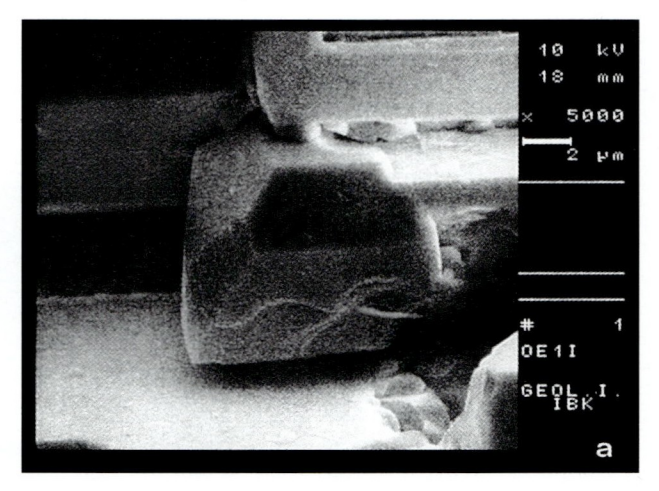

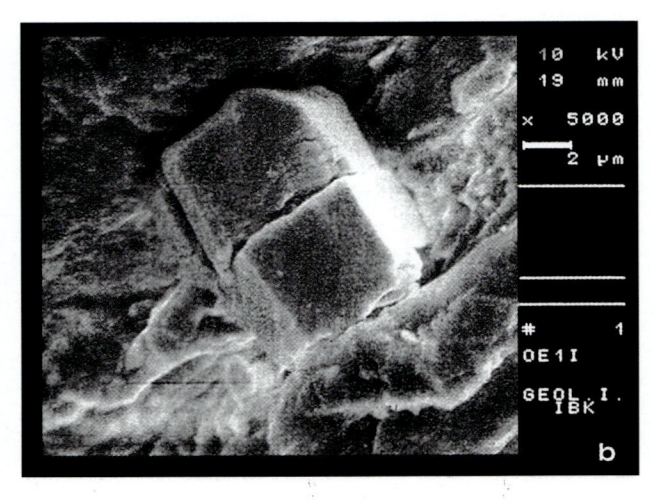

Fig. 2a, b. SEM pictures from vivianite on skin of the Iceman: (pictures from V. Stingl/Institute of Geology & Paleontology/University of Innsbruck)

of Innsbruck) showed corresponding patterns. A boolean search on the ICDD-data base with 6.73 Å and the colour "blue" gives only one match: vivianite.

The effect of vivianite oxidation was also observed: after some days of exposure to air the colour of the mineral changed from a light blue to a deeper blue (F. Tiefenbrunner, pers. comm.).

2.3. Chemistry of vivianite

The results of chemical analyses by TXRFA are presented in Table 1.

For comparative purposes 3 other natural vivianites (the same that have been used for XR-diffraction) have been analyzed. A total analytical error (precision and accuracy) of ±10% relative should be considered for all results.

Stoichiometric vivianite should have a composition of 33.4% Fe, 12.4% P, 51.0% O and 3.2% H; obviously this is not the case in all 4 investigated samples. However the stoichiometry is between ±10% relative,

which is a) quite normal within natural occurrences and b) within the estimated precision/accuracy-error of TXRFA.

Comparision of the four samples shows that all other elements have a wide variation (for example, manganese from 300 ppm in Tisenjoch vivianite to 30000 ppm in Hagendorf vivianite), which gives some hints to source rocks and/or source fluids from which vivianite was formed.

Generally the amounts of other elements in the Tisenjoch vivianite are rather low, compared to the three other vivianites; a fact which can be easily explained, since the formation of Tisenjoch vivianite is expected to be influenced solely by the chemistry of the mummy bones (human apatites normally have low concentrations in elements other than Ca and P) and/or by the chemistry of the surrounding rocks.

Table 2 shows that the rock composition of the surrounding rocks from the place where the mummy has been found, simply represents a normal composition of

Table 1. Chemistry of investigated vivianites; amounts in ppm (otherwise noted); Remark A: arrangement of elements according to ascending atomic number; below atomic number of silicon (a sample holder consisting of pure silicon was used) detection of the lighter elements was not possible. Remark B: fitting indices from the spectra deconvolution process have been very low (between 2.0 and 3.0), which gives an indication that all elements present in the samples have been analyzed. Furthermore it is suggested that all elements not involved in the deconvolution process are below detection limits (e.g. Sc, Co, Ge, Br, J, Ta, PGE, REE, Hg, Tl, Bi etc. the rest of the periodic table), since the excitation conditions used were high enough to excitate all elements; the detection limits for these elements can roughly be estimated between 3 and 10 ppm

	VIV 1 Waldsassen Germany A 2311	VIV 2 Hagendorf Germany A 2309	VIV 3 Luax Mine Colorado A 1965	**VIV 4** Hauslabjoch Mummy Ö1 I 9 30.3.95
P	12.6%	13.5%	13.2%	12.8%
S	<100	<100	<100	<100
Cl	<50	<50	<50	<50
K	44	7	42	272
Ca	220	156	8102	684
Ti	337	163	301	159
V	<10	<10	<10	<10
Cr	260	<10	218	176
Mn	1175	3.0%	9740	299
Fe	28.4%	28.8%	28.6%	27.4%
Ni	<10	<10	<10	<10
Cu	<5	<5	<5	<5
Zn	11	522	3.1%	87
As	5	3	12	2
Se	<5	<5	<5	<5
Rb	<3	<3	<3	<3
Sr	7	10	12	19
Y	<2	<2	<2	<2
Zr	<5	<5	<5	<5
Mo	<3	<3	<3	<3
Ag	<3	<3	<3	<3
Cd	<5	<5	<5	<5
Sn	<5	<5	<5	<5
Sb	<3	<10	<3	<3
Ba	<10	<10	<10	<10
Pb	<5	<5	<5	<5

Table 2. Major- and trace element chemistry of surrounding rocks; major elements in weight % oxides, Fe_{total} as Fe_2O_3, Loss on Ignition (LOI) at $1000\,°C/24\,h$; trace elements in ppm

	Unweathered gneiss (coarse) Nr. 200 18.8.92	Weathered gneiss (fine) Nr. 200 18.8.92		Unweathered gneiss (coarse) Nr. 200 18.8.92	Weathered gneiss (fine) Nr. 200 18.8.92
SiO_2	71.12	56.62	As	< 4	5
Al_2O_3	13.59	19.65	Ba	594	852
Fe_2O_3	5.68	7.71	Be	2	3
MnO	0.06	0.06	Ce	47	126
MgO	1.65	2.77	Cl	< 100	< 100
CaO	0.58	0.52	Co	15	18
Na_2O	1.62	1.35	Cr	67	95
K_2O	3.07	4.60	Cu	24	56
TiO_2	0.73	0.99	Ga	19	32
P_2O_5	0.16	0.25	La	85	151
LOI	1.77	5.23	Nb	14	18
			Ni	15	27
Total	100.03	99.75	Pb	19	39
			Rb	113	184
			S	802	1330
			Sc	13	20
			Sr	143	166
			Th	12	14
			U	< 3	6
			V	77	143
			Y	29	34
			Zn	46	104
			Zr	222	229

a para/ortho-gneiss, which is one of the dominant rock types in the Austroalpine Ötztal-Stubai Complex (Purtscheller, 1985).

Of course the weathered parts of this rock (the silt-clay fractions are more dominant) have a different major-element composition (less silica, enhanced alumina,- iron-, potassium- and water content) and trace composition (enrichment of almost all trace elements, as they are preferably bound in minerals which are more resistant to weathering).

However, all trace elements of the rocks are just in normal abundances, so that during vivianite formation no extraordinary concentration of trace elements in the mineral should be expected. This is not the case in the three other occurrences, since all represent vivianites from more or less poly-metallic ore deposits.

2.4. Chemistry of the skin from the Iceman

The results of chemical analyses by TXRFA of the skin in contact with the investigated vivianite are shown in Table 3 (together with a compilation of skin data extraced from Iyengar et al., 1978).

Comparison clearly indicates that there are no remarkable differences in composition between the "normal" skins and the Iceman's skin surrounding the vivianite. The only exception is the As-content, which is almost twice ($0.5\,ppm \pm 0.1\,ppm$) the highest reported value.

Abstract

A description of the unique occurrence of vivianite from the Iceman is given. Mineralogical-chemical data of the vivianite, the accompanying skin and the surrounding rocks are presented and discussed.

Zusammenfassung

Die außergewöhnliche Bildung von Vivianit auf der Mumie des Mannes im Eis wird beschrieben. Mineralogisch-chemische Daten des Vivianits, der anliegenden Haut und des umgebenden Gesteins werden gegeben und diskutiert.

Riassunto

Viene presentata una descrizione della singolare presenza divivianite sulla mummia della Tisenjoch, vengono presentati e commentati i dati chimico-mineralogici della vivianite e della roccia circostante.

Table 3. Trace element chemistry of investigated skin; amounts in ppm; Note: Remark B of Table 1 also holds true for these data except the estimated detection limits which are lower (0.1 to 1 ppm) because of the lighter organic matrix

	Compiled Skin Data (mostly whole skin data) (Iyengar et al., 1978)	**Skin** Hauslabjoch Mummy Ö1 I 9 30.3.95
P	300–1300	674
S	1380–1850	1510
Cl	–	6
K	880–11000	44
Ca	34–16000	1190
Ti	2–72	7
V	<1	<1
Cr	0.44–41	5
Mn	0.003–22	3
Fe	27–1900	592
Ni	0.1–47	3
Cu	7–143	8
Zn	6–1000	140
As	0.06–0.23	0.5
Se	0.06–1.20	0.1
Rb	–	<0.5
Sr	0.02–13	7
Y	–	<0.5
Zr	–	<0.5
Mo	0.06–1.70	<0.5
Ag	0.0004–4	<0.5
Cd	0.1–36	<0.5
Sn	–	<0.5
Sb	0.03–0.22	<0.5
Ba	14–17	<0.5
Pb	0.1–46	0.2

Résumé

La présence exceptionnelle de vivianite sur la momie du Tisenjoch fait l'objet de cette description qui présente et commente les caractéristiques minéralogico-chimiques de la vivianite, de la région cutanée intéressée et des roches environnantes.

Acknowledgements

The author thanks the company Atomika Instruments, Munich for permission to use their instruments, and U. Reus at the Institute of Physics, GKSS Forschungszentrum Geesthacht, Germany for helpful guidance during TXRF-Analysis. B. Baumgarten, Naturmuseum Bozen and V. Stingl, Institute of Geology & Paleontology, University of Innsbruck are acknowledged for providing the author with photographs. W. Platzer, O. Gaber, K.H. Künzel, F. Tiefenbrunner and H.J. Battista (all Medical Faculty at the University of Innsbruck) are thanked for providing the author with samples and/or information about the Iceman or general aspects to this work. V. Mair, Institute of Mineralogy & Petrography, University of Innsbruck is acknowledged for reading this paper carefully and giving corrections to a first version of the manuscript.

References

Edwards H. G. M., Williams A. C. and Barry B. W. (1996) Human skin: a Fourier transform Raman spectroscopic study of the Iceman: Spectroscopy Europe 8(1): 10–18.

Iyengar G. V., Kollmer W. E. and Bowen H. J. M. (1978) The Elemental Composition of Human Tissues and Body Fluids - A Compilation of Values for Adults. Verlag Chemie, Weinheim New York: 100–102.

Minsource 2.1. (1996) Ed. Howie R. A. (Abstracts) and Clark A. M. (Hey's Mineralogical Index), Chapman & Hall, London.

Krivan V., Schneider G., Baumann H. and Reus U. (1994) Multi-element characterization of tobacco smoke condensate: Fresenius J. Anal. Chem. 348: 218–225.

Purtscheller F. (1985) Geologischer Führer Ötztaler und Stubaier Alpen, 2. Aufl., Vol. 53. Bornträger, Berlin Stuttgart.

Seidler H., Bernhard M., Teschler-Nicola M., Platzer W., Zur Nedden D., Oberhauser A. and Sjøvold T. (1992) Some Anthropological Aspects of the Prehistoric Tyrolean Iceman. Science 258: 455–457.

Tessadri R., Reus U., Baumgarten B., Mair V., Stingl V., Platzer W. and Mirwald P. W. (1996) Vivianite from the Iceman of the Hauslabjoch (Tyrol, Austria): preliminary results. Mitt. Öst. Min. Ges. 141: 232–233.

Ethnomycological remarks on the Iceman's fungi

U. Peintner and R. Pöder

Institut für Mikrobiologie, Universität Innsbruck

1. Introduction

Three fungal objects were among the numerous items of the Iceman's equipment: the so called "Black Matter" filling up the major part of the "girdle bag" and two different shaped fungal fragments, each mounted separately on a leather thong. The "Black Matter" was shown to be a tinder material prepared from the "true tinder bracket" *Fomes fomentarius* (L. : Fr.) Fr. (Pöder, Pümpel and Peintner, 1995; Peintner, Pöder and Pümpel, 1998), the two objects on the leather thongs were identified as fruitbody fragments of the polypore *Piptoporus betulinus* (Bull. : Fr.) Karst (Pöder, Peintner and Pümpel, 1992; Peintner et al., 1998). Thus, the Iceman carried material of two species of polypores on his journey. From this arises the simple but important question: what significance did these fungi have for the prehistoric mountain traveller? The aim of the present article is to discuss the possible use of these two polypore species by giving a comprehensive overview of the available ethnomycological data.

2. The "Black Matter" (*Fomes fomentarius*) found in the "leather bag"

The so called "Black Matter" filled the major part of the girdle bag in which also several sharpened flint-stones, a small drill like a piece of flint and a slender bone-tool were found (Fig. 3). At first this material was thought to be a kind of resin, interpreted as part of a prehistoric repair kit (Lippert and Spindler, 1991). But this clotted, water-saturated mass became dull and fibrous while drying. In dry condition it could easily be loosened up to a wad-like consistency by using forceps. Finally, the "Black Matter" could be proven to consist nearly exclusively of context hyphae of *Fomes fomentarius* (Pöder et al., 1995). Apart from accidental impurities like a few animal hairs, traces of pyrite could be detected among the fungal hyphae (Sauter and Stachelberger, 1992). For a full description of the "Black Matter" see Pöder et al. (1995) and Peintner et al. (1998), for further figures Egg and Spindler (1993).

2.1. Ethnomycological implications

For primitive people the improvement of life conditions by controlling fire was enormous. Many species of polypores have been used as the primary tinder in the making of fire, but as "touchwood" or "punk" *Fomes fomentarius* has enjoyed primacy from the beginning (Wasson, 1968, p. 238f). There is evidence that this fungus has been used for millennia, as not only native fruitbodies of the "true tinder bracket", but also fruitbodies bearing traces of human handling have often been found in archaeological sites (Killermann, 1936, 1938; Kreisel, 1965/56, 1977; Neuweiler cited in Göpfert, 1982; Thoen, 1982), the oldest ones dating back to 11555±100 BP (Kreisel, 1996).

Various authors (for instance Buller, 1914; Herrmann, 1962; Champion, 1976; Weiner, 1981; Göpfert, 1982) describe in detail how tinder – also called "amadou" – was prepared and used: the context ("flesh") between the pileal crust and the spore-forming tube layer) of the fruitbodies was cut into slices and beaten until it became soft. Dipped into dung water [later on a solution of salpetre (KNO_3) was used], its inflammability was increased. Thereby it was rendered fit for use as tinder, for making fire with flint and steel before matches came into general use.

In ancient times dried and partly hollowed basidiomata were also used to transport fire (Cordier, 1870 cited in Thoen, 1982). The importance of tinder and fire for ancient people seems to be the background of a German Good Friday custom called "Weihfeuertragen", where people till today have been carrying the holy fire in partly hollowed fruitbodies of the true tinder bracket *F. fomentarius* (Walter, 1982).

One case came to our knowledge where *F. fomentarius* served for a criminal purpose: Ladurner (pers. comm., 1997) informed us about a case of incendiarism that happened in South Tyrol (Italy) at the turn of this century. A farm-hand, disappointed of his discharge, deposited a smouldering fruitbody in the wooden loft of the farm house of his former employers. The fire broke out a few days later and destroyed the whole building.

Apart from its widespread use as classical tinder since prehistoric times *F. fomentarius* has also been used

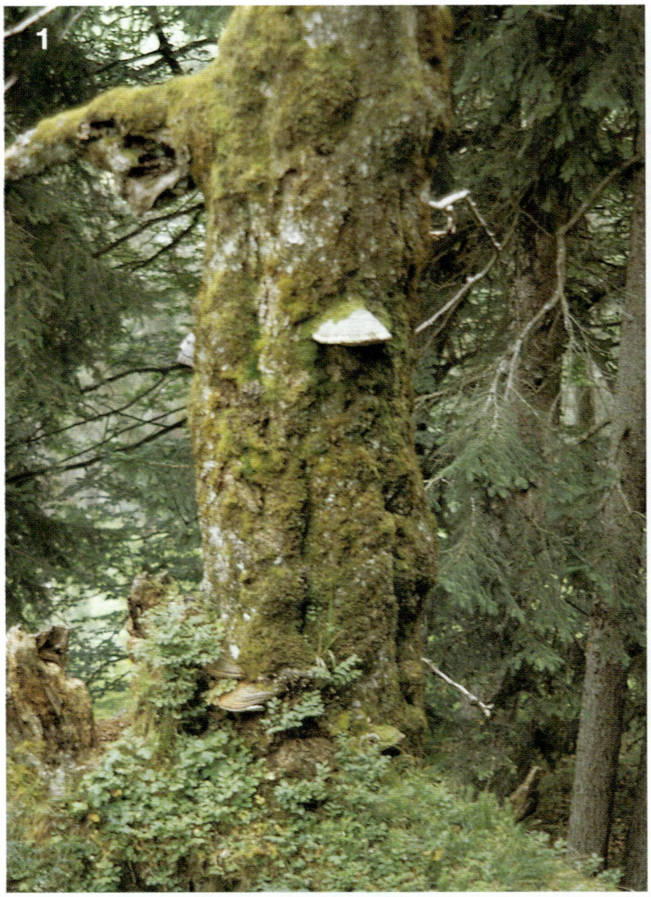
1

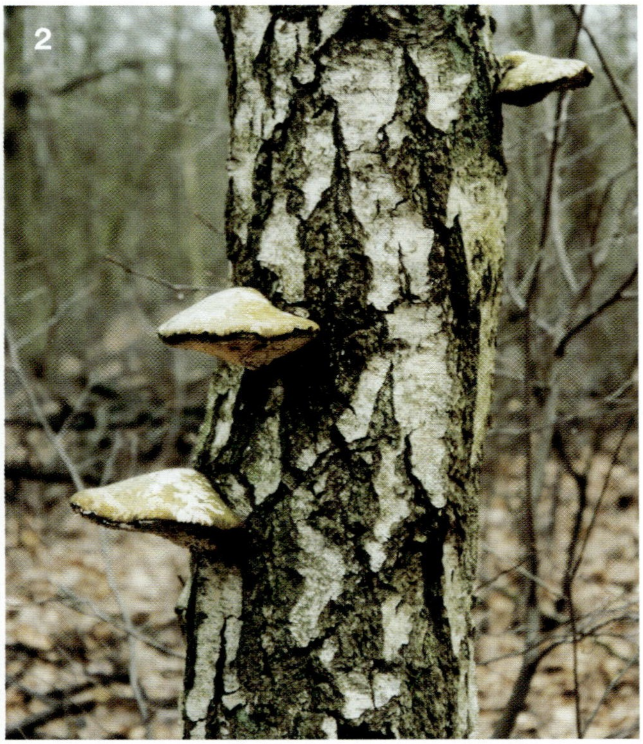
2

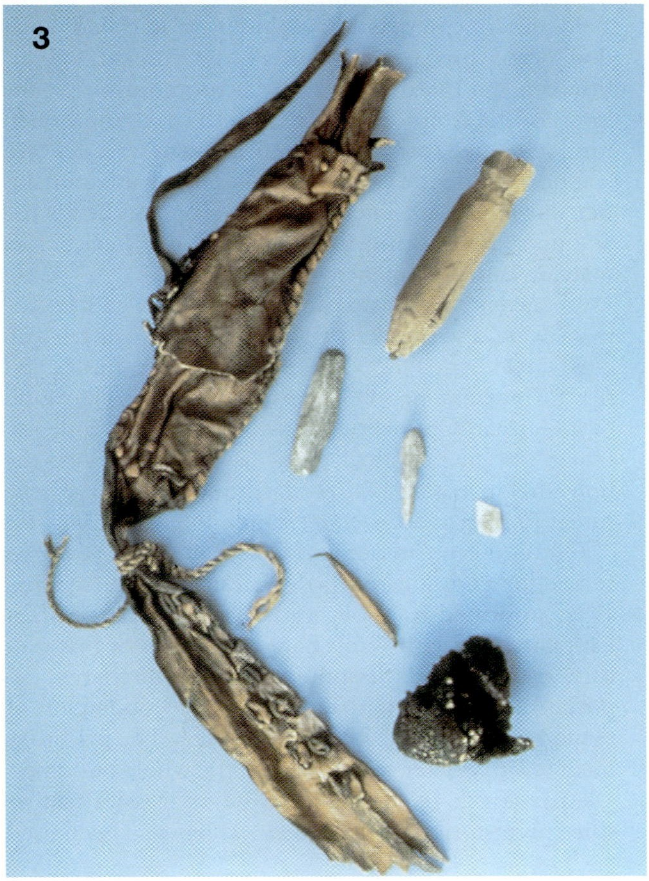
3

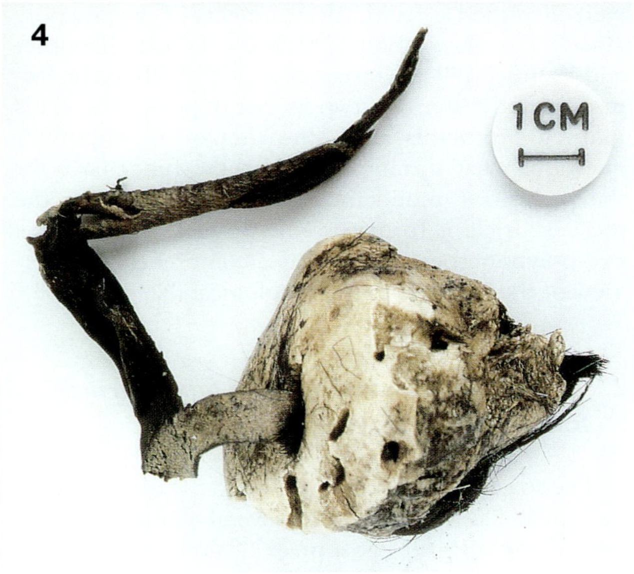
4
1CM

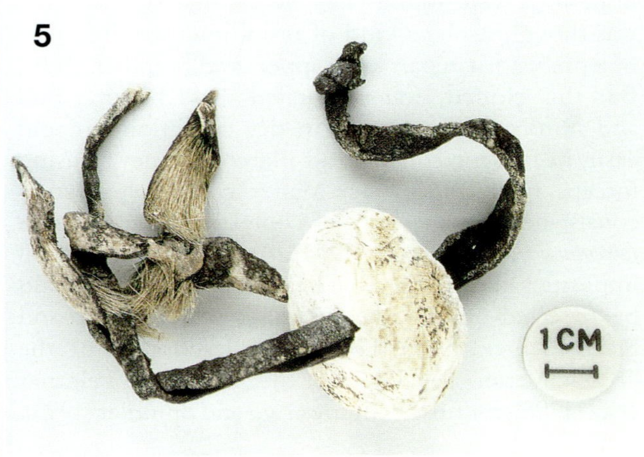
5
1CM

in various ways, for medical purposes, and even spiritual uses have been reported (Buller, 1914; Sponheimer, 1936; Herrmann, 1962; Champion, 1976; Thoen, 1982; Saar, 1991; Vaidya and Rabba, 1993). The first historical reference seems to come from Hippocrates in the fifth century B.C.. Hippocrates, writing on the practice in medicine, advised cauterization with amadou for cure of certain complaints. The cauterization was accomplished by lighting the tinder and applying it, as it was smouldering, to the skin on the outside of the affected organ (Buller, 1914; Rolfe and Rolfe, 1925). Maybe one fruitbody of *F. fomentarius* described and illustrated by Göpfert (1982) was used for cauterization: the fruitbody bore rubbing marks on the carbonized pore layer, and a twig of *Ulmus* stuck in a hole on the surface.

Rolfe and Rolfe (1925) write, that "cauterization by means of fungi appear to have survived among uncivilized races almost to the present day". They cite Rees (1819), who reports: "the Laplanders have a way of using funguses, or common toadstools, as we call them, as the Chinese and Japanese do the moxa, to cure pains. They collect the large funguses which they find on the bark of beech and other large trees, and dry them for use. Whenever they have pains in their limbs, they bruise some of the dried matter, and pulling it to pieces with their fingers, they lay a small heap of it on the part nearest to where the pain is situated, and set it on fire. In burning away, it blisters up the part, and the water discharge by this means generally carries off the pain."

Furthermore, *F. fomentarius*, was widely used as a styptic by surgeons, barbers and dentists and therefore even called "agaric of the chirurges" or "surgeon's agaric" (Cordier, 1870 cited in Thoen, 1982; Buller, 1914; Rolfe and Rolfe, 1925; Göpfert, 1982). Moreover, Killermann (1938) reports that the true tinder bracket can be used as a remedy against dysmenorrhoea, haemorrhoids and against bladder disorders, the active substance being "Fomitin".

In Indian folk medicine *F. fomentarius* was commonly known as "Gharikum" or "Chattri-Kiain" and used as a diuretic, a laxative and as a nervine tonic (Chopra, 1956 cited in Vaidya and Bhor, 1991). Furthermore, a kind of absorbing dressing made of tinder (it contains some iodine) was externally applied to wounds and burns (Vaidya and Rabba, 1993). Also the Kamchadales, a people living in the northern part of Kamchatka (Siberian Fareast), used it for pain relief (Wasson, 1968). Ying et al. (1987) describes different medical effects of *F. fomentarius*: As pure mushroom tea it can be used for the treatment of oesophagus, gastric and uterine carcinoma; if a decoction is made together with red rock lichen it relieves indigestion, and reduces stasis.

When natural people discover the medical power of polypores in preserving and improving human health, they often refer the origin of these properties to spiritual sources. Therefore, the Kanthy, a people living in western Siberia, used *F. fomentarius* in two ways: as a wound dressing or as protective smoke. The dressing was prepared like tinder, pounding the fungal context in a mortar until it was soft. It was pressed on wounds to stop bleeding, and it was also used as a warming compress for aching extremities but never on the body. When illnesses arose without evident external reasons supernatural beings were thought to be the cause. In such cases *F. fomentarius* was used as smoke. The smoke was obtained by burning the fruitbodies alone or together with silver fir bark in a chimney or in a tin pail. This ritual was also carried out when a person died, and it was continued until the deceased had been taken out of the house. Also the people coming from the funeral had to pass through this smoke. "The aim of the procedure was not to let the dead or supernatural beings have any influence on the living" (Saar, 1991). A very similar ritual is also reported from the Ainu people of Hokkaido (Japan), who, in the case of epidemics which they believed to be caused by demons, burned *F. fomentarius* around their houses the whole night through, with the aim of banishing the bad spirits or demons (Yokohama, 1975 cited in Thoen, 1982).

Furthermore, the fruitbodies of *F. fomentarius* were used for various commodities: as the soft material processed from the context has insulating and drying capacities, this polypore was often used in Germany, Hungary and parts of former Yugoslavia for making

Fig. 1. Old, mossy beech-tree with fruitbodies of *Fomes fomentarius* on the upper part of the stem. A second polypore species, *Ganoderma lipsiense* (Batsch) Atk. grows at the base of the stem (Location: Achenkirch, Tyrol, Austria; about 1000 m a.s.l.). **Fig. 2.** Birch-tree with adult fruitbodies of *Piptoporus betulinus*. The relatively old fruitbodies are loosing their thin, paper-like crust hence the underlying white context becomes visible (Location: Aldrans near Innsbruck, Tyrol, Austria; about 700 m a.s.l.). **Fig. 3.** The Iceman's girdle bag and its content: A part of the "Black Matter" is shown on the right hand below. This material, originally filling up the major part of the bag, represents a mass of clumped hyphae of the "true tinder bracket" *Fomes fomentarius*. Next to the sample of the "Black Matter" a 6.2 cm long bone-awl can be seen, above which three flint objects are depicted (from left to right: great silex blade, silex drill, small silex blade). The wooden tack right above was not part of the bag content (Photograph: Römisch-Germanisches Zentralmuseum, Mainz, Germany). **Figs. 4–5.** The two *Piptoporus betulinus* fruitbody fragments on leather thongs. **Fig. 4.** The somewhat cone-like formed piece officially labelled as object 91/133a in its original, water-saturated condition. **Fig. 5.** The sphaeroidal piece (91/133b) is mounted on a leather thong, which bears an elaborated three-lobed leather tassel. The picture was taken after lyophilization

caps, chest protectors and other articles of clothing (Buller, 1914; Ramsbottom, 1923; Rolfe and Rolfe, 1925; Broendegaard, 1981; Walter, 1982). For this purpose the fruitbodies were soaked in water and the outer crust and the porelayer were removed. The context was then made softer in a alkaline solution (using ashes) and rinsed thoroughly. Finally it was pounded with a wooden object, and nice "suede" was the end product that could be stretched to ten times its original size. In Swartzwald (Germany) huge fruitbodies were used to produce special gowns for the archbishop, but mostly the cloth was used for headwear, which was shaped on wooden moulds. In Hungary this industry still exists today. When sand is added, this tinder material can be used for sharpening razors (Broendegaard, 1981). For mounting such delicate insects as flies and mosquitoes for entomological collections, pith is often employed, into which the delicate silver entomological pin impailing the insect is inserted. The context material of the true tinder bracket, but also that of other polypores like *Piptoporus betulinus*, has been found to be an excellent material for such pith (Thoen, 1982). In addition, such material has been used as pin-cushions to prevent needles from rusting (Broendegaard, 1981). Moreover, Siberian peoples like the Ostiaks used powders ground from dried *F. fomentarius* and *Phellinus igniarius* (L. : Fr.) Quél. fruitbodies as snuff or added them to conventional snuff (Wasson, 1968; Thoen, 1982), while the Athapaska Indians, the Eyak, Tanaina and other Eskimo people of Eastern America used to smoke ashes of the true tinder bracket with or also without mixing it with tobacco (Thoen, 1982). Finally, whole fruitbodies were often used for decorative purposes and, hollowed out, as flower pots in Germany and former Bohemia (Cordier, 1870 cited in Thoen, 1982). A recent use of *F. fomentarius* concerns the art of fly fishing: fishing tackle manufacturers still offer pads of "amadou" as excellent absorbent to use for drying water-logged flies.

3. The two *Piptoporus betulinus* pieces on the leather thongs

The first *Piptoporus betulinus* fragment is more or less shaped like a Scots pine cone with a maximum dimension of 4.5 to 5 cm. It is perforated along its longitudinal axis and mounted on a decorated leather thong which is tied into a quite complicated knot at one end (Fig. 4). The other fragment is more or less globular with somewhat flattened poles; it is 3.5 – 4.7 cm in diameter and 2.4 cm high. It is also mounted on a leather thong, which bears an elaborate three lobed leather tassel. The surface and the context of both polypore fragments are whitish (Figs. 4, 5). A full description is given by Pöder et al. (1992), Peintner et al. (1998).

3.1. Ethnomycological implications

When screening the available ethnomycological and archaeological literature no hints could be found of archaeological findings or prehistoric uses of this distinctive fungus. However, various authors refer to a comparatively recent use of *Piptoporus betulinus* for food, in folk medicine and in other ways.

P. betulinus is one of the few edible polypores, at least when the fruitbodies are young. Wasson (1968, p. 238) mentions that the Kamchadal use a white polypore growing on birch trees for food. He cites Steller (1774), who reported that "the Kamchadal knock them off birches with sticks, break them up with axes, and eat them frozen." For Wasson (1968, p. 239) this fungus is *P. betulinus*, "a soft, white, rather spongy or rubbery growth on the birch, without much taste, which is sometimes eaten raw or cooked by mycophiles in Europe and the United States." In this context Wasson refers to a passage in Krasheninnikov (1755), where the latter speaks of the Kamchadal as "omnivorous creatures, for they pass by neither *zhagra* nor *mukhumor*, though the former has no taste and does not satisfy hunger, and the latter is obviously harmful". *Zhagra* means primarily "punk", "touchwood" but also "tree fungus" or "polypore" in general; referring to Krasheninnikov's text Wasson interprets *zhagra* as "the white polypore on birch trees", (= *P. betulinus*). *Mukhumor*, without doubt, means the hallucinogenic "fly agaric" *Amanita muscaria* (L. : Fr.) Hook.

Before modern medicine superseded many natural healing methods, various polypores were used for medical purposes (Thoen, 1982; Cochran, 1978; Ying et al., 1987): mushroom teas, for instance prepared of "Tchaga" (= *Inonotus obliquus* (Pers. : Fr.) Pilát), were very popular in Russian folk medicine as a remedy against cancer. The Russian author Solzhenitsyn (1968) even dedicated a whole chapter of his book "Cancer Ward" to "Tchaga". Species used for mushroom teas in Russian folklore include also *P. betulinus* (pers. comm. from Dr. R. Golubjatnikov, in Cochran, 1978). Stamets (1993) described the effect of *P. betulinus* teas as anti-fatiguing, immuno-enhancing and soothing.

People in southwest Surrey (Great Britain) used *P. betulinus* mainly externally. The context was cut into small stripes and used as a styptic; Such stripes with a perforation were used as corn pads, while charcoal from *P. betulinus* was much appreciated as an antiseptic. Such antiseptic charcoal was prepared by putting small pieces of the fungus into a casket of iron which was gently heated on a fire until it stopped smoking (Thoen, 1982). Hilton (1987), an Australian paediatrician, packed fruitbody pieces of *P. betulinus* behind ingrowing toenails which led to "excellent results".

In Tibetian traditional medicine a fungus closely related to the birch polypore, *Cryptoporus volvatus* (Peck) Hubbard, was used to cure sore throats: twice daily the patients keep 5 to 8 pieces of the fungus previously simmered in water in the mouth without chewing (Zang, 1984). Furthermore, *C. volvatus* was used against toothache, furuncles and intestinal bleeding (Bo and Yun-Sun, 1980 cited by Kreisel, 1994).

Besides the nutritional and medical purposes, *P. betulinus* was also used in other different ways: fruitbodies of *P. betulinus* have been used as razor strops, in England this polypore and also *Polyporus squamosus* (Huds.) Fr. were even called "razor-strop fungi" (Gilbertson, 1980). For this function, fruitbody strips were glued together to a strap, a practice known to the barbers of Caesalpinus'-day (Grant, 1993). *P. betulinus* fruitbodies have also been processed to make sweat pads in hats. Furthermore, this soft material was used by the Swiss watch industry for polishing the metal parts of their watches (Thoen, 1982). Entomologists use it to make mounting blocks for small insects (Grant, 1993). Noble (1973) reports that the fungi on dead birch trees, called "*Polyporus betulinus*", are "though and light like Polystyrene; they were used by Highlanders as packing for the back of their circular shield or targes." Beekeepers in England are reported to use smouldering *Piptoporus betulinus* and *Daedalea quercina* fruitbodies for anaesthetizing bees (Gilbertson, 1980; Tyler, 1977). Another interesting use is reported by Bondartsev (1986): After a special treatment fruitbodies of *P. betulinus* were used for the manufacture of drawing-charcoal. Moreover Thoen (1982) reported that the pretty white fruitbodies were used in Germany among other polypores like *Fomes fomentarius* as house decorations. In Austria (Styria) fruitbodies of *P. betulinus* were even used for decorative carvings (Lohwag, 1965). Finally, the tradition of dying wool with many fungal species including *Piptoporus betulinus* as well as *F. fomentarius* is still alive in Scandinavian countries (Sunderström and Sunderström, 1982).

4. Discussion

From an ethnomycological point of view one of the most striking facts is that the Iceman, on his long and fatiguing journey, carried two different species of fungi with him, which were treated and stored in different ways. As far as it is known this represents the only case where fungi obviously were part of a prehistoric person's equipment.

The greatest part of the Iceman's girdle bag was filled up with the "Black Matter", a mass of loosely interwoven context hyphae of the "true tinder bracket" (*Fomes fomentarius*). This material does not show the compactness and organisation of untreated or native fungal context: It had been treated mechanically in order to gain a material of a wad-like consistency. Furthermore, traces of pyrite could be found among the hyphae. These facts, namely the special consistency of the material and its safe storage in a bag keeping out dampness, strongly indicate its use as classical fire-starting tinder. However, it can never be proven whether or not it was additionally used for other purposes, for instance as a styptic wound compress.

The two fruitbody pieces of *Piptoporus betulinus*, each mounted separately on a leather thong, certainly served some purpose other than making fire. This polypore has never been reported to be a useful tinder material. On our own experience, material scraped off from fruitbodies of *P. betulinus* burns like paper, but it smoulders badly compared to *F. fomentarius*. Such simple experiments provide another argument against the assumption that *P. betulinus* is a good fire-starting tinder material. But *fide* Swanton (1916), it may be used for the renewing or duplication of fire, if the *Piptoporus* material is placed in a tin with restricted ventilation.

According to the available literature *P. betulinus* has occasionally been used for food, in folk medicine, and for the production of various commodities in historical times. No indications could be found regarding a prehistoric use of this remarkable polypore. On the other hand it is not surprising that the birch polypore has never been found in archaeological sites since the annual fruitbodies of *P. betulinus* are not as hard and durable as the pluriennal ones of *F. fomentarius* or others (e.g., *Phellinus igniarius*); under normal environmental conditions – that means, when they are not preserved in ice as in this case – the annual fruitbodies of *P. betulinus* are degraded by animals and micro-organisms within a short time. *P. betulinus* forms whitish, strikingly shaped fruitbodies exclusively on birch-trees, thus prehistoric man should have noticed this polypore. The fact that the Iceman obviously used *P. betulinus* supports this hypothesis. But what did he use this fungus for?

A merely alimentary use can be excluded for two reasons at least: Firstly, the corky consistency of this fungus, which is accompanied by a rather unpleasant taste, is not inviting. Secondly, who would store two small portions of simple food on such elaborated leather thongs? And what kind of commodity could these objects have been? Shape and size of these two fungal fragments do not give us any hint as to their possible practical application. Also a pure ornamentative or decorative function without any spiritual background seems not very likely: at least the more or less cone-like shaped piece with its irregular outline and its rough surface appears not to be very decorating to us, although we accept that *de gustibus non est disputandum*. But in contrast thereto, many parts of the Iceman's equipment, for instance the stone-disk, which is mounted on a tassel of nicely twisted leather straps, demonstrate the technical skill of this Neolithic people.

Summarising the above mentioned considerations it seems reasonable to us to associate the significance of these fruitbody pieces with a "medical-spiritual" use. The medical and spiritual properties of polypores have been described in the literature many times (e.g., Buller, 1914; Cochran, 1976; Thoen, 1982; Saar, 1991; Blanchette et al., 1992; Hoobs, 1996; Blanchette, 1997).

Concerning medical properties, it has been proven that *P. betulinus* produces pharmacologically active five-cyclic triterpens, the polyporenic acid A, B, C. Animal studies have shown that these substances have antimicrobial and antiphlogistic activity, and they inhibit the growth of malignant neoplastic cells; further, they are immuno-enhancing by acting as interferon inducers (Cross et al., 1940; Birkinshaw et al., 1952; Marcus, 1952; Wandokanty et al., 1958; Efimenko, 1961a; Efimenko et al., 1961b; Blumenberg and Kessel, 1962; Shibata et al., 1968; Cochran, 1976; Kawecki et al., 1978; Ying et al., 1987; Hoog, 1996).

Historically, cultures from tropical Amazonia to the extreme northern polar zones of Eurasia have discovered the power of polypores in preserving and improving human health (Stamets, 1993). Medical and spiritual properties were often mixed up, so it is not surprising that polypores have figured prominently in the cosmological view of native peoples, often being referred to as sources of eternal strength and wisdom (Wasson, 1968; Thoen, 1982; Saar, 1991; Blanchette et al., 1992; Blanchette, 1997).

Blanchette et al. (1992), for example, report on the importance of *Fomitopsis officinalis* (Vill. : Fr.) Bond. et Singer for the shamans of the American Pacific Northwest in the 19th century. *F. officinalis*, a polypore well-known for its medical properties, had an important spiritual as well as medical role in the Indian societies of this region. These peoples called *F. officinalis* "bread of ghosts", and the fruitbodies were used for carving spirit figures which were considered to have supernatural powers. Such carved figures were used by the shamans in rituals for curing the sick. After the death of the shaman, these important objects were placed at the head of the grave to guard the site and protect the shaman's spirit.

Equally interesting is a recent publication by Blanchette (1997): He reports that various North American indigenous people up to the 20th century used *Haploporus odorus* (Sommerf. : Fr.) Sing. – a pluriennal polypore with a fragrant anise-like scent – as a component of sacred objects, of medicine bundles or other religious materials. Medicine bundles are objects with "spirit power" or "magically protective" objects contained in wrappings. They were associated with rituals and considered to have special powers. Sacred robes were also symbols of spiritual powers and often considered to be medicine bundles themselves. "The frequent use of *H. odorus* basidiocarps on sacred robes and in shaman or other medicine bundles demonstrates the reverence and spiritual uses the Plains Indians attributed to this fungus". Older people wore fruitbody necklaces "to ward off illness", and fruitbodies were "burnt to produce a perfumed smoke in case of sickness". Furthermore, many robes were borne in order to receive protection from "supernatural strengths" through them during battles.

Moreover, Blanchette suggests a possible medical use of *H. odorus*. According to him *H. odorus* is identical to the "*Polyporus*" with medical properties mentioned by Hellson (1974). The latter reported, that this "*Polyporus*" was scraped with a knife and affixed to weasels robes when it could be used in emergency first-aid treatment. It was styptic on wounds and boiled with *Psoralea esculenta* to treat coughs. The infusion of the fungus was taken to stop diarrhoea and treat dysentery'.

The analogy of the Iceman's *P. betulinus* fragments on the leather thongs to the *H. odorus* objects illustrated and described by Blanchette (1997) is striking, thus providing additional evidence for their possible medical-spiritual use. If the Indigenous people of North America selected *H. odorus* for its fragrant anise-like scent and its medical properties, why should *P. betulinus* not have been selected for its medical virtues and its beautiful white fruitbodies five thousand years earlier? A further hint for a medical-spiritual use is the fact that *P. betulinus* grows exclusively on the birch-tree, which is regarded as the tree of life and fertility in many European and Siberian myths (e.g., Heeger, 1938; Wasson, 1968). Wasson (1968) for instance writes: "The birch is pre-eminently the tree of Siberian shamanism".

Finally, what can be said about the significance of the Iceman's fungi? Concerning the "Black Matter", its interpretation as classical fire-starting tinder seems well confirmed by the current body of evidence. Regarding the *P. betulinus* objects, it is much more difficult to find an adequate answer without leaving a firm scientific footing. Whatever the Iceman might really have done with these fungi will therefore probably remain his secret.

Abstract

The possible use of the Iceman's fungi is discussed on the basis of available ethnomycological data. There are strong indications that the "Black Matter" in his girdle bag, consisting of *Fomes fomentarius* hyphae, was used as classical fire-starting tinder, while a medical-spiritual use is discussed for the two *Piptoporus betulinus* fruitbody pieces on the leather thongs.

Zusammenfassung

Anhand der verfügbaren ethnomykologischen Daten werden mögliche Verwendungszwecke der beim Eismann gefundenen Pilze erörtert. Die "Schwarze Masse" aus *Fomes fometarius* Hyphen in seiner Gürteltasche diente höchstwahrscheinlich als klassisches Zundermaterial, während für die zwei Fruchtkörperfragmente von *Piptoporus betulinus* an den Lederstreifen medizinisch-kultische Zwecke diskutiert werden.

Résumé

Sur la base des données ethnomycologiques disponibles, les auteurs discutent l'usage possible des champignons retrouvés auprès de l'Homme des glaces: la "matière noire" composée d'hyphes de *Fomes fomentarius*, découverte dans son sac-ceinture, pourrait bien avoir été utilisée comme amadou classique, alors qu'un usage médico-religieux est supposé pour les deux fragments d'un carpophore de *Piptoporus betulinus* retrouvés sur les lanières de cuir.

Riassunto

Sulla base dei dati etnomicologici viene discusso il probabile uso dei funghi dell'uomo del ghiaccio. È moeto probabile che la "materia nera" del marsupio consistente di ife di *Fomes fomentarivs* sia stata usata come classica esca ignitera, mentre per i due pezzi del corpo fruttifero di Piptoporus betulinus, individuate su strisce di pelle, viene preso in considerazione un uso medico- spirituale.

Acknowledgements

This research project has been supported by the Austrian Science Fund (FWF P 10148-SOZ). For providing us with valuable literature or photographical material we are most grateful to F. Bellú, H. Kreisel, I. Krisai-Greilhuber, T. Laessø, M. Moser, H. Müller, P. Stamets, Römisch-Germanisches Zentralmuseum (Mainz, Germany), S. Wasser and E. Zissernig.

References

Birkinshaw J. H., Morgan E. N. and Findlay P. K. (1952) Biochemistry of the wood-rotting Fungi. Biochem. J. 50: 509–516.
Blanchette R. (1997) Haploporus odorus: A sacred fungus in traditional native American culture of the northern plains. Mycologia 98: 233–240.
Blanchette R. A., Compton B. D., Turner N. Y. and Gilbertson R. L. (1992) Nineteenth Century Shaman Grove Guardians ars carved Fomitopsis officinalis Sporophores. Mycologia 84: 119–124.
Blumenberg F. W. und Kessel F. J. (1962) Die Wachstumshemmung des Mäusesarkoms S37 durch den Birkenschwamm (Polyporus betulinus). Arzneimittel-Forschung 13: 198–200.
Broendegaard J. V. (1981) Svampe 3: 9–10.
Buller A. H. R. (1914) The fungus lore of the Greeks and Romans. Transactions of the British Mycological Society 5: 21–66.
Champion H. (1976) Feuermachen vor 5000 Jahren. Helv. Archeol. Archäologie in der Schweiz 7/1976–27/28: 70–74.
Chopra R. N. and Chopra I. C. (1956) Glossar of the Indian Medical Plants, C.S.I.R., New Delhi.
Cochran K. W. (1978) Medical effects of edible mushrooms. In The Biology and Cultivation of Edible Mushrooms. Part I. General Aspects. (eds. Chang S. T. and Hayes W. A.), pp 169–187. Academic Press, New York.
Cross L. C., Eliot C. G., Heilbron I. M. and Jones E. R. H. (1940) J. Chem. Soc. 632: Cited in Birkinshaw J. H. et al., 1951.
Efimenko O. M. (1961a) Polyporenice acid A – an antibiotic isolated from the tinder fungus *Polyporus betulinus*. Antibiotiki 6, 215–220. Cited in Hobbs C., 1996.
Efimenko O. M. (1961b) Triterpenoid acids from *Polyporus betulinus*. Kompleksn. Izuch. Fiziol. Aktivn. Veshchestv. Nizshikh- Rast. Akad. Nauk SSSR, Botan. Inst. Cited in Hobbs C., 1996.
Egg M. und Spindler K. (1993) Die Gletschermumie vom Ende der Steinzeit aus den Ötztaler Alpen. Vorbericht. Mit einem Beitrag von Roswitha Goedecker-Ciolek. Sonderdruck JbRGZM 39, Mainz.
Gilbertson R. L. (1980) Wood-rotting fungi of North America. Mycologia 72: 1–49.
Göpfert H. (1982) Pilze aus jungsteinzeitlichen Siedlungen. Schweizerische Zeitschrift für Pilzkunde. Übergangsheft 1982 B zu Mycologia Helvetica, 50–70.
Grant M. (1993) Letter to Science 260, 9 April 1993, 147.
Heeger F. (1936) Pfälzer Volksheilkunde. Ein Beitrag zur Volkskunde der Westmark. Verlag Daniel Meininger, Neustadt an der Weinstraße.
Hellson J. C. (1974) Ethnobotany of the Blackfood Indians. Canadian Ethnol. Serv. Pap. 19: 1–138. National Museums of Canada, Ottawa.
Herrmann M. (1962) Die Verwendung des Echten Zunderschwammes Fomes fomentarius (Fr.) Kickx, einst und jetzt. Mykologisches Mitteilungsblatt 6(3): 56–62.
Hilton R. N. (1987) Podiatric polypore. Mycologist 21: 121.
Hobbs C. (1996) Medicinal mushrooms. An exploration of Tradition, Healing & Culture. Interweave Press, Loveland, C0.
Kawecki Z., Kaczor J., Karpinska T., Sujak I. and Kandefer-Szerszen M. (1978) Studies of RNA isolation from Piptoporus betulinus as interferon inducers. Archivum Immunologiae et Therapiae Experimentalis 26: 517–522.
Killermann S. (1936) Die ältesten Pilzfunde und Berichte. Zeitschrift für Pilzkunde 22(4): 113–116.
Killermann S. (1938) Ehemalige Apothekenpilze. Zeitschrift für Pilzkunde 22: 11–13.
Krasheninnikov (1755) New edition of the Soviet Academy of Sciences. (ed. Berg L. S., 1946) Cited in Wasson G., 1968.
Kreisel H. (1956/57) Zunderschwämme, *Fomes fomentarius* L. ex Fr., aus dem Mesolithikum. Wissenschaftliche Zeischrift der Ernst Moritz Arndt-Universität Greifswald VI: 299–301.
Kreisel H. (1977) *Lenzites warneri* (Basidiomycetes) im Pleistozän von Thüringen. Feddes Repertorium 88(5–6): 365–373.

Kreisel H. (1996) Beitrag. In: Terberger T., 1996

Lippert A. und Spindler K. (1991) Die Auffindung einer frühbronzezeitlichen Gletschermumie am Hauslabjoch in den Ötztaler Alpen (Gem. Schnals). Archäologie Österreichs 2(2): 11–17.

Lohwag K. (1965) Birkenschwammschnitzereien. Mitt. Naturwiss. Ver. Steiermark 95: 136–139.

Marcus S. (1952) Antibacterial activity of the triterpenoid acid (polyporenic acid C) and of ungulinic acid, metabolic products of Polyporus benzoinus (Wahl.) Fr. Biochem. J. 50: 516–517.

Noble M. (1973) *Fomes fomentarius* in the Highlands of Scotland. Bulletin of the British Mycological Society 7: 35.

Ortner L. (1995) Von der Gletscherleiche zu unserem Urahn Ötzi. Zur Benennungspraxis in der Presse. In: Spindler K., Rastbichler-Zissernig E., Wilfing H., zur Nedded D. und Nothdurfter H. (eds.), Der Mann im Eis-Neue Funde und Ergebnisse. The Man in the Ice, vol. 2. Springer Verlag, Wien: pp 299–320.

Peintner U., Pöder R. and Pümpel T. (1998) The Iceman's Fungi. Mycological Research 102(10): 1153–1162.

Pöder R. (1993) Iceman's Fungi: Discussion Rekindled. Science 262: 24. December 1993, 1956.

Pöder R., Peintner U. und Pümpel T. (1992) Mykologische Untersuchungen an den Pilz-Beifunden der Gletschermumie vom Hauslabjoch. In: Höpfel F., Platzer W. und Spindler K., eds. Der Mann im Eis, Bd. 1. Veröffentlichungen der Universität Innsbruck 187: 311–320.

Pöder R., Pümpel T. und Peintner U. (1995) Mykologische Untersuchungen an der "Schwarzen Masse" vom Hauslabjoch. In: Spindler K., Rastbichler-Zissernig E., Wilfing H., zur Nedded D. und Nothdurfter H. (eds.), Der Mann im Eis - Neue Funde und Ergebnisse. The Man in the Ice, vol. 2. Springer Verlag, Wien: pp 71–76.

Ramsbottom J. (1923) Mushrooms and toadstools. Collins: London. Cited in Thoen D., 1982.

Rolfe R. T. and Rolfe F. W. (1925) The Romance of the Fungus World. Chapman & Hall, London.

Saar M. (1991) Fungi in Kanthy Folk Medicine. Journal of Ethnopharmacology 31: 175–180.

Sauter F. und Stachelberger H. (1992) Materialuntersuchungen an den Begleitfunden des "Mannes vom Hauslabjoch". Die "Schwarze Masse" aus dem "Täschchen". In: Der Mann im Eis, Bd. 1. (Höpfel F., Platzer W. und Spindler K., eds.) Veröffentlichungen der Universität Innsbruck 187: 442–453.

Shibata S., Nishikawa Y., Mai Fufuoka F. and Nahanishi M. (1968) Anti-tumor effects on some extracts of basidiomycetes. GANN 59: 159–161.

Solzhenitsyn A. (1968) Cancer Ward. Farrar, Straus and Giroux, New York.

Sponheimer B. (1936) Eine ungewöhnlich große Nebennutzung aus dem Baumschwamm. Zeitschrift für Pilzkunde 20(2): 77–79.

Stamets P. (1993) Growing gourmet and medical mushrooms. Ten Speed Press, Berkeley.

Steller G. W. (1774) Beschreibung von dem Lande Kamtschatka, dessen Einwohnern, deren Sitten, Namen, Lebensart und verschiedenen Gewohnheiten. Leipzig, pp 92–93. Cited in Wasson G., 1968.

Sunderström C. und Sunderström E. (1982) Färga med svampar. ICAA bokförlag Västeras, Stockholm.

Swanton E. W. (1916) Economic and Folk Lore Notes. Trans. Brit. Myc. Soc. V: 408–409.

Terberger T. (1996) Die "Riesenhirschfundstelle" von Endingen, Lkr. Nordvorpommern. Spätglaziale Besiedelungsspuren in Nordostdeutschland. Mit Beiträgen von Kloss K. und Kreisel H. Archäologisches Korrespondenzblatt 26: 13–32.

Thoen D. (1982) Usage et legendes lies aux Polypores. Note D'Ethnomycology. Bulletin Trimestriel de la Societe Mycologique de France 98: 289–318.

Tyler V. E. (1977) Folk uses of Mushrooms – Medicoreligious aspects. In: Mushrooms and Man, 29–46. Linn Benton Community College. Cited in Thoen D., 1982.

Vaidya J. G. and Bhor G. L. (1991) Medically important wood-rotting fungi with special emphasis to Phansomba. Deerghayu International VII(2): 14–16.

Vaidya J. G. and Rabba A. S. (1993) Fungi in Folk Medicine. The Mycologist 7(3): 131–133.

Walter J. (1982) Trachtenstücke aus dem Buchenschwamm. Da schau her. Beiträge aus dem Kulturleben des Bezirkes Liezen. 3/1982/Juli, 8–10.

Wandokanty F., Utzig J. and Klotz J. (1958) The action of Polyporus betulinus and Poria obliquus on spontaneous tumors in dogs including cancers of the papilla. Medycyna weterynaryjna 3: 148–151.

Wasson R. G. (1968) Soma. Divine mushroom of immortality. Ethno-mycological studies 1. Harcourt Brace Jovanovich.

Weiner J. (1981) Mit Stahl, Stein und Zunder. Die in Vergessenheit geratene Technik des Feuerschlagens. Pulheimer Beiträge zur Geschichte und Heimatkunde 5. Jahresgabe des Vereins für Geschichte und Heimatkunde E. V: 13–18.

Ying J., Mao X., Ma Q., Zong Y. and Wen H. (1987) Icones of medical fungi from China. (Wang H. and Fan S., eds.) Science Press, Beijing.

Zang M. (1984) Mushroom distribution and the diversity of habitats in Tibet, China. McIlvainea 6(2): 14–20.

Contribution to insect remains from the accompanying equipment of the Iceman

W. Schedl

Institut für Zoologie und Limnologie, Universität Innsbruck

1. Introduction

The recovery of the Late Neolithic mummy found in September 1991 in glacial ice of Similaun at the Tisenjoch (3210 m NN) in the Ötztaler Alpen very near to the border of South Tyrol (Italy) afforded an unique opportunity to explore insects remains in the accompanying equipment of the Iceman.

The whole surface of the frozen body of the mummy was inspected by Aspöck und Auer (1992), but no ectoparasites of the class of insects were found. Later on Gothe and Schöl (1992, 1994) have isolated from hair samples numerous remains of Deer keds, *Lipoptena cervi* (L.) (Hippoboscidae, Diptera). Insects of the accompanying equipment of prehistoric Central European human beings scarcely have been documented in the scientific literature, but no doubt without a human connection f.i. in Lendahl (1988), Osborne (1988), and Geiser (1998).

2. Material and methods

The author received a first part of insects remains, approximately a tenth of all, from Prof. Klaus Oeggl (Institute of Botany, University of Innsbruck) at the beginning of May 1994. The numbers of the material belong to a sorting out of the Römisch Germanisches Zentralmuseum, Mainz, Germany. The insect remains found in different material of the Iceman clothing and equipment as in leather, fur, hairs, string, grass and texture have been conserved in a mixture of alcohol, glycerin and distilled water (1:1:1). 36 vials were cleaned, analyzed and inspected under a stereomicroscop of WILD M 5 by spreading in a small petri dish. Some of greater remains were mounted on little labels, some as usual microscopical slides for further photographic interpretation made with a WILD Photomakroskop M 400.

3. Results

In the investigated samples the insect remains are not well preserved, often they were soiled with minerals, they are often fragile touching them with a needle or a fine hair brush, but nevertheless several important taxonomic features could be proved and particularly demonstrated. No specimen has been complete. Only one small insect, a drosophilid fly, has shown all three body parts, but without their wings and legs.

In most cases I could save of the sample material only few typical sclerites of the caput, thorax, abdomen, the legs or wings. Some of these remains are possible to coordinate to a distinct order of insects, in some cases also to a lower category of the system (Table 1). Therefore it has been of great advantage, that the author has worked in a wider entomological interest from 1966 to 1974 in the subalpine up to the higher alpine belt of the Ötztaler Alpen (Schedl, 1970, 1972, 1976, 1982).

From remains of Coleoptera I could recognize often elytra of 1,2–4,0 mm in length, the fine-structures were lost, but parallel stripes or rows of punctures are visible. Some elytra and double-folded hind wings belong to small Staphylinidae. Three times I have found the sixth tergite with characteristical distal spines (Fig. 7) of the genus *Tachinus* (see Lohse, 1964). Other elytra belong to small Curculionidae, perhaps to *Polydrosus* and/or *Dichotrachelus* (?), other three elytra belong to Coccinellidae, one fits exactly to the Scolytidae *Pityokteines spinidens* Reitter, which is boring in the bark of *Abies alba*, seldom also in *Picea abies, Pinus sylvestris* and *Larix decidua* (Schedl, K. 1981) (Fig. 2). This barkbeetle the author has not expected in these samples.

Some femora and tibiae from the leg I with special end-spines are characteristically for the Scarabaeidae genus *Aphodius* (Fig. 6), also one caput and pronotum has been found in the same sample. Species of *Aphodius* are abundant in the grass-heath of the subalpine and alpine belt, also driffted on perpetuated snow-fields and on glaciers.

Of the order hymenoptera I could recognize some small capites without antennae (Formicoidea), parts of thoraces and tergites (also from Vespoidea), two bases of the gaster (from an aculeate and a parasitic hymenoptera), and a part of a hind wing perhaps of a winged Formicidae (?).

Table 1. Evidences of taxa based on recognizable parts of insect remains in different categories of studied specimens

	order	superfamily	family	genus	species
Coleoptera	42		Staphylinidae	*Tachinus*	
			Scarabaeidae		
			Coccinellidae		
			Scolytidae	*Pityokteines*	*spinidens* Reitter
Hymenoptera	11	Formicoidea			
		Ichneumonoidea			
		Vespoidea	Vespidae		
Heteroptera	11		Lygaeidae	*Oxycarenus*	*modestus* (Fallén)
Diptera	9		Drosophilidae		
			cf. Syrphidae		
Siphonaptera	2		Pulicidae	*Pulex*	cf. *irritans* L.
Homoptera	2	"leaf hopper"			
		Psylloidea	Triozidae	*Bactericera*	cf. *striola* (Flor)

Some fore wings of small Diptera (Fig. 4) belong sensu Colyer (1951), and Bächli und Burla (1985) to Drosophilidae, one probably to Syrphidae (Fig. 5). One slightly decayed body of a Drosophilidae (head with antenna basis and arista visible) I have mentioned above. There exists also one haltere of 0,5 mm length.

The order Homoptera is represented only by one end of an abdomen with typical ovipositor of a leaf-hopper (cf. Cicadelloidea) and of a fore-wing of a Psylloidea, Triozidae, cf. *Bactericera striola* (Flor), which is living on *Salix* spp. normally in lower belts (in litt. 2.8.94 Burckhardt). From the order Heteroptera are conserved in some samples several clavi of the Lygaeidae of *Oxycarenus modestus* (Fallén), in one sample (Nr.91/123) also the coria and membrane of a right and left fore wing and two capites partly with prothorax of the same species (Fig. 1). This is a boreomontane species living on leaves of *Alnus incana* and *glutinosa* known also from North and South Tyrol (Heiss, 1973; Tamanini, 1982). One small hind wing (Fig. 3) I try to associate with a species of the order Psocoptera. Even from the order Siphonaptera are isolated from sewed leather of the Iceman two capites, not complete, but with the typical blood-sucking mouthparts (Fig. 8).

The shape of the head without pecten, the position of the complex-eye, the antennal cavity and parts of the tentorium leads to *Pulex irritans* L. (Smit, 1957). There exist also no other species without pectines and such a head-shape in this altitude. Perhaps further remains of this flea species will bring taxonomical clearness. In order to this finding it is demonstrated that the Iceman has had probably at least 2 human fleas in his dress.

The greatest part of the insects remains (80–90%) are untypical, only parts of sclerites and unpossible to place them to an insect order. In the sample number 91/123 are found two small mites of the family of Oribatidae (2 species!). Although all hitherto existing samples for this study came directly from the accompanying equipment of the Iceman in September 1991. It is not out of question, that some remains are less than 5000 years old. The bad condition of the remains and the presented material let me believe, that these remains are Neolithic. Till now I could not recognize any remain of an insect larva and no very typical insect remain for the case, that the Iceman has been exposed a longer time at sunny weather for attack of typical dead body insects sensu Smith (1986).

Abstract

A first part of insect remains found in different materials of leather, fur, hairs, string, grass and texture of the famous Iceman from 3210 m NN in September 1991 has been conserved in 36 vials in a special solution. Most of the remains are small (under 5 mm in length) and in poor condition. Only more sclerotizised parts of the ectosceleton have survived 5000 years in the surroundings of the mummy. Relativ numerous are remains of Coleoptera (f.i. of small Staphylinidae, Scarabaeidae, Coccinellidae), of Hymenoptera (f.i. Formicoidea), of Diptera and small Heteroptera (*Oxycarenus modestus*). Very interesting is the probable evidence of two capites of *Pulex irritans* L. found in sewed leather, a real ectoparasite of human being.

Zusammenfassung

Ein erster Teil der Insektenreste aus den verschiedenen Materialien – Leder, Fell, Haare, Schnüre, Gras und Gewebe – des berühmten Eismannfundes vom September 1991 auf 3210 m.ü.M. wurde in 36 Probenröhrchen in einer Speziallösung konserviert. Die meisten dieser Reste sind klein (unter 5 mm lang) und in schlechtem Zustand. Nur stärker sklerotisierte Teile des Außenskeletts haben die 5000 Jahre in der Umgebung der Mumie überlebt. Relativ zahlreich sind Reste von Käfern (z.B. kleine *Staphylinidae, Scarabaeidae, Coccinellidae*), Hautflüglern (z.B. *Formicoidea*), von Zweiflüglern und kleinen Wanzen (*Oxycarenus modestus*). Sehr interessant sind die in einem Stück

Fig. 1. Pair of fore-wings of *Oxycarenus modestus* (Fallén) without the basical cavi (nat. length of the left wing 3,6 mm) MPr.7.
Fig. 2. Left elytron of the Coleoptera *Pityokteines spinidens* Reitter (nat. length 2,0 mm) TPr.4. **Fig. 3.** Fore-wing of a Triozidae, cf. *Bactericera striola* (Flor) (nat. length 3,05 mm) MPr.6. **Fig. 4.** Fore-wing of a Drosophilidae (nat. length 2,75 mm) MPr.5.
Fig. 5. Apical part of a Diptera fore-wing (Syrphidae ?) (nat. length 5,9 mm) MPr.8

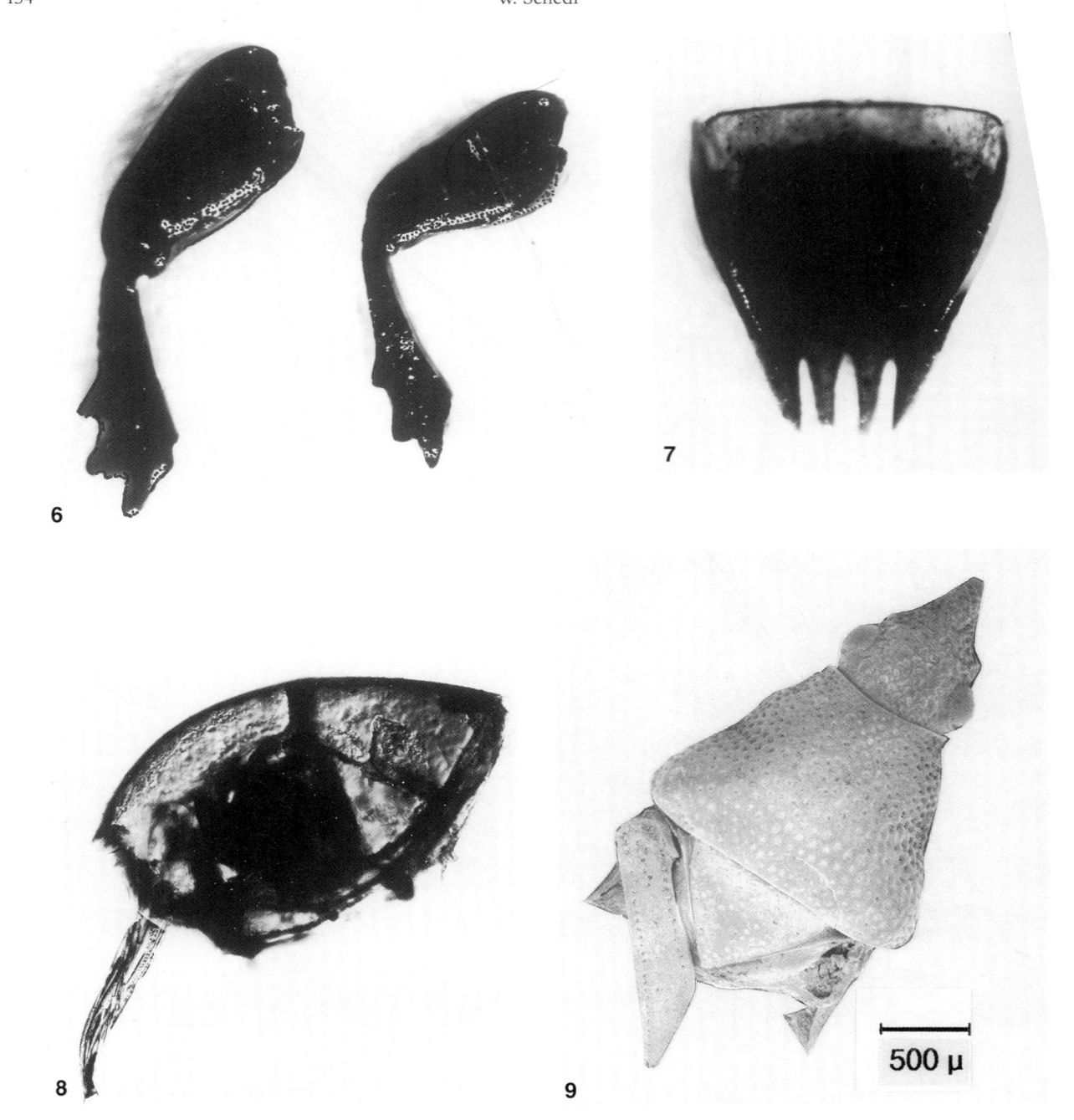

Fig. 6. Femur/Tibia I of *Aphodius* sp. (nat. length 2,0 mm) TPr.2. **Fig. 7.** 6th tergite of the coleoptera *Tachinus* sp. (nat. length 1,3 mm) TPr.3. **Fig. 8.** Main part of a caput of *Pulex irritans* L. (?) with the blood-sucking mouthparts epipharynx and two laciniae, without antenna, sand-grain before the compound eye (nat. length 0,65 mm) MPr.10. **Fig. 9.** Caput, prothorax and part of the mesothorax with left clavus of the Heteroptera *Oxycarenus modestus* (Fallén) (nat. length 2,4 mm) TPr.1, REM-foto

genähtem Leder gefundenen Hinweise auf zwei Exemplare von *Pulex irritans* L., einem echten Ektoparasiten des Menschen.

Résumé

Une première partie des résidus d'insectes détectés dans les différents matériaux–cuir, fourrure, poils, cordes, herbe et tissus–retrouvés sur le célèbre Homme des glaces, découvert en septembre 1991 à 3 210 m d'altitude, a été conservée dans 36 éprouvettes remplies d'une solution spéciale. La plupart des résidus sont très petits (c.-à-d. d'une longueur inférieure à 5 mm) et en très mauvais état. Seules les parties sclérosées de l'ectosquelette se sont conservées pendant 5000 ans auprès de la momie. Parmi ces résidus, assez nombreux sont ceux de coléoptères (ex.: *Staphylinidae, Scarabaeidae, Coccinellidae* de petite taille), d'hyménoptères (ex.: *Formicoidea*), de diptères et de petits hétéroptères (*Oxycarenus modestus*). A signaler la présence probable, dans un morceau de cuir cousu, de deux exemplaires de *Pulex irritans* L., authentique ectoparasite de l'homme.

Riassunto

Una prima parte di relitti di insetti individuati nei diversi materiali quali cuoio, pellicce, capelli, corde, erba, tessuti dell'uomo del Similaun, ritrovato a 3210 m s.l.m. nel settembre del 1991, sono stati conservati in 36 fiale in una soluzione speciale. La maggior perte dei relitti sono di dimensioni ridotte (meno di 5 mm di lunghezza), ed in cattive condizioni, soltanto le parti più sclerotizzate dell'Ectoskeleton sono sopravissute da 5000 anni nell'immediata vicinanza della mummia. Relativamente numerosi sono i resti di coleotteri (p.e di piccole Staphylinidae, Scaraboeidae, Coccinellidae) di Hymenoptera (Oxycarenus modestus).

Molto interessante è il ritrovamento su pelle cucita di due esemplari classificabili probabilmente quali Pulex irritans L., un vero e proprio ectoparassita umano.

Acknowledgements

The determination of the Lygaeidae down to the species level from Prof. Dr. Ernst Heiss (Innsbruck) and of the Triozidae from Dr. Daneil Burckhardt (Muséum d'Histoire naturelle, Genève) is grateful acknowledged. I also like to thank Univ. Doz. Dr. G. Haszprunar for his support at the POLYVAR and Mr. K. Eller for making one picture at the ZEISS DSM 950 (both from the Institute of Zoology, University of Innsbruck).

References

Aspöck H. und Auer H. (1992) Zur parasitologischen Untersuchung des Mannes vom Hauslabjoch. In: Höpfel F. et al. (eds.) Der Mann im Eis, Bd. 1. Veröffentlichungen der Universität Innsbruck 187: 214–217.

Bächli G. und Burla H. (1985) Drosophilidae. Insecta helvetica. Fauna 7: 1–116.

Colyer C. N. (1951) Flies of the British Isles. London New York, pp 384.

Geiser E. (1998) 8000 Jahre alte Reste des Bergblattkäfers *Oreina cacaliae* (Schrank) von der Pasterze. Wiss. Mitt. Nationapark Hohe Tauern 4: 41–46.

Gothe R. und Schöl H. (1992) Hirschlausfliegen (Diptera, Hippoboscidae: hussiv) in den Beifunden der Leiche vom Hauslabjoch. In: Höpfel F. et al. (eds.) Der Mann im Eis, Bd. 1. Veröffentlichungen der Universität Innsbruck 187: 299–306.

Gothe R. and Schöl H.: Deer keds (*Lipoptena cervi*) in the accompanying equipment of the Late Neolithic human mummy from the Similaun, South Tyrol. Parasit. Res., Heidelberg 80: 80–83.

Heiss E. (1973) Zur Heteropterenfauna Nordtirols (Insecta: Heteroptera) III. Lygaeoidea. Veröff. Mus. Ferd. Innsbruck 53: 125–158.

Lendahl G. (1988) Quarternary fossil insects, a source material for faunal history. Entomol. Tidskr., Stockholm, 109: 1–13.

Lohse G. M. (1964) Fam. Staphylinidae I (Micropeplinae bis Tachyporinae). In: Freude H. et al. (eds.) Die Käfer Mitteleuropas, Krefeld, Bd. 4: 1–264.

Osborne P. J. (1988) A late Bronze Age insect fauna from the river Avon, Warwickshire, England. J. Archaeol. Sci. 15: 715–728.

Schedl K. E. (1988) Fam. Scolytidae. In: Freude H. et al. (eds.) Die Käfer Mitteleuropas. Krefeld, Bd. 10: 34–99.

Schedl W. (1970) Planipennia (Insecta: Neuroptera) der subalpinen und höheren Stufen der Ötztaler Alpen (Tirol, Österreich). Ber. nat. med. Ver. Innsbruck 58: 305–312.

Schedl W. (1972) Bockkäfer (Insecta: Coleoptera, Cerambycidae) aus der subalpinen Stufe der Ötztaler Alpen (Tirol, Österreich). Ber. nat. med. Ver. Innsbruck 69: 87–93.

Schedl W. (1976) Untersuchungen an Pflanzenwespen (Hymenoptera: Symphyta) in der subalpinen bis alpinen Stufe der zentralen Ötztaler Alpen (Tirol, Österreich). Veröffentlichungen der Universität Innsbruck, Alpin-Biol. Studien VIII: 1–85.

Schedl W. (1982) Über aculeate Hymenopteren der zentralen Ötztaler Alpen (Tirol, Österreich) (Insecta: Hymenoptera). Ber. nat. med. Ver. Innsbruck 69: 87–93.

Smit F. G. A. M. (1957) Siphonaptera. In: Handbooks Identific. Brit. Insects, London (16): 1–94.

Smith K. V. (1986) A Manual of Forensic Entomology. London, pp 205.

Tamanini L. (1982) Gli Eteroteri dell'Alto Adige (Insecta: Heteroptera), Studi trent. Sci. nat., Acta Biol. 59: 65–194.

Compilation of DNA sequences from the Iceman's grass clothing

S. Antonini[1], S. Luciani[1], I. Marota[1], M. Ubaldi[1], and F. Rollo[1,2]

[1] Dipartimento di Biologia molecolare, cellulare e animale, Università di Camerino
[2] Scuola di Specializzazione in Biochimica e Chimica clinica, Università di Camerino

1. Introduction

In winter 1993, in our laboratory, the isolation of *Gramineae* DNA from a sample of radiocarbon dated (3362–3136 B.C.) grass (Rollo et al., 1994a; Rollo et al., 1994b) found at the Iceman's site during the 1992 archaeological campaign (Bagolini et al., 1995) demonstrated for the first time that the particular environment of an Alpine glacier had allowed the "molecule of life" to survive for more than five thousand years (Rollo et al., 1994a) thus opening the way to the palaeomolecular characterization of the biological remains (Rollo et al., 1994b). This finding was later confirmed by the isolation of the DNA of the mummy itself in Munich and Oxford (Handt et al., 1994).

Together with the original DNA of the grass, however, we also isolated other DNAs belonging to different organisms: algae, fungi, yeasts and most curiously, two dicotyledonous plants.

The origin of the fungal and algal DNAs became evident as soon as the samples were analyzed by scanning electron microscopy: the grass was covered by fungal hyphae and algal cysts (Rollo et al., 1994a; Rollo et al., 1994b; Haselwandter and Ebner, 1994; Rollo et al., 1995b; Rollo et al., 1995c). On the other hand the provenance of the docotyledonous plant DNAs was, and still is, puzzling. To explain this finding, one can only tentatively suggest that the grass cloak became stained with sap from a dicotyledonous plant (Rollo et al., 1994b).

The question is now whether the molecular data can offer us some clue about the Iceman's days: was, for example, the Iceman's grass stained with plant sap due to the fact that the man was lying on top of the undergrowth or, rather, was it the consequence of some sort of manual activity such as stripping a branch? We can also wonder if the microorganisms found on the grass became associated with the clothing when that was worn by the Iceman's or, conversely, whether the fungi and the algae colonized the remains of the equipment (and the body) following the death of the man.

For what concerns the colonization of the equipment by microorganisms, we can identify five possible stages (Table 1): a first stage, encompassing the growth of the grass in the wild till its harvesting by the Iceman or by someone for him; a second, following the harvesting of the grass and during the Iceman's everyday life; a third stage, from the death of the man till the covering of the body by the snow; a fourth, the longest, between the covering of the mummified remains by the snow and their recovery in 1991 and 1992 by the archaeologists; the last one, following the recovery of the remains from the glacier. However, due to the precautions taken, it does not seem probable that the grass has undergone a relevant colonization by modern microorganisms following its recovery from the glacier.

Whatever the time of the colonization of the grass by fungi and algae may be, it seems reasonable to affirm that knowledge of the ancient microorganisms could add significantly to the knowledge of the environment in which the Iceman lived, and/or to the understanding of the physical, chemical and biological factors that have contributed to the exceptionally good preservation of the remains.

As the correct interpretation of the data requires the contribution of many different specialists from a number of diverse disciplines such as microbiology, algology, botany, ecology, taphonomy, etc. we have made every effort for the database to be easily approachable by the broadest readership.

2. Technical notes

The DNA sequences were obtained using polymerase chain reaction amplification of the DNA extracted from grass samples T (= Tisenjoch) 27, 44 and 182 (Bagolini et al., 1995) followed by cloning of the amplicons and sequencing as previously described (Rollo et al., 1994a; Rollo et al., 1994b; Rollo et al., 1995b; Rollo et al., 1995c).

Search for homologous sequences in the EMBL and GenBank databases was performed using FAST-A.

Table 1. Hypothetical chronology of the colonization of the Iceman's grass by microorganisms (according to Rollo et al., 1995c)

Phase 1	In the wild: the grass hosts parasitic and/or epiphytic microorganisms
Phase 2	The grass is harvested and knotted: mesophilic/thermophilic saprophytes colonize the hay
Phase 3	The man dies: additional/different saprophytes thrive on the grass clothing and on the decomposing body
Phase 4	The corpse is covered by layers of snow: psychrophilic microorganisms colonize the remains[a]
Phase 5	The site is excavated by the archaeologists: the grass is colonized by opportunistic saprophytes from the present-day environment[b]

[a] In principle one cannot exclude that the remains underwent repeated cycles of freezing and thawing.
[b] Experimental evidence in this sense indicates that the contamination of the remains, if any, has been very low.

DNA sequences were aligned using the Higgins-Sharp function (Clustal 4) on the MAC DNASIS package (Hitachi). Evolutionary distances were corrected for superimposed mutations and the trees were constructed according, respectively, to the Tajima and Nei (1984) and the neighbor-joining (Saitou and Nei, 1987) options in the TREECON (Van de Peer and DeWachter, 1993) program.

3. Availability of the sequence data

Sequence data from the Iceman have been published elsewhere (Rollo et al., 1995a; Rollo et al., 1995b; Rollo et al., 1995c). The accession numbers of the reference sequences from EMBL and GenBank databases, are reported in the present paper.

4. Land plants

T2720 and T2717 (T2717 being identical to T2720) correspond to a 483 bp long portion of the chloroplast gene for the large subunit of the ribulose 1,5-bisphosphate carboxylase (*rbc*L). Figure 1 shows a phylogenetic tree of the grasses, including T2720. The accession

Fig. 1. *rbc*L phylogeny of the grasses including the T2720 (Iceman's grass) sequence. Bootstrap values for 100 replicates are shown. The scale indicates the branch length corresponding to 10 changes per 100 nucleotide positions

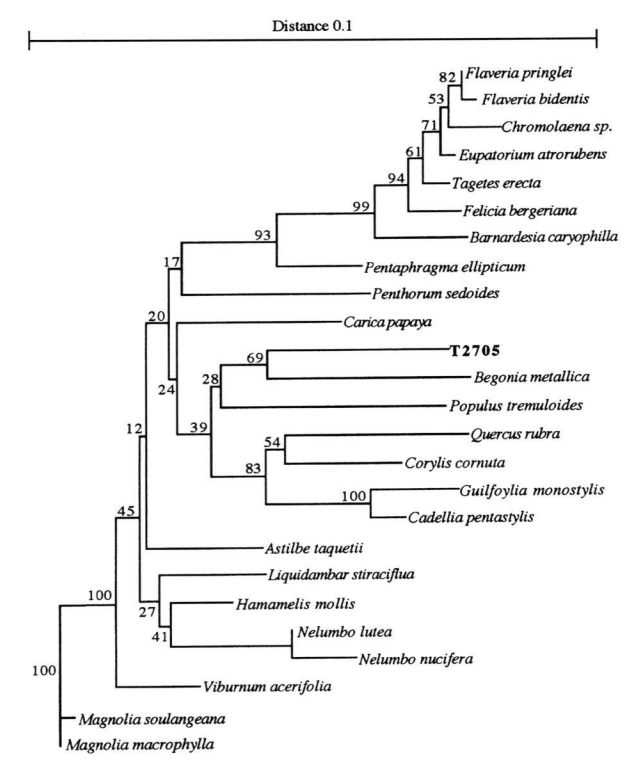

Fig. 2. *rbc*L phylogeny of dicotyledonous plants showing relatedness to the T2705 (contaminant of the Iceman's grass) sequence. Bootstrap values for 100 replicates are shown. The scale indicates the branch length corresponding to 10 changes per 100 nucleotide positions

codes of the reference sequences from databases used to construct the tree are reported in Table 2.

T2705 and T4420 correspond to a 483 bp long portion of the chloroplast gene for the large subunit of the ribulose 1,5-bisphosphate carboxylase (rbcL). Figures 2 and 3 show phylogenetic trees of dicotyledonous plant sequences related, respectively to the T2705 and T4420 sequences. The accession codes of the

Table 2. Sources of sequence data

Organism	EMBL/GenBank accession No.
Aegilops crassa	X62118
Avena sativa	L15300
Bambusa glaucesens	M91626
Cenchrus setigerus	L14622
Hordeum vulgare	X00630
Oryza sativa	X15901
Neurachne munroi	X55828
Neurachne tenuifolia	X55827
Pennisetum glaucum	L14623
Puccinellia distans	L14621
Sorghum bicolor	–
Sparganium sp.	M91633
Triticum aestivum	D00206
Zea mays	V00171

Table 3. Sources of sequence data

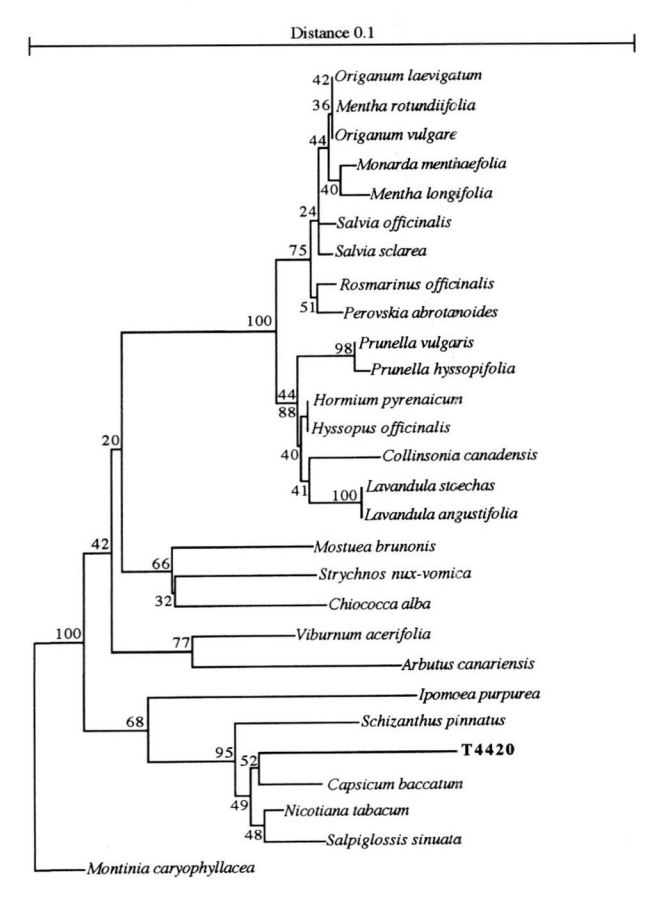

Organism	EMBL/GenBank accession No.
Astilbe taquetii	L11173
Barnardesia caryophilla	L01887
Begonia metallica	L01888
Cadellia pentastylis	L29491
Carica papaya	M95671
Chromolaena sp.	L13640
Corylis cornuta	X56619
Eupatorium atrorubens	L13649
Felicia bergeriana	L13639
Flaveria bidentis	X55830
Flaveria pringlei	X55829
Guilfoylia monostylis	L29494
Hamamelis mollis	L01922
Llquidambar stiraciflua	M58394
Magnolia macrophylla	X54345
Magnolia soulanngeana	M58393
Nelumbo lutea	H77032
Nelumbo nucifera	H77033
Pentaphragma ellipticum	L18794
Penthorum sedoides	L11197
Populus tremuloides	M58392
Quercus rubra	M58391
Tagetes erecta	L13637
Viburnum acerifolia	L01959

Fig. 3. rbcL phylogeny of dicotyledonous plants showing relatedness to the T4420 (contaminant of the Iceman's grass) sequence. Bootstrap values for 100 replicates are shown. The scale indicates the branch length corresponding to 10 changes per 100 nucleotide positions

reference sequences from databases used to construct the tree are reported in Tables 3 and 4.

5. Algae

T2725 is a 483 bp long portion of the chloroplast gene for the large subunit of the ribulose 1,5-bisphosphate carboxylase (rbcL). Figure 4 shows a phylogenetic tree of green algae (Chlorophyta) protists (Euglenophyta) and land plants (Bryophyta, Filicophyta, Psilotophyta and Pinophyta) including T2725. The accession codes of the reference sequences from databases used to construct the tree are reported in Table 5.

6. Fungi

T18217 is a 158 bp long sequence spanning the entire nuclear gene for the 5.8s ribosomal RNA (5.8s rRNA)

Table 4. Sources of sequence data

Organism	EMBL/GenBank accession No.
Arbutus canariensis	L12597
Capsicum baccatum	U08610
Chiococca alba	L14394
Collinsonia canadensis	Z37387
Hormium pyrenaicum	Z37393
Hyssopus officinalis	Z37394
Ipomoea purpurea	X60663
Lavandula stoechas	Z37408
Lavandula angustifolia	Z37404
Mentha longifolia	Z37416
Mentha rotundiifolia	Z37417
Monarda menthaefolia	Z37420
Montinia caryophyllacea	L11194
Mostuea brunonis	L14404
Nicotiana tabacum	Z00044
Origanum laevigatum	Z37426
Origanum vulgare	Z37427
Perovskia abrotanoides	Z37428
Prunella hyssopifolia	Z37432
Prunella vulgaris	Z37433
Rosmarinus officinalis	Z37434
Salpiglossis sinuata	U08618
Salvia officinalis	Z37446
Salvia sclarea	Z37450
Schizanthus pinnatus	U08619
Strychnos nux-vomica	L14410
Viburnum acerifolia	L01959

Table 5. Sources of sequence data

Organism	EMBL/GenBank accession No.
Adiantum raddianum	U05906
Anabaena sp.	L02521
Andreaea rupestris	L13473
Chara connivens	L13476
Chlamydomonas moewusii	M15842
Chlamydomonas reinhardtii	J01399
Chlorella ellipsoidea	M20655
Codium fragile	M67453
Coleochaete orbicularis	L13477
Euglena gracilis	M12109
Marchantia polymorpha	X04465
Microbiota decussata	L12575
Nitella translucens	L13482
Physcomitrella patens	X74156
Psilotum nudum	L11059
Spyrogira maxima	L11057
Sirogonium melanosporum	L13484

Other fungal sequences from the Iceman's grass are the T2709, the T44NS4, and the T44NS7. They have been described elsewhere; in particular, the T44NS7 sequence has been shown to originate from a microorganism related to the psychrophilic basidiomycetous yeast *Leucosporidium scottii* (Rollo et al., 1995c). Figure 5

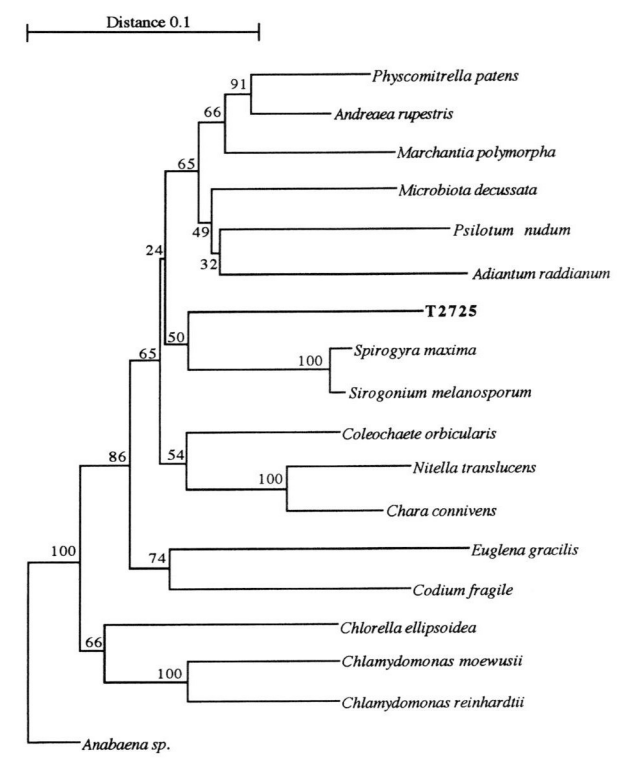

Fig. 4. *rbc*L phylogeny of green algae (Chlorophytes) protists (Euglenophyta) and land plants (Bryophyta, Filicophyta, Psilotophyta and Pinophyta) showing relatedness to the T2725 (contaminant of the Iceman's grass) sequence. Bootstrap values for 100 replicates are shown. The scale indicates the branch length corresponding to 10 changes per 100 nucleotide positions

Table 6. Sources of sequence data

Organism	EMBL/GenBank accession No.
Acremonium uncinatum	L20305
Alternaria alternata	X17454
Candida albicans	X71088
Cephalosporium acremonium	X06574
Chaetomium globosum	–
Cladosporium fulvum	L25430
Cladosporium sphaerospermum	L25433
Epichloe typhina	L07136
Epicoccum nigrum	–
Fusarium sambucinum	X65480
Leptosphaeria maculans	L07735
Neurospora crassa	X02447
Ophiosphaerella korrae	U04862
Penicillium variabile	L14507
Saccharomyces cerevisiae	U09327
Sclerotinia sclerotiorum	M96382
Septoria nodorum	U04237
Talaromyces mimosinus	L14526
Thermomyces lanuginosus	M10392

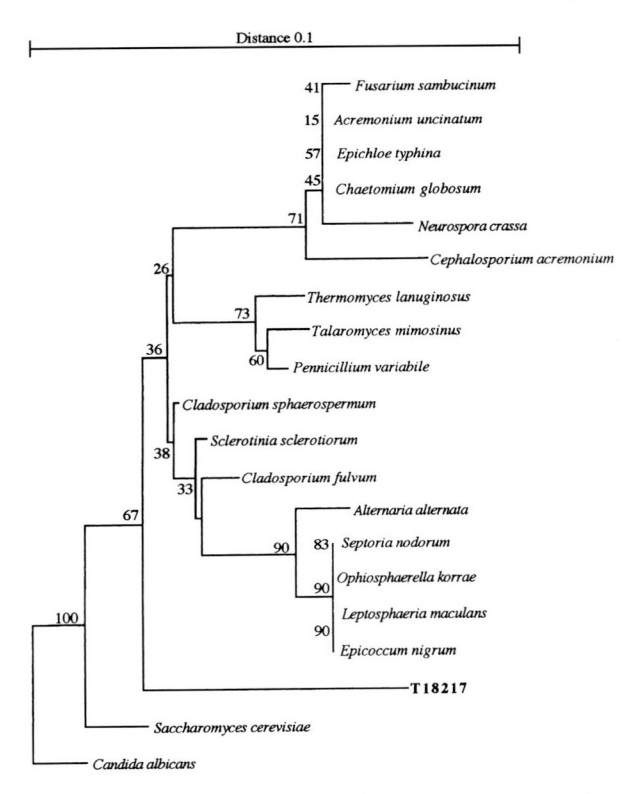

Fig. 5. 5.8s rRNA phylogeny of the ascomycetous fungi including the T18217 (fungus associated to the Iceman's grass) sequence. Bootstrap values for 100 replicates are shown. The scale indicates the branch length corresponding to 10 changes per 100 nucleotide positions

shows a 5.8s rRNA phylogeny of the Ascomycetes, which includes the T18217 sequence. The accession codes of the reference sequences from databases used to construct the tree are reported in Table 6.

Abstract

The paper presents a collection of DNA sequences of different origin (plant, algal, fungal sequences) isolated from three samples of grass collected during the August 1992 archaeological survey of the Iceman's site. The grass has been radiocarbon dated to 3362–3136 B.C. (Rollo et al., 1995a) and is believed to come from the cloak or the boots worn by the neolithic shepherd/hunter. In order to make the compilation as "friendly" as possible to consult, the DNA sequences have been compared with the corresponding sequences from EMBL and GenBank databases and used to construct phylogenetic trees of the relevant *taxa*. The compilation is intended as a reference tool and, possibly, as a stimulating subject of speculation to all those – microbiologists, botanists, ecologists, ethnologists, anthropologists and forensic scientists – presently working on the Iceman.

Zusammenfassung

Aus drei Grasproben, die während der archäologischen Ausgrabung auf dem Tisenjoch im August 1992 gesammelt wurden, hat man wurden DNA-Sequenzen verschiedener Herkunft (Algen, Pilze, Sproßpflanzen) isoliert. Das Gras wurde auf 3362–3136 v. Chr. radiokarbon-datiert (Rollo et al., 1995a) und stammt wahrscheinlich vom Grasumhang oder aus der Schuhfüllung des neolithischen Hirten/Jägers. Um die Zusammenstellung möglichst benutzerfreundlich zu gestalten, wurden die DNA-Sequenzen mit den korrespondierenden Sequenzen von EMBL und GenBank Datenbasen verglichen und verwendet, um phylogenetische Verbindungen der relevanten Taxa zu konstruieren. Die Zusammenstellung versteht sich als ein Hilfsinstrument und möglicherweise als ein stimulierendes Spekulationssubjekt für alle Mikrobiologen, Botaniker, Ökologen, Ethnologen, Anthropologen und forensischen Wissenschafter, die im Moment am Mann im Eis forschen.

Riassunto

L'articolo presenta una raccolta di sequenze di DNA di origine differente (vegetale, algale, fungina) determinate a partire da tre campioni di erba raccolti durante l'esplorazione archeologica compiuta nell'agosto del 1992, presso il sito archeologico dell'Uomo venuto dai ghiacci. I campioni sono stati datati tra il 3362 e il 3136 a.C. con il metodo del radiocarbonio (Rollo et al., 1995a) e, si ritiene, provengano dalla mantella o dai calzari indossati dal pastore/cacciatore neolitico. Al fine di rendere la compilazione quanto più possibile di agevole consultazione, le sequenze di DNA sono state paragonate con quelle conservate nelle banche dati EMBL e GenBank e inserite in altrettanti alberi filogenetici. La raccolta è offerta come uno strumento di riferimento e, magari come spunto per interessanti speculazioni teoriche a quanti – microbiologi, botanici, ecologi, etnologi, antropologi e medici legali, sono attualmente al lavoro intorno all' Uomo venuto dai ghiacci.

Acknowledgements

Our research was supported by Ministero per l'Università e la Ricerca Scientifica e Tecnologica (M.U.R.S.T.), by Ministero per le Risorse Agricole, Alimentari e Forestali (M.R.A.A.F.) and by Scuola di Specializzazione in Biochimica e Chimica clinica, University of Camerino.

References

Bagolini B., Dal Ri L., Lippert A. und Nothdurfter H. (1995) Der Mann im Eis: Die Fundbergung 1992 am Tisenjoch, Gem. Schnals, Südtirol. In: Spindler K., Rastbichler-Zissernig E., Wilfing H., Zur Nedden Z. und Nothdurfter H. (eds): Der Mann im Eis, Bd. 2. Springer-Verlag, Wien New York, pp. 3–22.

Handt O., Richards M., Trommsdorff M., Kilger C., Simanainen J., Georgiev O., Bauer K., Stone A., Hedges R., Schaffner W., Utermann G., Sykes B. and Pääbo S. (1994)

Molecular genetic analyses of the Tyrolean Iceman. Science 264: 1775–1778.

Haselwandter K. und Ebner M. R. (1994) Microorganisms surviving for 5300 years. FEMS Microbiol. Lett. 116: 189–194.

Rollo F., Asci W. and Marota I. (1994a) Neolithic plant DNA from the Iceman's site: rare, long and nicked. Ancient DNA Newsletter 2: 21–23.

Rollo F., Asci W., Antonini S., Marota I. and Ubaldi M. (1994b) Molecular ecology of a neolithic meadow: the DNA of the grass remains from the archaeological site of the Tyrolean Iceman. Experientia 5: 576–584.

Rollo F., Asci W., Marota I. and Sassaroli S. (1995a) DNA analysis of grass remains found at the Iceman's archaeological site. In: Spindler K., Rastbichler-Zissernig E., Wilfing H., Zur Nedden Z. and Nothdurfter H. (eds): Der Mann im Eis, Bd. 2. Springer-Verlag, Wien New York, pp. 91–105.

Rollo F., Antonini S., Ubaldi M. and Asci W. (1995b) The "neolithic" microbial flora of the Iceman's grass: morphological description and DNA analysis. In: Spindler K., Rastbichler-Zissernig E., Wilfing H., Zur Nedden Z. and Nothdurfter H. (eds): Der Mann im Eis, Bd. 2. Springer-Verlag, Wien New York, pp. 107–114.

Rollo F., Sassaroli S. and Ubaldi M. (1995c) Molecular phylogeny of the fungi of the Iceman's grass clothing. Curr. Genet. 28: 289–297.

Saitou N. and Nei M. (1987) The neighbor-joining method: a new method for reconstructing phylogenetic trees. Mol. Biol. Evol. 4: 406–425.

Tajima F. and Nei M.: Estimation of evolutionary distance between nucleotide sequences. Mol. Biol. Evol. 1: 269–285.

Van de Peer Y. and De Wachter R. (1993) TREECON: a software package for the construction and drawing of evolutionary trees. Comp. Applic. Biosc. 9: 177–182.

Epilogue: The search for explanations and future developments

K. Oeggl, J. H. Dickson, and S. Bortenschlager

In the initial phase of research into the Iceman, four hypotheses – the hunter, shaman, metal prospector and shepherd theories – were proposed to explain the find in its entirety (Egg et al., 1993). On the basis of the detailed scientific investigations conducted in the meantime, however, the assumption that the Iceman was in some way involved in an early form of transhumance has now gained general acceptance.

For many years research at the Innsbruck Department of Botany has concentrated on the postglacial development of the climate and timber line, with the upper Ötztal a main location for the fieldwork. The data obtained from this research provided the first indications that Neolithic man started to play a role in shaping the vegetation cover in addition to climatic impacts (Bortenschlager, 1984). Demonstrating anthropogenic changes to the vegetation is a complex undertaking as humans does not create a completely new habitat at altitudes above the timber line (Oeggl, 1994). Further interdisciplinary palaeoecological studies were subsequently undertaken to determine the human impacts on the alpine vegetation, taking the intensity of different forms of human activity into account. For the Mesolithic hunter and gatherer economies, in which the high mountains played an important role for hunting, it was found that the great diversity of species and locations to be found at altitudes between 1900 and 2500 m above sea-level attracted both man and fauna to those altitudes in the Early Holocene (Broglio and Lanzinger, 1996; Oeggl and Wahlmüller, 1994 a,b). In the subsequent Neolithic it would seem from the overall picture of the finds made in Upper Italy that prehistoric man had lost interest in the high mountains (Fedele, 1981). A more detailed analysis of the individual sites, however, clearly shows that human activity in the alpine stage was as common as ever (Bagolini and Pedrotti, 1992). Hunting alone, however, would not seem to be a sufficiently powerful motive. The discovery of the Iceman accelerated the process of research into the history of the vegetation of the upper Ötztal. Analysis of over twenty peat profiles presented a detailed biostratigraphic picture of the region. One striking feature of the diagrams obtained is the increase in pasture weeds and indicators of soil enrichment in the alpine stage starting in the Middle Neolithic. The extensive investigations conducted revealed changes in the vegetation cover as a result of Neolithic pastoral farming for the whole of the upper Ötztal area (Bortenschlager, this volume). That in turn proves that Neolithic man had an economic interest in the extensive areas of grassland located above the timber line in addition to their suitability for hunting. Even as early as the Neolithic, because of the absence of large areas of pasture in the valleys, livestock was driven to the higher elevations for grazing. That confirms the hypothesis put forward by scientists from the start that Neolithic man made use of these altitudes for primarily agricultural reasons, namely as high-level pasture.

The unique nature of the Iceman find in terms of completeness and state of preservation initially suggested that the Iceman's presence at that altitude must have been a singular event. But the existence of several mesolithic dwelling places in the vicinity of the Tisenjoch (Leitner, 1996; Niederwanger, pers. comm.) and in addition to pollen analysis, the results of radiocarbon dating show that the use of the Tisenjoch route across the main alpine ridge cannot have been exceptional. Among the otherwise highly consistent dating results for the vegetable and wood remains found at the site, which are also in agreement with the age of the mummy reported by Bonani et al. (1992), there are two items which do not fit in with the general pattern (Kutschera et al., this volume). In both cases the items are pieces of wood which must have arrived at the site of the find through human agency. The one, the remains of a pine twig (specimen no. 92/292, Oeggl and Schoch, this volume, Plate 6) is from the Early Neolithic, while the second is an artefact with clear traces of having been worked in the Iron Age (specimen no. 92/275, Oeggl and Schoch, this volume, Plate 8). It would therefore be a fallacy to treat the discovery of the Iceman at such an altitude as the result of a unique event. This conclusion is also confirmed by the recent discovery of the haft of a Neolithic axe in the same area (Oeggl and Spindler, in preparation). On the contrary, the results of scientific research prove that humankind has made continuous use of these altitudes from the Mesolithic to the present day.

For those researchers already familiar with the results of the palaeoecological studies the discovery of an ancient human body at a great height in the Ötztal Alps was not a total surprise because people must have crossed these inhospitable mountains. Before the discovery of the mummy, individual pollen profiles from the high-elevation moors of the upper Ötztal had already indicated an increase since the Middle Neolithic in those alpine taxa that benefit from pastoral farming. This was initially an isolated finding, as little was known about Neolithic settlement of the inner Alpine area. But even then, subsequent research with pollen analysis conducted along the Ötztal (Bortenschlager, 1984; Wahlmüller, unpublished data) suggested that human activity must have developed from the south. The result of this research also offers the first clue with regard to the Iceman's origins and the location of his home. It also supports the standard archaeological argument that the Iceman had his roots in the Remedello culture to the south of the main Alpine ridge (Spindler, 1996). It is for the archaeologists to provide the final answer with regard to the precise location of the Iceman's settlement, but analysis of the plant remains found with the mummy already permits his home to be narrowed down considerably. The numerous specimens recovered from the site of the find derive from a variety of sources. A small number of them are from plants that grew in and around the rocky depression where the Iceman lay, and they reflect the local flora of the nival zone since the Neolithic. Most of the vegetable material, however, was carried by the Iceman wittingly or unwittingly and entombed in the ice with him. The natural locations of the plants concerned vary between the valley floor and the timber line in the subalpine stage, and reflect the environments through which the Iceman passed before his death. The analytical significance of the individual plants varies. Analysis has shown that some of the wood used by the Iceman to make his weapons and tools was cut during the vegetation period and some was not. To that extent the spectrum of the types of wood involved does not give a picture of the situation immediately prior to his death but is rather the product of a longer period of time. In the Neolithic, wood was an important raw material for making all kinds of utensils, and therefore had to be easily available in adequate amounts. Phytogeographic evaluation of the wooden specimens found reflects the Iceman's habitat to a limited extent. Most of them derive from the montane zone. Their ecological requirements indicate a transitional zone between thermophilous oak forests (*Quercetalia pubescenti-petreae*) and the montane spruce forests (*Piceetum montanum*). Norway maple (*Acer platanoides*), yew (*Taxus baccata*), ash (*Fraxinus*), lime (*Tilia*) and elm (*Ulmus*) are indicative of a humid habitat with fresh soils rich in minerals, and a climate characterised by mild winters such as is typical of the steep forested slopes to be found at the beginning of Schnalstal and Vinschgau in South Tyrol (Oeggl and Schoch, this volume). These locations are only 25 km from the site of the find.

Separate analysis of the mosses recovered from the Iceman's clothes produced the same result. Here, too, the crucial mosses occur in abundance only 10–20 km south of the main Alpine ridge in Schnals Valley and Vinschgau. To the north of the Alps, on the other hand, the same spectrum of mosses is only to be found at distances of 40 km and more (Dickson, this volume).

The investigations carried out in an attempt to identify the Iceman's home also include analysis of the remains of insects found at the site. Here again the habitats of the species encountered range from the montane to the Alpine stage. One interesting find is the bark beetle (*Pityokteines spinidens*), which lives in firs (*Abies alba*) and occasionally in spruce (*Picea abies*), pine (*Pinus sylvestris*) and larch (*Larix decidua*). Two other insects from the montane zone were also found. There was a greater prevalence, however, of species from the subalpine-alpine grass heaths (Schedl, this volume). They confirm the results of the botanical investigations and show that the Iceman had spent time both in the valley and at higher elevations. The final piece of evidence relating to the Iceman's origins is provided by pollen analysis of the contents of the mummy's intestines, where the presence of pollen grains from the Hop hornbeam (*Ostrya carpinifolia*) offers final confirmation of the Iceman's southern home. Vinschgau and the lower part of Schnals Valley represent the northern limit of the natural spread of the Hop hornbeam (*Ostrya carpinifolia*). The Ötztal or the Inntal to the north of the main alpine ridge can therefore be discounted as the Iceman's original territory (Oeggl, this volume). Information on the Iceman's physical environment is provided by analysis of the stomata of dwarf willow leaves extracted from the sediment found in the rock cleft in which the Iceman lay. The stomata density of the dwarf willow (*Salix herbacea*) is inversely proportional to the carbon dioxide content of the air. With the help of a modern calibration data set it was possible to determine the carbon dioxide content of the air the Iceman breathed. The results show that there was no significant change in carbon dioxide concentrations from the prehistoric to the pre-industrial period (Birks, this volume). One astonishing feature of the Iceman's equipment is the make of his tools and weapons, and his careful choice of materials. He made exclusive use of those types of wood that are best suited for the intended purpose in the individual case. This applies to the bow, for example, where the Iceman selected yew

(*Taxus baccata*) of good quality and worked it with due regard for the physical characteristics of the piece. For the shafts of his arrows, too, the Iceman chose the best materials. The rigidity of the wood of the wayfaring tree (*Viburnum lantana*) makes it the best suited of all indigenous types of wood for that purpose. Only in one case, namely the repaired shaft of no. 12 arrow, did he make use of a second-best material, i.e., the wood of the cornel tree (*Cornus* sp.) (Oeggl and Schoch, this volume). Similarly, for vegetable binding material the Iceman employed only bast from the lime (*Tilia*) (Pfeifer and Oeggl, this volume). For fire-lighting he used a piece of true tinder fungus (*Fomes fomentarius*), while a piece of birch fungus (*Piptoporus betulinus*), which grows only on birch trees, served combined medical and ritual purposes (Peintner and Pöder, this volume).

The Iceman's last meal was reconstructed through pollen analysis and macrofossil analysis of the contents of the colon. The identified bran of the einkorn (*Triticum monococcum*) is typical of the cereals grown in the inner Alpine area in the late Neolithic. The results of the pollen analysis, showing cereal pollen grains of the *Triticum*-type, confirm the macrofossil analysis. But the very exciting discovery is the predominance of pollen of the hop hornbeam (*Ostrya carpinifolia*). This rare instance of the preservation of the microgametophyte plus the spectrum of the other pollen taxa indicate late spring as the time of death, as do the stemless leaves of the Norway maple (*Acer platanoides*), which still contain chlorophyll (Oeggl and Schoch, this volume). The fine state of preservation of the hop hornbeam pollen means that it was ingested by the Iceman immediately after flowering and that exposure to the digestive enzymes was brief. If the Iceman died at Hauslabjoch he must have made a rapid ascent from the valley to the site of the find. Together with the spectrum of the wood used for his equipment, this fact indicates that the Iceman belonged to a local community, and the theory of a chance wanderer in the mountains can be excluded.

An insight into hygienic conditions in the Neolithic is provided through the discovery of a human ectoparasite (Schedl, this volume) in the form of two fleas (*Pulex irritans* L.) and an endoparasite, namely the whipworm (*Trichuris trichiuria*) (Aspöck et al., this volume). The Iceman apparently suffered from the latter. Together with pathogenic changes to the mucous membrane of the intestine (Platzer, personal communication), the large number of eggs found in the pollen analysis specimens is indicative of a general infestation. Such a condition leads to accelerated peristaltik, which is also a factor in the good state of preservation of the pollen grains. Indirectly, this also supports the theory of the Iceman's rapid ascent to the site of the find.

Some questions still remain unanswered, however. There is a contradiction between the complete array of the Iceman's equipment on the one hand and the non-functional state of certain items on the other. The bow is unfinished, and few of the arrows are ready for use. The theory of a deliberate act of deposition of the body or a gentle death through exhaustion (Spindler, 1996) is inconsistent with the fact that a number of artefacts are badly damaged. The torn quiver cap, the broken quiver strut, the fractured arrow shafts, and the remote location of some of the pieces well away from the quiver suggest that considerable mechanical force was involved. It is true that conditions for excavation at such altitudes are difficult and that the first attempts at recovery involved the use of force, but there can be no doubt that the above artefacts were fragmented on deposition and were recovered in that damaged condition from the undisturbed ice (Zissernig, 1992; Lippert, 1992; Egg et al., 1992; Bagolini et al., 1996). The above fractures can hardly be attributed to forces at work in the glacial ice. At present the distribution of the pieces in the rocky hollow can best be explained as deposition by water in a melt water pool. That means either that the Iceman must have lain in water immediately after his death or that the ice in the rock cleft melted at least once between his death and final discovery in 1991.

This raises a number of questions, including questions relating to the chronology of the deposition of the various items. In particular, what is the reason for the widely scattered location of the remains in the rock cleft? This must be explained on the basis of systematic analysis of the macrocomponents found there. The distribution of the corresponding vegetable remains will shed light on the original location of the various items and clarify the question of their possible translocation after the Iceman's death. If translocation has not occurred after death, the distribution pattern of the remains is that of their original deposition. That in turn would mean that, at the moment of the Iceman's deposition in the rocky hollow, powerful forces were at work which caused fragmentation of the various artefacts. The fact that the Iceman must have lain in water for some length of time at least once after his death can be seen not only from the macrofossil analysis but also from the results of the medical investigations performed on the body (Bereuter et al., 1996) and also from the fact that some of the Iceman's equipment is missing, e.g. parts of the birch-bark containers, fragments of the wooden frame for the backpack, pieces of leather from the quiver cap, etc.

These conclusions in turn raise questions of palaeoclimatology. Research to date has shown that climatic cooling has occurred between approx. 4500 BC and the modern period (Lamb, 1977). That was the argument in support of the hypothesis that the body of the Iceman had been locked in ice ever since his death. Investigations in the high mountains confirm this long-term climatic pattern (neoglaciation), although the gradual drop in temperature has been repeatedly interrupted by warmer episodes (Patzelt and Bortenschlager, 1973; Bortenschlager, 1992; Oeggl and Wahlmüller, 1994; Wick and Tinner, 1997). If the Iceman thawed out once or several times and subsequently lay in a meltwater pool, there must also have been warmer periods at Tisenjoch similar to the summer of 1991. When those periods occurred and how often is the subject of a separate interdisciplinary research programme. All together, the investigations performed on the Iceman and related finds offer a detailed insight into a brief period of prehistory, which has not previously been achieved through pollen analysis for high-elevation moors. Fine stratigraphic research has produced new findings into the genesis of the alpine grass heaths and early pastoral farming there, and will in future provide more detailed information on the climatic and cultural history of the high mountains.

References

Bagolini und Pedrotti (1992) Vorgeschichtliche Höhenfunde in Trentino-Südtirol und im Dolomitenraum vom Spätpaläolithikum bis zu den Anfängen der Metallurgie. In: Höpfel F., Platzer W. und Spindler K. (eds.): Der Mann im Eis, Bd. 1. Veröffentlichungen der Universität Innsbruck 187: 359–377.

Bagolini B., Dal Ri L., Lippert A. und Nothdurfter H. (1996) Der Mann im Eis: Die Fundbergung 1992 am Tisenjoch, Gem. Schnals, Südtirol. In: Spindler K., Rastbichler-Zissernigg E., Wilfling H., zur Nedden D. und Nothdurfter H. (eds.): Der Mann im Eis-Neue Funde und Ergebnisse. The Man in the Ice, vol. 2. Springer Verlag, Wien: 3–23.

Bereuter T. L., Reiter C., Seidler H. and Platzer W. (1996) Postmortem alterations of human lipids – part II: lipid composition of a skin sample from the Iceman. In: Spindler K., Wilfling H., Rastbichler-Zissernig E., zur Nedden D. and Nothdurfter H. (eds.): Human Mummies. The Man in the Ice, vol. 3: 2275–278.

Bonani G., Ivy S., Niklaus Tn. R., Suter M., Housley R. A., Bronk C. R., Van Klinken G. J. und Hedges R. E. M. (1992) Altersbestimmung von Milligrammproben der Ötztaler Gletscherleiche mit der Beschleuniger-Massenspektrometrie- Methode (AMS). In: Höpfel F., Platzer W. und Spindler K. (eds.): Der Mann im Eis. Veröffentlichungen der Universität Innsbruck 187, Bd. 1: 108–116.

Bortenschlager S. (1992) Die Waldgrenze im Postglazial. In: Eder-Kovar J. (ed.): Palaeovegetational development in Europe and regions relevant to its palaeofloristic evolution. Museum of Natural History Vienna: 9–14.

Broglio A. and Lanzinger M. (1996) The human population of the southern slopes of the Eastern Alps in the Würm Lateglacial and early Postglacial. Il Quaternario 9: 499–508.

Egg M., Goedecker-Ciolek R., Groenman-Van-Wateringe W. und Spindler K. (1993) Die Gletschermumie vom Ende der Steinzeit aus den Ötztaler Alpen. Jb RGZM 39: 128pp.

Fedele F. (1981) Il popolamento delle Alpi nel Palaeolithico. Le Scienze 160: 22–39.

Lamb H. H. (1977) Climate: present, past and future. Climatic history and the future. London.

Leitner W. (1996) Der "Hohle Stein" – eine steinzeitliche Jägerstation im hinteren Ötztal. (Archaeologische Sondagen 1992/93). In: Spindler K., Wilfling H., Rastbichler-Zissernig E., zur Nedden D. und Nothdurfter H. (eds.): Human Mummies. The Man in the Ice, vol. 3: 209–213.

Lippert A. (1992) Die erste archäologische Nachuntersuchung am Tisenjoch. In: Höpfel F., Platzer W. und Spindler K. (eds.): Der Mann im Eis, Bd. 1. Veröffentlichungen der Universität Innsbruck 187: 245–253.

Oeggl K. und Wahlmüller N. (1994a) Holozäne Vegetationsentwicklung an der Waldgrenze der Ostalpen: die Plancklacke 2150 m, Sankt Jakob im Defreggental (Osttirol). Diss. Botanicae (Festschrift Lang) 234: 389–411.

Oeggl K. and Wahlmüller N. (1994b) The Environment of a High Alpine Mesolithic Camp Site in Austria. American Association of Stratigraphic Palynologists, Contribution Series 29: 147–160.

Oeggl K. und Spindler K. (2000) Ein weiterer neolithischer Beilhorn vom Hauslabjoch. Archäologisches Korrespondenzlatt (in press).

Oeggl K. (1994) Palynological record of human impact on alpine ecosystems. In: Biagi P. and Nandris J. (eds.): Highland Zone Exploitation in Southern Europe. Monografie di "Natura Bresciana" 20: 107–122.

Spindler K. (1996) Iceman's last weeks. In: Spindler K., Wilfling H., Rastbichler-Zissernig E., zur Nedden D. and Nothdurfter H. (eds.): Human Mummies. The Man in the Ice, vol. 3: 252–263.

Wick L. and Tinner W. (1997) Vegetation changes and timberline fluctuations in the Central Alps as indicators of Holocene Climatic Oscillations. Arctic and Alpine Research 29: 445–458.

Zissernig E. (1992) Der Mann vom Hauslabjoch: Von der Entdeckung bis zur Bergung. In: Höpfel F., Platzer W. und Spindler K. (eds.): Der Mann im Eis, Bd. 1. Veröffentlichungen der Universität Innsbruck 187: 234–244.

SpringerHumanBiology

Konrad Spindler et al. (eds.)

Human Mummies

A Global Survey of their Status
and the Techniques of Conservation

1996. VIII, 294 pages. 226 partly coloured figures.
Hardcover DM 168,–, öS 1176,–
(recommended retail price)
ISBN 3-211-82659-9
The Man in the Ice, Volume 3

The contributions to the volume have their origin in a spontaneously organized meeting of experts at Innsbruck 1991 for finding the best measures for the conservation of the newly found "Man in the Ice".

Due to their target the techniques of conservation of corpses all over the world are described: Smoke-drying (Ecuador), sun-drying (Canary Islands), natural mummification by combination of low temperatures and dry air (Eskimo mummies of Qilakitsoq); the reader also will find descriptions of permafrost mummies of Greenland and "bog-mummies" of North-West-Europe including new methods of investigation in the fields of roentgenological, microbiological, microscopical, and microchemical methods and the latest imaging techniques in medical archaeology.

„Die zahlreichen gekonnt zusammengestellten Forschungsarbeiten von international renommierten Fachleuten eröffnen dem Anatomen eine sehr interessant gestaltete Sicht auf die moderne Mumienforschung. Das Buch stellt für jeden Anatomen nicht nur eine Abwechslung und Wissenserweiterung dar, in der nicht etwa Fakten von dem Spezialisten für den Spezialisten aneinandergereiht sind, sondern eine fesselnde und umfassende Darlegung eines anatomisch faszinierenden Forschungsgebietes wird aufgetan".
                                                                                    Annals of Anatomy

SpringerWienNewYork

A-1201 Wien, Sachsenplatz 4–6, P.O.Box 89, Fax +43.1.330 24 26, e-mail: books@springer.at, Internet: **www.springer.at**
D-69126 Heidelberg, Haberstraße 7, Fax +49.6221.345-229, e-mail: orders@springer.de
USA, Secaucus, NJ 07096-2485, P.O. Box 2485, Fax +1.201.348-4505, e-mail: orders@springer-ny.com
Eastern Book Service, Japan, Tokyo 113, 3–13, Hongo 3-chome, Bunkyo-ku, Fax +81.3.38 18 08 64, e-mail: orders@svt-ebs.co.jp

SpringerHumanbiologie

Konrad Spindler et al. (Hrsg.)

Der Mann im Eis

Neue Funde und Ergebnisse

1995. X, 320 Seiten.
231 z.T. farb. Abbildungen und 1 Beilage.
Gebunden DM 154,–, öS 1078,–
ISBN 3-211-82626-2
The Man in the Ice, Volume 2

Die Ergebnisse der Nachgrabung 1992 am Hauslabjoch durch ein italienisch-österreichisches Archäologenteam werden ausführlich (mit detaillierter Fundliste und steingerecht gezeichnetem Fundplan) vorgestellt. Weiters: Schädel und Gewebe, Fingernägel, Haare und Tätowierungen, Überlegungen zur Herkunft des Mannes, detaillierte DNA-Analysen zu Grasfunden, tierische Rohstoffe (Leder/Felle), die „Schwarze Masse" aus der Gürteltasche, Federkeratine der Pfeilbewehrungen, Rohstoffe der Steininstrumente, absolute Datierung.

„Die vorliegenden Berichte über neue Funde und Ergebnisse dieser als ‚Sternstunde der Archäologie' eingestuften Entdeckung sind die ersten wissenschaftlich fundierten Annäherungen an einen einmalig konservierten Menschen aus der späten Jungsteinzeit ... Eine Fülle von Farb- und Schwarzweißbildern, Grafiken und Zeichnungen ergänzen die Beiträge. Eine Stärke dieser Sammlung machen die Aufsätze über die kulturhistorischen Hintergründe aus ... Der ‚bunte' Eindruck dieser Aufsatzsammlung wird verstärkt durch die linguistische Analyse der in den Medien verwendeten Personenbezeichnungen für den ‚Mann im Eis'. Vom ‚urzeitlichen Alpenwanderer' bis zum ‚frozen Fritz' der Amerikaner sind alle Facetten medienwirksamer Phantasie aufgeführt".

Freiburger Universitätsblätter

SpringerWienNewYork

SpringerBiology

Kurt W. Alt,
Friedrich W. Rösing,
Maria Teschler-Nicola (eds.)

Dental Anthropology

Fundamentals, Limits, and Prospects

1998. XXVI, 564 pages. 224 figures.
Hardcover DM 138,–, öS 966,–
(recommended retail price)
ISBN 3-211-82974-1

The study of teeth is an important key to the biology and culture of past and living populations. Their extreme hardness, their numerous and highly heritable characteristics and their susceptibility to environmental responses allow inferences of human evolution, adaptation, living conditions, environments, group affinity, kinship and the life and death of single persons. This has been gradually discovered during the last thirty years, and it lead to the formation of the new field of dental anthropology. It is rooted in biology and has strong ties to dentistry, forensics and the humanities.

This book gives a comprehensive view of this new field of dental anthropology. It provides a basic introduction as well as a reference for the specialist in anthropology, forensics, ecology, paleontology and dentistry. Most aspects are completely new, particularly the syntheses. The basic experience and literature is of a broad international and multilingual origin.

"... an excellent overview of the multidisciplinary nature of dental anthropology including some of the limitations of present approaches, as well as future directions ... a valuable addition to the literature in dental anthropology ... recommended to teachers, researchers and postgraduate students in the field."

Anthropological Science

SpringerWienNewYork

A-1201 Wien, Sachsenplatz 4–6, P.O.Box 89, Fax +43.1.330 24 26, e-mail: books@springer.at, Internet: **www.springer.at**
D-69126 Heidelberg, Haberstraße 7, Fax +49.6221.345-229, e-mail: orders@springer.de
USA, Secaucus, NJ 07096-2485, P.O. Box 2485, Fax +1.201.348-4505, e-mail: orders@springer-ny.com
Eastern Book Service, Japan, Tokyo 113, 3–13, Hongo 3-chome, Bunkyo-ku, Fax +81.3.38 18 08 64, e-mail: orders@svt-ebs.co.jp

SpringerBiology

H. Preuschoft, D. J. Chivers (eds.)

Hands of Primates

1993. IX, 421 pages. 224 figures.
Hardcover DM 218,–, öS 1525,–
(recommended retail price)
ISBN 3-211-82385-9

The hand commonly is considered to have exerted great influence on the evolution of typically human features, like upright posture, stereoscopic vision, "manipulative" handling of parts of the environment. The hands of the other primates are not less closely related to the necessities of life than in humans. But beyond this general statement, only few satisfying and precise analyses of their functions exist. Most considerations begin and end up with Napier's discrimination and definition of power grip and precision grip – which has turned out to be very useful in surgery – and the restating of man's distinctiveness.

The characteristic features of the human hand are to a large extent shared by the hands of other primates, and therefore it seems logical to approach the human hand by looking into the details of hand function and hand morphology in non-human primates.

This book presents a well-integrated series of articles which deepen our knowledge regarding the problems mentioned above.

SpringerWienNewYork

A-1201 Wien, Sachsenplatz 4–6, P.O.Box 89, Fax +43.1.330 24 26, e-mail: books@springer.at, Internet: **www.springer.at**
D-69126 Heidelberg, Haberstraße 7, Fax +49.6221.345-229, e-mail: orders@springer.de
USA, Secaucus, NJ 07096-2485, P.O. Box 2485, Fax +1.201.348-4505, e-mail: orders@springer-ny.com
Eastern Book Service, Japan, Tokyo 113, 3–13, Hongo 3-chome, Bunkyo-ku, Fax +81.3.38 18 08 64, e-mail: orders@svt-ebs.co.jp